RADAR
CROSS SECTION

The Artech House Radar Library

Radar System Analysis by David K. Barton

Electronic Intelligence: The Analysis of Radar Signals by Richard G. Wiley

Electronic Intelligence: The Interception of Radar Signals by Richard G. Wiley

Principles of Secure Communication Systems by Don J. Torrieri

Shipboard Antennas by Preston E. Law, Jr.

Radar Propagation at Low Altitudes by M.L. Meeks

Interference Suppression Techniques for Microwave Antennas and Transmitters by Ernest R. Freeman

Radar Cross Section by Eugene F. Knott, John F. Shaeffer, and Michael T. Tuley

Radar Anti-Jamming Techniques by M.V. Maksimov et al.

Synthetic Array and Imaging Radars by S.A. Hovanessian

Radar Detection and Tracking Systems by S.A. Hovanessian

Radar System Design and Analysis by S.A. Hovanessian

Radar Calculations Using the TI-59 Programmable Calculator by William A. Skillman

Radar Calculations Using Personal Computers by William A. Skillman

Techniques of Radar Reflectivity Measurement, N.C. Currie, ed.

Monopulse Principles and Techniques by Samuel M. Sherman

Receiving Systems Design by Stephen J. Erst

Signal Theory and Random Processes by Harry Urkowitz

Radar Reflectivity of Land and Sea by M.W. Long

High Resolution Radar Imaging by Dean L. Mensa

Introduction to Monopulse by Donald R. Rhodes

Probability and Information Theory, with Applications to Radar by P.M. Woodward

Radar Detection by J.V. DiFranco and W.L. Rubin

RF Radiometer Handbook by C.W. McLeish and G. Evans

Synthetic Aperture Radar, John J. Kovaly, ed.

Infrared-To-Millimeter Wavelength Detectors, Frank R. Arams, ed.

Significant Phased Array Papers, R.C. Hansen, ed.

Phased Array Antennas, A. Oliner and G. Knittel, eds.

Handbook of Radar Measurement by David K. Barton and Harold R. Ward

Statistical Theory of Extended Radar Targets by R.V. Ostrovityanov and F.A. Basalov

Antennas by Lamont V. Blake

Radars, 7 vol., David K. Barton, ed.

RADAR
CROSS SECTION

Its Prediction,
Measurement and
Reduction

Eugene F. Knott
John F. Shaeffer
Michael T. Tuley

International Standard Book Number: 0-89006-174-2
Library of Congress Catalog Card Number: 85-047750

Printed and manufactured in the United States by Book-mart Press, Inc., North Bergen, N.J.

CONTENTS

PREFACE

The term radar cross section, and its acronym RCS, is familiar to most scientists and engineers involved in radar systems, but to others it may seem to be an obscure characteristic of a body or target scanned by a radar beam. They have somehow learned or heard that this strange area is not necessarily the geometrical cross section of the body; if not, then what is it? The name radar itself may even evoke memories of difficult electromagnetic courses taken in college and thus, for some, the concept of radar cross section is mysterious and elusive

In order to acquaint scientists and engineers who may be competent in their own disciplines with a seemingly new and unfamiliar technology, Georgia Tech introduced a short course on radar cross section reduction in January 1983. The course focused not only on what radar cross section is, but how to reduce it as well. Accompanying the formal lectures was a set of course notes numbering over 700 pages. This book is an outgrowth of those notes.

As the reader will perceive, the book is not intended as an exhaustive survey or treatise. Such would defeat our purpose of exposing the novice to what can quickly become a very complicated subject. For those who wish to pursue the subject in more detail, several books are available which delve into the intricacies of prediction and measurement techniques. In addition, many of these books are classics, giving far more information than is included here. The reader will find many of these books and papers listed in the references at the end of each chapter.

In addressing our purpose of presenting the flavor of RCS, if not its intricacies, we have organized the book into five groups of chapters. Chapters 1 and 2 contain background information; Chapters 3 through 5 introduce the concept of scattering and present useful RCS prediction techniques; Chapter 6 displays examples of RCS behavior for simple and complex bodies; Chapters 7 through 9 address radar cross section reduction methods; Chapters 10 through 13 discuss techniques for measuring absorber properties and the scattering characteristics of test targets; and Chapter 14 examines practical ways to identify scattering mechanisms on complex targets.

The reader should appreciate that technological advances have been steady in this field, and will continue to be so, and that even during publication of this book measurement and prediction techniques are being improved and modified. Moreover, in the interest of national security, some aspects of radar cross section measurement, prediction, and reduction cannot be presented here. Therefore, the book is not as detailed nor as complete as the authors would have preferred. Nevertheless, we think that it fills a need in the engineering community. We hope the reader agrees.

Eugene F. Knott
John F. Shaeffer
Michael T. Tuley

June, 1985

ACKNOWLEDGMENT

This book could not have been published without the assistance of many people. The authors are grateful to Dr. E. K. Reedy, Director of the Radar and Instrumentation Laboratory of the Georgia Tech Research Institute, for urging us to proceed with publication, and for furnishing moral and other support. In addition, we are indebted to our clerical staff, Cecelia Edwards, Helen Williams, and Patricia Winn, for converting our disorganized prose into legible text by passing it through their word processors several times.

CHAPTER 1

INTRODUCTION

E. F. Knott

1.1 OVERVIEW

This book is an introduction to the rather broad field of the echo characteristics of radar targets. It is intended to acquaint engineers, scientists, and managers with what may be a new and unfamiliar discipline, even though a great body of knowledge has existed since before the widespread use of radar in World War II. Modern weapons systems often carry RCS performance specifications in addition to other, more conventional requirements, such as speed, weight, and payload. Integrating these RCS specifications into a new or existing system requires that engineers of several disciplines interact with each other, and our intention is to promote that interaction and to show why certain procedures or features are important in system design from the standpoint of electromagnetics.

The book is structured around five major topic areas:
- background information
- electromagnetics
- RCS phenomenology
- absorbing materials
- measurements

The background information is contained in this chapter (Ch. 1) and Chapter 2. Since the radar echo mechanism is strictly an electromagnetic phenomenon, three chapters (Ch. 3-5) are devoted to the development of electromagnetic theory and RCS prediction techniques; examples of echoes from simple targets are given in Chapter 6. The control of the echo properties from man-made targets is of great practical and tactical importance, and the two primary ways to control it are by shaping and through the use of radar absorbing materials. Shaping is discussed in Chapter 7, and the analysis and design of radar absorbers is contained in Chapters 8 and 9. The measurement of materials and RCS are discussed in Chapters 11 through 13, and Chapter 14 discusses data analysis and display.

1.2 RADAR SYSTEMS

The acronym RADAR was coined during World War II, and was so new that many 1945 dictionaries did not list the word. It stands for "radio detection and ranging," and since its introduction in the 1940s, it has become a common household word. Radar was initially developed to replace visual target detection for several reasons. Radio waves suffer much less attenuation through the atmosphere than light waves, and signals in the lower frequency ranges actually propagate over the visible horizon. This makes it possible to detect targets long before they are visible optically. Radars also work well at night when there is little or no ambient light to illuminate the target.

A radar emits its own energy and does not rely on the illumination of the target by other sources. As such, it is an active device, rather than a passive sensor. In fact, a major advantage of radar is that it is an active system which can clock the time it takes for energy to travel to the target and back again.

The radar uses the known velocity of propagation of an electromagnetic wave to determine the distance to the target. The velocity of light is 186,282 miles per second, but a more convenient number to remember is 11.8 inches per nanosecond. Within a margin for error of only 1.6%, this is one foot per nanosecond. Accounting for the two-way propagation (out to the target and back again), the distance R to the target is simply

$$R = c \, \Delta t / 2$$

where c is the speed of light and Δt is the time interval between the transmission of an energy pulse and the reception of the radar echo. Thus, the radar is a timing device.

However, distance alone does not reveal the target location. Two direction angles must be measured, and there are a variety of ways to do so. Fortunately, because of the nature of most radar antennas, it can be done simply by measuring the direction in which the antenna is pointed. If the antenna is mounted on a pedestal having two axes of rotation (azimuth and elevation, for example), the two angles can be measured by devices mounted on the rotation shafts. The distance and the two angular directions then serve to locate the target.

In many instances the target is low enough on the horizon that an elevation angle measurement is unnecessary. Morever, there are other cases when the waves propagating between the target and the radar curve gently downward with increasing distance, hence the elevation angle of the antenna does not indicate the true elevation angle to the target. This curve or bending is due to a gradual reduction in the atmospheric index of refraction with increasing height; the bending itself is called refraction.

Like the eye and other optical systems, the radar antenna is deliberately designed to be more sensitive in a given direction than in other directions. This serves to concentrate more energy on the target upon transmission, and to increase the receiver sensitivity upon reception. The directivity increases as the antenna becomes physically larger, and energy concentrations of 1000 to 100,000 times the omnidirectional value are not uncommon.

Phased arrays are another common form of antenna. They consist of a large collection of elementary antennas, such as dipoles or waveguide slots, each of which is excited or fed from a distribution network. Depending on the spacing between elements and the relative phase of their excitation, the net radiation from this collection can be swept from one direction to another without any physical motion of the array itself. Since this scanning can be done electronically, very high scan rates can be achieved.

Whether or not a radar is caused to scan depends on its mission and deployment. Low frequency radars are typically long-range systems used for surveillance and detection well beyond the horizon. Low frequency antennas are usually fixed, permanent installations because of their large size. Shorter range surveillance systems operate at higher frequencies and are typically scanned in azimuth; some scan in elevation as well as azimuth with an up and down nodding motion, superposed on a slower azimuth scan. Scanning radars can be found in land-based systems, on ships, and in aircarft.

Fire control radars track their targets in space. The tracking systems, therefore, must include angle tracking as well as range tracking. A computer is often an integral part of the radar, extracting trajectory information from the track history and feeding pointing information to the gun or other firing system. Terminal homing radars are small and light, because they must fit inside a relatively confined area in a missile. There has been a tendency to exploit shorter wavelength instrumentation for these systems because of the weight and space constraints.

The great variety of radars and their missions make it impossible to list and discuss the many kinds. Consequently, the systems discussed in Chapter 2 are generic, and only a few examples are given. The simplest system is the moving target indicator (MTI) radar, whose output is an indication of only targets that move; an example is a police radar. The radar emits a continuous stream of energy, and a sample of the transmitted signal is used as a local oscillator signal, to be mixed with the received target signal. This "mixing" action generates a signal whose frequency is the difference between the transmitted frequency and the target reflected frequency. The only way the two frequencies can be different is if the target is in motion, in which case the difference frequency is proportional to the component of the target velocity toward or away from the radar.

Except for continuous wave (CW) radar systems, radars are pulsed because of the need to emit a burst of energy and then remain quiet while "listening" for the echo. In addition, pulsed operation allows high-power transmission for short intervals of time. Typically, a conventional pulsed radar is "on" for only 0.1% to 0.5% of the time, and consequently it spends most of its time listening for the echo.

1.3 ELECTROMAGNETICS

We live in an environment of electromagnetic (EM) waves, and the microwave frequencies of radar systems occupy only a small region of the EM spectrum. Nevertheless, these waves, whether light, radio, or microwaves, all obey the same physical laws. The laws governing the behavior of EM waves were established by the year 1900, thanks to the pioneering efforts of men like Hertz, Faraday, Ampere, Coulomb, and Maxwell.

Early experimenters were downright curious, and inventors like Bell and Edison exploited known phenomena, as well as undertaking their own profit-oriented research. Each made his own contribution, but it was James Clerk Maxwell who assembled the results of the experimenters and formulated the basis of modern electromagnetic theory. His famous four equations, augmented by a handful of others, are the foundations of the theory, much as the Navier-Stokes equations are the foundations of fluid mechanics.

The EM waves of concern to us are harmonic in both time and space, for which Maxwell's equations take on a specific form. Remote from any source of radiation and reflection, the electric and magnetic fields remain at right angles to each other as well as to the direction of propagation. The electric and magnetic fields both attain their maximum values at the same time and place, and in the course of their harmonic variation they also go to zero at the same time and place.

When an EM wave impinges on a body, it induces oscillating charges and currents inside that body and on its surface. For the special case of a perfectly conducting body, the induced charges and currents are confined to the surface and, at microwave frequencies, even a poorly conducting metal like steel approximates the behavior of a perfect conductor. Therefore, even if the body is a thin steel shell, there are no induced fields or charges inside, provided the shell is a complete surface.

Even if the body is a dielectric material capable of supporting induced charge and currents in its interior, attention can be confined to the charges and currents induced on its surface. As shown in Chapter 3, the electric and magnetic fields at any point in space can be expressed in terms of a surface integral of the charges and currents induced on the surface of a body. Although this is a well known result for source-free regions, it is rather remarkable that we do not need to know the interior distributions of charge and current in order to calculate the scattered fields.

The total field at an observation point due to radiation by induced fields over the surface of the target obstacle is comprised of the incident and scattered fields. Presumably, the incident field is known, and all that need be done is to subtract it from the total field in order to obtain the field scattered by the body. This is trivial.

What is not so trivial is estimating the induced charges and currents. Indeed, this is the crux of the problem, for if the charge and current distributions are known, the integral can be evaluated, numerically if not analytically. Finding the induced fields is the problem, but in some cases the integral equation can be broken into a collection of homogeneous linear equations, which are then amenable to solution.

However, such a solution is restricted by computer memory to bodies not much more than a wavelength in size. If some particular feature of the target can be exploited, such as roll symmetry or the two-dimensional nature of infinite bodies, the size that can be treated approaches about 10 wavelengths. This form of solution for the induced currents and charges is called the method of moments, and the fields on the body must be sampled at intervals of around $\lambda/10$. The interaction of each segment with every other segment creates a square matrix which, when inverted, yields the solution.

Another approach to finding the fields scattered by the body is to restrict our attention to the far field and to make what is known as the tangent plane approximation. It results in a prescription known as physical optics, which yields a great deal of useful information. In applying the tangent plane approximation, we first identify the local patch of integration over the body surface. An infinite plane must be tangent to the body there, and the local tangential fields are assigned the identical values that they would have had, if the body had been truly flat and infinite in extent at the tangent point.

These local tangential fields are available from the theory of geometric optics, which simply invokes the known boundary conditions for an infinite plane separating two different media. Usually the body is taken to be metallic, but the approach also works for dielectric materials. It has even been applied successfully to thin membranes, such as found in soap bubbles and the leaves of trees.

Having invoked the tangent plane approximation, we can evaluate the surface integral. Note that the problem of solving for the induced surface fields has been completely bypassed; in essence, the local radius of curvature has been assumed to be large enough that the induced fields are negligibly different than those of an infinite sheet viewed from the same angle.

There are only a few cases in which the surface integral can be evaluated exactly, and even then we must bear in mind that the integral itself is now only approximate. The integral is almost trivial if the body is a flat plate, and rapidly becomes complicated even for a cylinder, as shown in Chapter 5. Generally, there is no exact evaluation for doubly curved surfaces except for very symmetrical bodies, and the method of steepest descents must be used.

1.4 RCS PHENOMENOLOGY

Whatever the case, evaluation of the physical optics integral yields important and useful results. In the case of a flat plate, it yields the correct result when the plate is viewed at broadside incidence, and the answer is reasonably accurate several degrees away from the specular (broadside) orientation. In fact, the angular sector over which the theory gives acceptable results depends on the electrical size of the plate.

However, another analytical procedure can be used: the geometrical theory of diffraction (GTD). In formulating his GTD, J. B. Keller extended the notion of geometric optics for the single ray reflected by a surface to a cone of rays diffracted by an edge. The cone of diffraction still contains the geometric optics feature whereby diffracted rays leave the edge at the same angle made by the incident ray. However, the theory can be exploited to postulate induced edge currents, which expand the angular (non-specular) coverage of the diffracted fields. More importantly, GTD has a built-in polarization dependence that is absent in physical optics and geometric optics.

Examination of the results for a flat plate gives some insight into a heirarchy of scattering types. The broadside return from the plate is proportional to the square of the area of the plate and the square of the frequency. For this particular orientation, the plate behaves like a dihedral or trihedral corner reflector, whose returns also increase with the square of the area and the square of the frequency. Thus, the trihedral corner, the dihedral corner viewed in a direction perpendicular to its axis, and the flat plate viewed at normal incidence, all constitute one class of scatterer.

If the plate is oriented off the specular angle, but in such a way as to maintain a pair of edges perpendicular to the radar line of sight, the radar return is no longer proportional to the square of its area, but to the square of the edge length. The return is also independent of frequency, a characteristic shared by the doubly curved surfaces, of which the sphere is an example. Thus, spheroids and edges presented normal to the line of sight have frequency-independent echo strengths.

If the plate is angled so that no edge is perpendicular to the radar line of sight, the radar simply senses the returns from the four corners. The echo of a corner has no dependence on length because a corner or vertex has no physical dimension: it is a point. Thus, the returns from vertices vary inversely with the square of the frequency.

Consequently, a rather simple structure like a flat plate exhibits at least three kinds of scattering behavior, depending on its orientation. A cylinder also has more than one scattering behavior: at broadside incidence, the return from a cylinder varies as the cube of some body dimension and directly with the frequency. When the cylinder is tilted with respect to the line of sight, its return degenerates to the contributions from a pair of curved edges, one at either end. The echo from a curved edge varies inversely with the frequency, and therefore lies somewhere between the characteristic of a straight edge and the characteristic of a vertex.

These characteristics do not include the creeping wave mechanism, which is a phenomenon generally associated with smooth bodies that are not large with respect to the wavelength. The creeping wave is launched at a shadow boundary and traverses the shadowed side of a smooth body, eventually emerging at the shadow boundary on the other side of the body. It loses energy as it propagates around the rear curved surface, and the longer the electrical path length, the more energy it loses. The creeping wave phenomenon becomes negligible for most bodies that are larger than a dozen or so wavelengths.

Even if the target does not have a smooth termination, edges that also form shadow boundaries can diffract energy across shadowed regions. Right circular cones are examples, and patterns presented in Chapter 8 demonstrate how a pad of absorber changed the nose-on cross section, even though the pad was cemented to the shadowed base.

A wave often confused with the creeping wave is the surface traveling wave launched along surfaces at small grazing angles. The phenomenon occurs only when there is a component of the electric polarization parallel to the surface along the direction of incidence, when the angle of incidence is within a few degrees of grazing, and when there is a discontinuity at the far end of the body that reflects the wave back toward the radar. The difference between this mechanism and the creeping wave mechanism is that the wave travels along an illuminated surface in the one case and over a shadowed surface in the other. The surface wave contribution can be reduced by smoothing out the discontinuity at the rear of the body, which urges the wave forward into the shadow region where it becomes a creeping wave.

1.5 ABSORBING MATERIALS

One way of reducing the radar echo from a body is to soak up the incident electromagnetic energy, thereby reducing the net energy available for reflection back to the target. In order to absorb energy, materials must be found in which the the induced currents are in phase with the incident fields, as are the currents in a resistor. In fact, many absorbing materials are manufactured with carbon providing the loss mechanism, and the dissipation of energy takes place by the conversion of electromagnetic energy into heat.

Energy absorption, however, does not necessarily require carbon. There are dielectric materials whose indices of refraction are complex numbers, and it is the imaginary part that gives rise to the loss. The molecules in the material are essentially small dipoles that try to orient themselves along the incident field. If the field changes too fast, or, if the dipoles lag the impressed field variations, torque is exerted and energy is deposited in the material. The dipoles experience a kind of molecular friction as they try to follow the oscillations in the fields.

Conduction losses are lumped together with the imaginary part of the index of refraction for engineering convenience. Moreover, magnetic losses can occur as well as electric losses, because the index of refraction contains the product of the magnetic permeability and the electric permittivity. Thus, lossy materials may also include ferrites or carbonyl iron in addition to, or in the place of, carbon.

The earliest form of radar absorber, the Salisbury screen, was a sheet of porous material impregnated with graphite and spaced a quarter-wavelength off a metallic backing plate. We know from transmission line theory that a short circuit (the metal plate) placed a quarter-wavelength behind a load effectively creates an open circuit at the load itself. Thus, the incident wave "sees" free space and there is no reflection. In fact, all of the power in the incident wave is delivered to the resistive sheet and none is reflected.

If the spacing between the sheet and the metal backing plate is not a quarter-wavelength, the wave sees a finite impedance in parallel with the impedance of the resistive sheet, producing a net complex impedance different from that of free space. Consequently, part of the incident power in the wave is reflected, with the reflection being greater the more the frequency of the incident wave moves away from $\lambda/4$ frequency spacing. Thus, the Salisbury screen is a narrowband device.

The bandwidth can be increased by cascading several sheets, one behind the other and separated by spacers. The resistance should decrease from sheet to sheet, with the lower values being used for the sheets closer to the metal backing plate. This collection of sheets is called a Jaumann absorber, and the bandwidth rises with each additional sheet used in the design. However, additional sheets require additional spacers, and the thickness of the absorber increases.

However, the resistive sheet is only one form of absorber; there are bulk materials which are loaded with carbon as well. Although the conductivity of carbon is much lower than metals such as aluminum and steel, it is still too high to be used as a solid material. The bulk conductivity must be reduced, typically by spreading a thin coating on the fibers of a fibrous mat, or by infusing carbon in a matrix of spongy urethane foam.

Graded dielectric materials are manufactured by bonding together two or more layers of these spongy sheets. The carbon loading density must increase from layer to layer, with the highest loading (highest conductivity) being in the layer closest to the metal backing whose reflectivity is to be reduced. Unfortunately, these materials are bulky and fragile, and are not suitable for applications involving exposure to weather and wind blast.

Much thinner absorbers can be fabricated by the use of materials having magnetic losses. Magnetic substances of value are iron oxides (ferrites) and carbonyl iron. The ferrites can be sintered in solid form, and the carbonyl dust can be mixed with an organic binder to form a matrix of the appropriate loss density. Although for a given level of performance the magnetic materials are typically thinner than ordinary dielectric absorbers, they are heavy because of their high concentrations of iron.

Another kind of material, called circuit analog RAM, utilizes lossy material deposited in geometric patterns on a thin lossless film. The thickness of the deposit, which is like aquadag, controls the effective resistance of the layer, and the geometry of the pattern controls the effective inductance and capacitance. Thus, the layer can be tailored to specific values of inductance, capacitance, and resistance, and its performance can be analyzed in terms of lumped elements. Typically, an absorber panel is comprised of more than one layer and the layers are separated by light low-density spacers. Absorber panels can be made that are much more durable than the foam materials mentioned above.

Finally, hybrid absorbers can be designed and fabricated using combinations of the materials mentioned above. The design requires a knowledge or tabulation of the electromagnetic properties of the materials, including those of the spacers and the bonding agents used to hold the layers together. Typically, we seek a broadband design, and, consequently, these electromagnetic properties should be available over the intended range of frequencies.

In addition to absorber materials, only one other method has been found practical for radar cross section control: shaping. Shaping is a more useful tool for systems in the design stage, but only if RCS threat sectors can be identified. The technique is virtually useless for retrofitting to existing vehicles because of the cost and probable impairment of the vehicle's mission. For example, moving the jet intakes of an aircraft from below the wings to above the fuselage is out of the question for operational aircraft.

If all viewing angles are equally likely, shaping offers no advantage. This is because of a rule of thumb which states that a reduction in the RCS at one aspect angle is always accompanied by an enhancement at another. For most airborne systems, however, RCS control is of more importance in the angular forward cone than in the broadside sectors. Therefore, shaping can be used to shift large returns from the nose-on region to the broadside sectors. Highly swept wings are an example, but we should appreciate that shaping is best performed in the system design stages.

1.6 MEASUREMENTS

As in any other design requirement, RCS measurements are necessary to verify anticipated performance as well as to evaluate design approaches. In addition, measurements are required for the evaluation of absorber designs. Consequently, several chapters are devoted to the discussion of measurement tools and facilities. Chapter 10 focuses on material measurements, and Chapters 11 through 13 describe RCS measurement techniques and facilities.

Absorber materials can be tested in several ways, one of which is with a coaxial or wavequide transmission line. This technique is advantageous in that the equipment can be set up on a laboratory bench and the fields to be measured are totally confined within the transmission line itself. Basic measurement techniques require that the voltage reflection coefficient of a material sample be measured, and this requires either multiple measurements with different terminations or a slotted section (a length of transmission line with a small probe inserted into the line to sample the field). The amplitude of the internal standing wave pattern gives the magnitude of the reflection, and the shift of the pattern one way or the other (when the sample is inserted in the line) gives the phase angle.

In the days when equipment was not as advanced as it is now, the measurements had to be performed manually at each frequency of interest. The introduction of time domain reflectometry and swept signal sources has made the task much easier and much faster to perform. In either event, the magnitude and phase of the reflection coefficient can be used, along with the sample thickness, to calculate the elctromagnetic properties of the sample.

Transmission line techniques also have disadvantages: samples must be very carefully machined to fit inside the waveguide or coaxial line, and there is a risk of creating higher-order field modes within the sample. Improperly fitting samples and higher-order modes lead to erroneous results. Moreover, if the material is not homogenous, there is the risk of characterizing a sheet or panel of material on the basis of inhomogenities existing in the small sample.

These errors can be reduced by testing or measuring large panels of absorbing materials, but it cannot be done in a transmission line. In one technique, the famous NRL arch method, the absorber panel is illuminated by a small horn. The system is calibrated by replacing the absorber panel with a metallic plate. Two horns, one for transmission and one for reception, are mounted on an arch centered over the test panel, and measurements can be conducted for a variety of angles of incidence and reflection.

It is difficult to measure the relative phase of the reflected signal and, consequently, the arch technique is typically used only for characterizing the amplitude of the reflection. However, the natural isolation between the transmitting and receiving horns makes it easy to use swept frequency sources and, consequently, the measurements can be made rapidly and the results displayed as a function of frequency minutes after the measurements.

However, the NRL arch technique may place the test panel in the near field, yielding reflectivity values that are lower than might be measured under far field conditions. An alternative is to utilize an RCS measurements range where far field conditions can usually be met. The material sample is mounted on a flat plate which is in turn installed upon a target support column in a vertical position. The plate is rotated through 360 degrees in azimuth, and the broadside flash from the back side of the metal plate serves to calibrate the broadside flash from the absorber covered side. Unless special instrumentation is available, the measurements must be repeated for each frequency of interest.

An RCS range serves a much more valuable purpose than merely evaluating absorber samples. The RCS range is a tool for engineering and scientific studies, as well as testing the performances of various design approaches, or simply constructing a data base for a collection of targets or target conditions. Such measurements are useful in characterizing the actual radar signatures of operational systems in dynamic flight scenarios.

As with any other engineering tool, RCS ranges have their particular virtues and shortcomings and they come in all sizes, shapes, and geometries. Early RCS measurement facilities were indoor anechoic chambers, although some measurements were made by training a radar on an outdoor target. Currently, there are a large number of indoor and outdoor ranges in operation throughout the US and some of them are described in Chapters 12 and 13.

Indoor ranges suffer limitations in the sizes of the targets that can be measured, while outdoor ranges suffer as much as 35% down-time due to unfavorable weather conditions. Although the indoor ranges offer protection against the weather and unauthorized exposure of the target, outdoor ranges can often measure full-scale targets under far field conditons. Most, but not all, outdoor ranges take advantage of the ground plane effect, which actually enhances system sensitivity.

A problem common to both indoor and outdoor ranges is how to expose the target to the incident radar beam on an "invisible" target support. As shown in Chapter 11, no target support mechanism is actually invisible, although some are acceptably so. Plastic foam has been a traditional material for target supports, but a recent improvement is the absorber-covered metal pylon. Nonetheless, the pylon is not without its limitations, one of which is the necessity to machine a hole in the target for mounting attachments. The hole itself can introduce undesired target reflections.

Outdoor ranges invariably use pulsed radar instrumentation, while indoor ranges use CW and FM/CW systems. The measured patterns are essentially the same in either case, provided that the pulse width of the outdoor system is long enough to comfortably bracket the target. In each case it is possible with stepped frequency synthesizers to collect coherent signature data that can be processed to create radar imagery. As we shall see, radar imagery is of great diagnostic value and is another tool for use in RCS control.

1.7 SUMMARY

In this introductory chapter, we have outlined some of the major topics that will be covered in subsequent chapters. This introductory material will continue with the background information on radar systems presented in Chapter 2. The remainder of this book focuses on four major areas:

- electromagnetics: Ch.3-5
- RCS phenomenology: Ch. 6,7
- radar absorbers: Ch. 8-10
- RCS measurements: Ch. 11-13

CHAPTER 2

RADAR FUNDAMENTALS

M. T. Tuley

2.1 INTRODUCTION

While the study of radar cross section and RCSR is interesting from a purely academic point of view, we assume that the reader has a practical application in mind. That application is typically the prediction of a radar's performance against a given target, and then modification of the target RCS to reduce the performance of the radar.

The purpose of this chapter is to provide a brief survey of radar fundamentals which can be used to put the RCS and RCSR problems in context. Obviously, only some of the most basic points concerning radar can be covered in a single chapter, and the reader who seeks more detail can find it in any one of a number of books on the general topic of radar [1, 2, 3, 4, 5]. Here, after a brief history, several types of radars are described, with emphasis on operational rather than instrumentation systems. The radar range equation is developed, and its implication for RCSR is examined. Detection theory is then briefly discussed, and, finally, electronic countermeasures (ECM) techniques are outlined, with an emphasis on the effects of RCS control on ECM effectiveness.

2.2 HISTORY OF RADAR DEVELOPMENT

Radar did not come to the forefront until World War II, when tremendous strides were made in both the theory and practice of radar technology. The earliest account of the reflection of radio frequency waves from metallic and dielectric bodies was given by Hertz, who in 1886 used a 450 MHz spark-gap transmitter and receiver to test Maxwell's theories. The first detection of what might be called a military target by radar was achieved by Christian Hulsmeyer, a German engineer, who demonstrated a ship detection device to the German Navy in 1903. However, the range of the device was so limited that little interest was generated.

The earliest US radar detection work was carried out by Taylor and Young of the Naval Research Laboratory (NRL). In late 1922 they used a continuous wave (CW) interference radar operating at 60 MHz to detect a wooden ship. The first aircraft detection, also made using a CW interference radar, was made in 1930 by Hyland of NRL.

Early radar work, both in the US and abroad, often used a CW transmitter. These radars detected targets based on modulation of the received signal caused by the Doppler shifted reflection from the target beating with the direct signal from the transmitter. Low frequencies were used because high power transmitters were not available at the higher frequencies. These CW radars detected the presence of a target, but no range information could be extracted. Also, at the low frequencies used, little angular information was available. To allow range measurement, pulsed systems were required.

A pulsed radar operating at 28 MHz and using 5 μs pulses was developed at NRL and tested in early 1935. The tests were unsuccessful, but the radar was subsequently modified, and it detected its first target echo in 1936. Shortly after that, in late 1936, the US Army Signal Corps tested its first pulsed radar. The Army also developed the first operational anti-aircraft fire control radar, the SCR-268, which was fielded in 1938.

The British successfully demonstrated a pulsed radar at 12 MHz in 1935, obtaining detection ranges of more than 40 miles against a bomber aircraft. Both the US and Britain were aware of the reduction of the physical size of equipment and angular resolution advantages available at higher frequencies, and groups in both countries were working at 200 MHz by the late 1930s.

The pivotal event allowing the practical development of microwave radar was the invention of the cavity magnetron by Randall and Boot in Britain. The magnetron is a self-excited crossed-field (i.e., the magnetic and electric fields are perpendicular) oscillator whose frequency of operation is determined by the dimensions of a regular series of holes and slots cut into a cylindrical anode structure surrounding a cylindrical cathode. The first cavity magnetron produced a peak pulse power of 100 kW at 3 GHz, a power level much greater than had previously been achieved at those frequencies. In late 1940, Britain and the US began cooperative efforts in the radar area. The focal point for US radar development efforts during World War II was the Radiation Laboratory, established at MIT in November 1940. The initial staff of 40 had grown to about 4,000 by mid-1945, and their activities are documented in a 28-volume set of books (commonly known as the Rad Lab Series [6]), which even now, 40 years later, provide a valuable reference on radar fundamentals.

Improvements in radar technology have been enormous since World War II. In the transmitter area, the development of the high power traveling wave tube (TWT), millimeter wave power tubes, and solid-state microwave sources have been important. In the receiver area, solid-state technology has improved mixers and allowed development of low noise amplifiers. In the antenna area, large scale phased arrays have become practical. In signal processing, and in every other area of radar, the advent of the small, fast digital computer has made practical radar techniques which could not otherwise have been considered.

2.3 RADAR FREQUENCY BANDS

A radar may be defined as a device which transmits an electromagnetic wave and detects objects by virtue of the energy scattered from them in the direction of the receiver. In that regard, a radar could operate at any frequency. Practically speaking, the devices which we normally designate as radars are limited to a relatively narrow slice of the electromagnetic spectrum for reasons of availability of components, propagation effects, angular resolution requirements, target scattering charateristics, and a host of other concerns. Generally, radars fall within the frequency limits given by the standard radar bands in Table 2-1. These bands cover a frequency range from 3 MHz to 300 GHz, but the greatest number of operational radars fall within what are commonly called the microwave frequency bands, designated in the table as L, S, C, X, and K_u.

The radar band designations began in World War II as a security measure, but they have since come into standard usage. In general, the bands are divided so that the power sources, propagation effects, and target reflectivity are similar for frequencies within a band, but may be very different from one band to another. Table 2-2, adapted from reference [7], provides typical uses for the most commonly employed bands. VHF and UHF frequencies are generally used for very long range surveillance, because of their ability to provide over-the-horizon coverage, and due to favorable target and clutter characteristics. However, angular accuracy at these low frequencies requires very large antennas, so these bands are only used when their unique propagation characteristics are required. Search radars and tracking radars are most often found in one of the higher radar bands, with S, C, and X bands the most prevalent. K_u, K_a, and higher frequencies are finding increasing use in mapping, fire control, and missile guidance applications, because small beamwidths can be obtained using reasonably sized antennas.

Table 2-1
Radar and ECM Bands

STANDARD RADAR BANDS*		ELECTRONIC COUNTERMEASURES BANDS**	
Band Designation (1)	Frequency Range (MHz)	Band Designation	Frequency Range (MHz)
HF	3 – 30	A	0 – 250
VHF (2)	30 – 300	B	250 – 500
		C	500 – 1000
UHF (2)	300 – 1000	D	1000 – 2000
		E	2000 – 3000
L	1000 – 2000	F	3000 – 4000
S	2000 – 4000	G	4000 – 6000
		H	6000 – 8000
C	4000 – 8000	I	8000 – 10,000
X	8000 – 12,000	J	10,000 – 20,000
K_u	12,000 – 18,000	K	20,000 – 40,000
		L	40,000 – 60,000
K	18,000 – 27,000	M	60,000 – 100,000
K_a	27,000 – 40,000		
Millimeter (3)	40,000 – 300,000		

Notes: (1) British usage in the past has corresponded generally, but not exactly, with the letter-designated bands.

(2) The following *approximate* lower frequency ranges are sometimes given letter designations: P band (225 – 390 MHz); G band (150 – 225 MHz); and I band (100 – 150 MHz).

(3) The following *approximate* higher frequency ranges are sometimes given letter designations: Q band (36 – 46 GHz); V band (46 – 56 GHz); and W band (56 – 100 GHz).

*From IEEE Standard 521-1976, 30 November 1976.
**From AFR-55-44(AR105-86, OPNAVINST 3430.9B, MEO 3430.1), 27 October 1964.

Table 2-2
Radar Frequency Bands and General Usages

Band Designation	Frequency Range	General Usage
VHF	50 – 300 MHz	Very Long Range Surveillance
UHF	300 – 1000 MHz	Very Long Range Surveillance
L	1 – 2 GHz	Long Range Surveillance, Enroute Traffic Control
S	2 – 4 GHz	Moderate Range Surveillance, Terminal Traffic Control, Long Range Weather
C	4 – 8 GHz	Long Range Tracking, Airborne Weather Detection
X	8 – 12 GHz	Short Range Tracking, Missile Guidance, Mapping, Marine Radar, Airborne Intercept
K_u	12 – 18 GHz	High Resolution Mapping, Satellite Altimetry
K	18 – 27 GHz	Little Used (Water Vapor Absorption)
K_a	27 – 40 GHz	Very High Resolution Mapping, Airport Surveillance
Millimeter	40 – 100+ GHz	Experimental

Table 2-1 also provides the electronic countermeasures (ECM) band designations listed in Department of Defense instructions. While IEEE Standard 521-1976 specifically notes that the ECM band designations are not consistent with radar practice, use of the ECM band descriptors in radar work has become pervasive. Thus, a designer of a radar absorber would be well advised, when asked about the performance of his product at C band, to be sure the questioner means 4 to 8 GHz and not 500 to 1000 MHz.

2.4 BASIC RADAR SYSTEM ELEMENTS

Any functional radar system must contain at least four basic elements. These are a transmitter, an antenna or antennas, a receiver, and an indicator. As illustrated in Fig. 2-1, the transmitter produces a radio frequency (RF) signal which is delivered to an antenna to be beamed toward the target. A portion of the transmitted electromagnetic wave is scattered by the target in the direction of the receiving antenna, where it is collected and delivered to a receiver. The final component of the radar system is an indicator, which conveys target information to the radar operator. Modern radars often contain sophisticated signal processors which enhance performance. However, such signal processors can normally be lumped within the receiver or indicator subsystems.

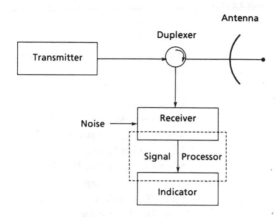

Figure 2-1 Basic radar system elements

2.4.1 Transmitters

Because radar is an active device, it must have a means of generating the desired RF signal, and signal generation is the function of the transmitter. The configuration chosen for the transmitter is always driven by the details of the radar application. Power level (peak and average), coherent *versus* noncoherent operation, duty cycle, size and weight limitations, frequency, and instantaneous bandwidth requirements are all factors which must be considered in transmitter design. In addition to discussions of transmitters in the general radar references listed previously, Ewell[8] provides an excellent treatment on the design of radar transmitters.

The history of radar development has been strongly influenced by the availability and capability of RF sources. The advent of microwave radar in World War II was a direct result of the invention of the cavity magnetron. Advanced coherent radar techniques could not be seriously considered before the development of high power traveling wave tubes. Continued improvement in extended interaction oscillators (EIOs) and amplifiers (EIAs), along with novel magnetron configurations appear to be key in future millimeter wave systems development. At the other end of the radar spectrum, solid-state source development has made the implementation of extremely large phased array radars at frequencies below S band practical.

The transmitter type and the source selected for a radar depend on the power required, the allowable cost, and the details of the application. The required output power is often derived directly from a power oscillator, and the magnetron is the prime example of such a device. Pulsed magnetrons can be built with output powers in excess of 10 MW. They require fairly simple modulators, are efficient, and are relatively inexpensive. Magnetron disadvantages include a low duty cycle, narrow bandwidth, and difficulty in coherent operation.

Coherent radars are more often implemented using a master-oscillator-power-amplifier (MOPA) chain. The master oscillator is often crystal controlled to provide maximum frequency stability. TWT amplifiers are typically used as the low and medium power amplifiers in a MOPA chain, with either a high power TWT, crossed-field amplifier (CFA), or klystron used in the final amplifier stage. The major advantages of a MOPA transmitter are coherent signal generation and extreme flexibility in signal design and coding. Disadvantages compared to magnetron transmitters include size, weight, cost, and transmitter complexity.

2.4.2 Antennas

The basic function of the radar antenna is to couple energy between free space and the transmission lines inside the radar. The transmitting antenna must take the RF energy from the transmitter and convert it into a beam of the desired shape illuminating the desired volume of space. The receiving antenna must receive electromagnetic energy from specific directions and guide that energy into the receiver. In radar, the same antenna is commonly used for both transmission and reception. In that case a duplexer must also be used to switch the antenna between the transmitter and receiver, and to protect the receiver from the transmitted pulse.

Radar antennas have taken many forms over the years. There are a number of excellent general antenna texts which also cover antennas suitable for radar use [9, 10, 11], and references [12, 13] provide complete treatments of radar antenna design and analysis. Perhaps the most commonly used radar antenna has been the paraboloidal reflector antenna. A number of parabolic configurations can be used, depending on the desired beam shape. A paraboloid (circular parabola or dish antenna) produces a pencil beam, which is often used in tracking applications. Sections of parabolas, or of parabolic cylinders, can be used to produce asymmetrical beams.

Antenna arrays, consisting of a number of individual elements, also have a long history of use in radar. Early, low frequency, search antennas often used large arrays of dipoles to provide a large enough aperture to obtain satisfactory angular discrimination. More recently, phased arrays have become a significant trend in combination search and track antennas. Phased arrays offer the advantage of rapid electronic beam steering, high power capability, and adaptive control of the radiation pattern [14]. These advantages are obtained, however, with the attendant disadvantages of higher cost and a more complex system. Radiating elements in phased arrays are typically waveguides, waveguide slots, or dipoles. Beam steering can be obtained by changing the transmitter frequency in frequency scanned arrays (this technique is often used in arrays which can scan mechanically in azimuth and electronically in elevation), by using phase shifters to change the relative phase between the array elements, or by control of the amplitude and phase of individual transmitting modules at each element for some of the modern solid-state transmitters.

Whatever the method used to obtain the desired antenna pattern, the achieved pattern and antenna gain (i.e., the ratio of the power radiated in a particular direction by an antenna to that radiated in the same direction by a perfectly efficient isotropic antenna) are a function of the size of the antenna aperture and the illumination function. Typically, beamwidth and gain must be traded off against allowable sidelobe levels. Table 2-3 illustrates the relative gain, half-power beamwidth, and first sidelobe intensity for a number of common aperture illumination functions, listed in order of decreasing level of the first sidelobe. Note that uniform illumination of the aperture provides the most efficient distribution [15], with the smallest half-power beamwidth and highest relative gain, but a uniformly illuminated aperture suffers from very high first sidelobes (only 13.2 dB below the mainlobe for a rectangular aperture, and 17.6 dB for a circular aperture).

Table 2-3 provides a good method of estimating antenna beamwidths, but it is also convenient to be able to estimate antenna gain based on beamwidth. One rule of thumb [16] for relatively narrow beam reflector antennas of typical efficiency estimates the gain G as

$$G = \frac{27,000}{\theta_1 \theta_2} \tag{2-1}$$

where θ_1 and θ_2 are the two principal plane half-power beamwidths (i.e., angular width between the points 3 dB down from the peak) in degrees. Gain is often expressed in dB as $10 \log_{10} G$.

2.4.3 Receivers

The purpose of the radar receiver is to take the RF energy delivered by the antenna and to process it into a form suitable for the chosen indicator. Radar receivers vary in complexity from simple crystal detectors to multiple conversion superheterodyne receivers with cryogenically cooled low-noise RF preamplifiers.

However, by far the most common receiver is the single conversion superheterodyne system, shown in Fig. 2-2. While an RF amplifier is indicated as

Table 2-3
Antenna Pattern Parameters for Various Aperture Distributions
(λ = wavelength, D = aperture width)

Aperture Shape	Illumination Type ($x \leq 1$)	Relative Gain	Half-Power Beamwidth (deg)	First Sidelobe (dB below maximum)
Rect.	$f(x) = 1$	1.000	$51 \lambda/D$	13.2
Rect.	$f(x) = 1 - (1 - 0.8)x^2$	0.994	$52.7 \lambda/D$	15.8
Rect.	$f(x) = 1 - (1 - 0.5)x^2$	0.970	$55.6 \lambda/D$	17.1
Circ.	$f(r) = 1$	0.865	$58.4 \lambda/D$	17.6
Rect.	$f(x) = 1 - x^2$	0.833	$66 \lambda/D$	20.6
Rect.	$f(x) = \cos(\pi x/2)$	0.810	$69 \lambda/D$	23.0
Rect.	$f(x) = \cos^2(\pi x/2)$	0.667	$83.1 \lambda/D$	32.0
Rect.	$f(x) = \cos^3(\pi x/2)$	0.575	$95 \lambda/D$	40.0

the input stage, in practical systems the mixer often acts as the first stage. In noncoherent systems, a reflex klystron or solid-state source is typically used as the local oscillator. Coherent systems use a stable local oscillator (STALO) locked to the master oscillator. The mixer stage (also called the first detector) translates the RF signal to an intermediate frequency (IF) for further amplification and processing. The IF chosen is dependent on signal bandwidth and the requirements for further processing. For narrowband systems, 30 MHz or 60 MHz are typical intermediate frequencies. Wideband systems (e.g., the AN/APS-116 radar which frequency chirps over a 500 MHz bandwidth[17]) may have a first IF in L band or above.

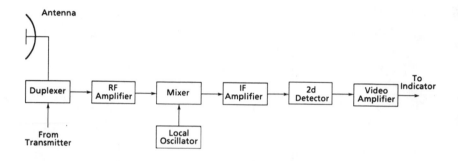

Figure 2-2 Superheterodyne Receiver

The advantage of the superheterodyne receiver in radar applications is that amplification and filtering can be more easily performed at IF than at RF. Since the RF bandwidth is typically wide, the IF bandwidth normally sets the noise bandwidth of the radar receiver. If a matched filter receiver is to be employed, it is normally implemented at IF. It is also in the IF section that automatic frequency control (AFC), instantaneous automatic gain control (IAGC), or sensitivity time control (STC), which is a commonly used method to prevent close-in clutter return from saturating the display, are usually implemented.

A second detector and video amplifier typically make up the final receiver stages. The second detector extracts modulation at video frequencies from the carrier. The video amplifier is used to provide a sufficient signal level to drive the chosen display.

2.4.4 Indicators

The indicator used on most radar receivers is a cathode ray tube (CRT) which displays target position in some coordinate system, but this is not

always the case. For instance, police speed radars typically provide only audio and a digital speed readout as an indicator. An area surveillance radar might use an alarm as an indicator which is triggered whenever a target is detected within the search perimeter. Nevertheless, most radars employ CRTs as indicators, and the remaining discussion focuses on the types of CRT presentation normally employed.

There are two basic parameters which can be controlled for information display on a CRT, beam deflection and beam intensity. On a deflection modulated display (an A-scope, which displays received signal level on the vertical axis *versus* time on the horizontal axis is one example), a target is indicated by the deflection of the electron beam as it writes across the tube. An intensity modulated display, such as the plan position indicator (PPI), indicates a target by a luminous spot on the face of the tube, while non-target areas remain dark. Figure 2-3 provides a number of the formats which have commonly been used in radar displays.

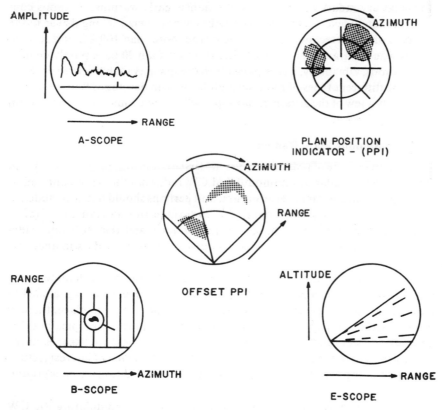

Figure 2-3 Various Display Types

Strides in signal processing and components have significantly altered what can be done in the display area. Highly synthetic presentations, heads up displays, graphics, color, touch panels or track balls, and other features have greatly improved the presentation to, and control by the radar operator. The ability to computer control and generate displays has added significant flexibility to formatting and presenting radar data.

2.5 EXAMPLES OF RADAR SYSTEMS

Radars are built in a wide variety of types and for a large number of different applications. The early areas of use for radars in civilian applications were in navigation and traffic control (air and marine). However, as radar became more widespread, and as technology improved, weather detection and tracking, automobile speed detection, collision avoidance, and buried object detection also became important as civilian radar applications.

Most of the civilian applications also apply in the military arena. In addition, military radars are used for surveillance, early warning, weapons control, and intelligence functions. The military has been a strong driver in the advance of radar technology. In fact, of the more than 100 US and foreign radars whose parameters are listed in [4], fewer than 20 have purely civilian designations. Obviously, it is impossible in the space available here to describe all of the different types of radars which have been developed and deployed. Thus, only a few of the generic radar types which have found wide application are briefly discussed.

2.5.1 Continuous Wave Radars

The simplest type of radar, and the type first demonstrated by Hulsmeyer in 1903, is the CW radar. An unmodulated CW radar has the significant limitation of providing no range information, and perhaps should not be included in a class of objects whose name stands for "radio detection and ranging." Nevertheless, CW systems are of historical importance, and they still find widespread use in police speed radar and as instrumentation radars in anechoic chambers.

Figure 13-11, in the chapter on indoor RCS measurement chambers, is a block diagram of a CW instrumentation radar. The nulling loop is included so that the background level can be cancelled without a target in place, thereby increasing the sensitivity of the system. A police radar, which is typically a CW homodyne system and uses transmitter leakage as the local oscillator signal to the mixer, does not require a nulling loop because the signal of interest (return from a speeding vehicle) is Doppler shifted from the stationary background echo.

The lack of range information can be overcome by modulating the CW signal, and thus providing a "timing mark." Frequency and phase coding are the two techniques normally employed in modulating CW radars.

Frequency modulated continuous wave (FM-CW) radars vary the frequency of the transmitted signal and measure the range based on the frequency difference between the instantaneous transmitted and received signals. Typical modulations include triangular, saw-tooth, and sinusoidal. Current applications of FM-CW include altimeters, tracking radars, and high resolution instrumentation systems.

Another method of providing a timing mark is by phase coding the CW signal. Either a binary or a polyphase code can be used. In binary coding the signal is divided into segments, with a +1 represented by 0° phase (in-phase) and a 0 represented by a 180° phase shift (out-of-phase), with respect to the carrier. The return signal can then be correlated with a delayed version of the transmitted code to produce range discrimination approximately equal to half the bit length.

2.5.2 Pulsed Radar

As indicated above, obtaining range information requires that the transmitted signal be coded to allow an elapsed time measurement to be made between some reference time and the time a signal is received back from the target. Frequency and phase coding in CW systems have been discussed, but by far the most common coding technique is the amplitude modulation employed in the pulsed radar, a block diagram for which is shown in Fig. 2-4. A timing device (clock) triggers the transmitter and simultaneously begins a range strobe on the indicator. The transmitted pulse leaves the antenna and propagates at the speed of light. When the pulse reaches a target, it is reflected and a portion of the energy returns to the radar, where it is detected and displayed. The range R to the target is given by the two-way transit time Δt and the speed of light c as

$$R = \frac{c\Delta t}{2} \tag{2-2}$$

A radar typically transmits a continuous train of pulses at some pulse repetition frequency (PRF). A return from an earlier pulse received after a subsequent pulse was transmitted would be interpreted to be coming from a target at a much shorter range, producing a range ambiguity. Because the primary task of a search radar is to provide the range and angular coordinates of a target, ambiguities in range are undesirable. In addition, a range ambiguous target will be displayed at a shorter range, and may thereby be masked by close-in clutter return. The maximum unambiguous range which a radar can measure is given by

$$R_{unamb} = \frac{c}{2f_r} \tag{2-3}$$

where f_r is the PRF in Hz.

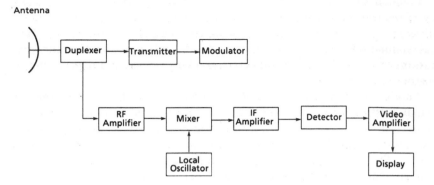

Figure 2-4 Pulse radar block diagram

In addition to their wide operational employment, pulsed radars are often used as instrumentation radars, both in field situations and on outdoor ranges. One major advantage of the pulsed radar is that returns over a specific range interval can be selected and recorded, and unwanted signals from other range intervals can be discriminated against based upon their time of arrival. Simple pulsed radars are more difficult to use in indoor measurements because the short ranges involved require very short receiver recovery times.

2.5.3 Pulse Compression

In a sense, the simple pulsed radar provides an amplitude code on a CW signal which allows range discrimination. The range resolution provided is a function of the pulse width and is given by

$$\Delta R = \frac{c\tau}{2} \tag{2-4}$$

where τ is the pulsewidth. Thus, a one nanosecond (1 ns) pulse would provide a 6 in range resolution.

For a given unambiguous range and a given peak power, shorter pulses imply a lower average power, and therefore shorter detection ranges in noise. One way to overcome this difficulty and provide better range resolution, while maintaining a high average power, is to further code the pulses, and a number of techniques are used to do so under the general name of pulse compression.

One of the common methods of pulse compression is to apply a linear frequency sweep or "chirp" to each pulse as it is transmitted. Figure 2-5 is the block diagram of the AN/APS-116 radar, which uses high range resolution for discrimination against sea clutter [17]. A 0.5 μs pulse is transmitted which contains a 500 MHz linear frequency sweep. In the receiver, a dispersive filter

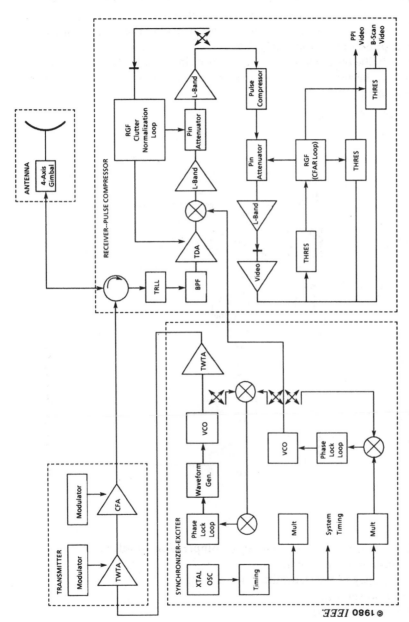

Figure 2-5 Block diagram of AN/APS-116 pulse compression radar system (adapted from Smith and Logan [17])

© 1980 IEEE.

(i.e., a delay line with a variable delay *versus* frequency which causes the different portions of the received pulse to reach its output at nearly the same time) with a delay characteristic which is the inverse of the chirp, compresses the output pulse to approximately 3 ns.

Figure 2-6 illustrates the waveforms present in a linear FM pulse compression system. The achievable compressed pulsewidth τ_c, is related to the swept bandwidth B by [18]:

$$\tau_c \approx 1/(f_2 - f_1) = 1/B \tag{2-5}$$

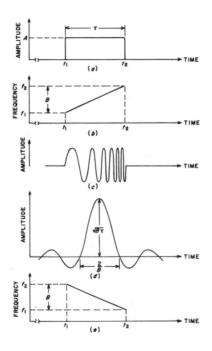

Figure 2-6 Frequency-modulation pulse-compression waveforms:
  (a) Transmitted waveform;
  (b) Frequency of the transmitted waveform;
  (c) Time waveform;
  (d) Compressed pulse;
  (e) Output characteristics of the pulse compression filter. (adapted from Skonik [1])

Note from the figure that the compressed pulse has an amplitude which is proportional to $\sin(\pi Bt)/\pi Bt$, and thus has 13.2 dB first sidelobes in the time (or range) dimension. Often, amplitude weighting of the received pulse is done to reduce the sidelobe levels.

Any technique which effectively increases the bandwidth of the radar output can be used to enhance the range resolution. FM-CW provides the high duty cycle and high average power of the CW radar with the added advantage of range resolution. In fact, radar altimeters are often FM-CW systems because of the ease of obtaining the required range resolution with a low power transmitter (a few watts output is typical).

Other methods of pulse compression include intrapulse phase coding and, for instrumentation systems, frequency stepping. Stepped frequency systems can obtain extremely high resolution by taking data over very wide total bandwidths. For instance, the Georgia Tech Compact Range is now instrumented to take stepped frequency data over a 2-18 GHz band. The frequency domain data (phase and amplitude) can be stored and then transformed to the time (range) domain using the fast Fourier transform (FFT). Theoretically, the 16 GHz bandwidth could give about three-eighths of an inch range resolution, although a Hanning amplitude weighting, which is used to reduce the time sidelobes, somewhat degrades the actual resolution achieved.

2.5.4 Moving Target Indication (MTI) and Pulse Doppler Radar

The Doppler shift caused by the radial motion of a target can be used as an additional discriminant for moving targets in a fixed target and clutter background, and MTI and Pulse Doppler systems make use of the Doppler discriminant. The designation of a radar as an MTI system rather than a Pulse Doppler system is rather arbitrary, but, historically, Doppler radars which are ambiguous in range but unambiguous in velocity have been called Pulse Doppler radars, while radars which are unambiguous in range are designated as MTI systems.

The major requirement for a Doppler radar is a coherent receiver. This implies that the local oscillator can maintain phase coherence with the transmitted pulse, so that the Doppler shift can be detected. Coherence is normally obtained by use of a MOPA, as described in sec. 2.4.1, but it can be obtained with a coherent-on-receive system using a coherent local oscillator (COHO) phase locked to the magnetron output pulse.

A simple method of implementing MTI is by use of a delay line canceller. The simplest version of the delay line canceller utilizes a single delay line and single loop, as shown in Fig. 2-7. The coherent video signal is split, one-half the signal is delayed by one pulse repetition interval, then subtracted from the undelayed signal. Targets at a constant range will experience no phase shift pulse-to-pulse, and therefore will cancel in the subtractor. The returns from

moving targets, on the other hand, will experience a phase shift due to changing range and will not cancel. A disadvantage of the delay line canceller is that
radial velocities which produce a 360-degree phase shift between pulses (called
blind speeds) will also cancel, so that at velocities given by

$$V_n = \frac{n\lambda f_r}{2} \quad n = 1,2,3\ldots \tag{2-6}$$

no output will occur. In addition, some signal loss will be incurred through the
canceller for all velocities except those half-way between the blind speeds, as is
also illustrated in Fig. 2-7. Multiple loop cancellers may be used to give a
sharper notch at the blind speeds, and PRF jitter can be used to eliminate
blind speeds.

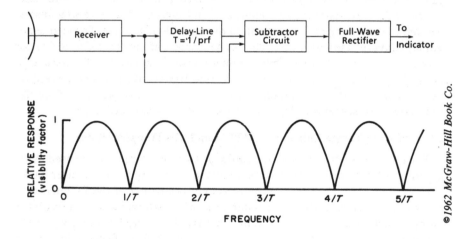

Figure 2-7 Single pulse delay line canceller (adapted from Skolnik [1])

Modern Pulse Doppler systems, which operate at PRFs high enough so
that blind speeds in the velocity range of interest are eliminated, typically
employ digital processors rather than delay line cancellers. Usually, a pulse
burst (generally at least eight pulses) is transmitted, and the return is Fourier
transformed in the receiver signal processor. The FFT output provides signal
levels as a function of frequency and allows discrimination between targets
and clutter. One advantage of the digital systems is that computer processing
in airborne radars allows much more effective suppression of the non-zero
Doppler clutter caused by radar motion than can be achieved in analog systems.

The improvement of MTI and Pulse Doppler radars is an area of continuing research. The advent of digital processors capable of providing the required FFT calculations in real time, phased arrays which allow electronic beam stepping to be accomplished, and MOPA chains providing vastly improved transmitter stabilities have allowed orders of magnitude improvement in the ability, particularly of airborne look-down radars, to suppress clutter. Thus, targets can be detected which without Doppler processing would be buried in the background clutter. For further information on MTI and Pulse Doppler systems, reference [19] provides an excellent compilation of important papers in the area.

2.5.5 Tracking Radars

No real distinction has been made in the above discussion concerning radar function, but in most cases the implicit assumption has been that the radar is used for detection. However, another important radar function is target tracking, and a number of special techniques have been developed to allow radars to automatically track a target. Generally, tracking radars are separated into two groups, continuous trackers and track-while-scan (TWS) systems.

TWS systems have become more popular with the advent of enhanced computing capabilities. Typically, continuous trackers and TWS systems both implement automatic range tracking electronically with a split-gate tracker. Two adjacent signal samples of a chosen width, called an early gate and a late gate, are generated, and the voltages in the two gates are subtracted to provide an error signal which drives the center of the gates to the range centroid of the target return. Similar tracking must also be done in angle for TWS systems, and several methods are available to compute the target angular position. However, the fact that the target is not continuously illuminated means that some means of predicting target position for the next scan must be provided, and α-β or Kalman filters are often used. The application of modern estimation theory to the smoothing and prediction of radar tracks, combined with improved computing techniques, has greatly enhanced TWS system performance.

However, even with the increasing capability of TWS systems, continuous tracking systems are still used most often for weapons guidance or homing applications. Continuous trackers depend on the generation of an angular error signal which is used to move the antenna beam to the perceived center of the target. Tracking radars have many forms, but the major types of continuous trackers in use today are conical scan and monopulse systems. Each is discussed briefly below.

A simple method of obtaining the direction of a target in one dimension, relative to some reference, is to switch the beam position symmetrically to either side of the reference and to measure the amplitude of the target return in each position. Such a tracking technique is called sequential lobing, and the target is centered when the outputs at the two positions are equal. The technique can be extended to two dimensions by using two sets of beam positions orthogonal to one another.

A logical extension of sequential lobing, called conical scan, rotates the radar beam in a circle about the reference direction and continuously measures the target return. The low frequency portion of the detected output of such a system will be a sinusoid at the rotation frequency. The amplitude of the sine wave is proportional to the total angular error, and the phase, relative to the scan position, indicates the direction of the error. Conical scan systems are relatively easy to implement, but they suffer several disadvantages. The first is that pulses over the complete conical scan period must be processed to determine the track error, and changes in target RCS over the scan period will induce track errors. A second disadvantage, related to the first, is that a jammer which is modulated at the conical scan frequency will also induce error signals into the receiver and can cause loss of track. One technique to counter jamming is to deny knowledge of the scan frequency. This is accomplished with conical-scan-on-receive-only (COSRO) systems, for which the illuminating beam does not scan and only the receiving pattern rotates.

A technique to reduce tracking sensitivity to target amplitude fluctuations and jamming is to provide an error signal based on a single pulse, and such a technique is called monopulse. In monopulse systems, simultaneous lobes are created and a comparison is made, either in amplitude or phase, between them. Figure 2-8 provides a block diagram of an amplitude monopulse radar, and Fig. 2-9 illustrates the beams used and the resulting error voltage. As can be noted, the individual beams are squinted relative to each other. A sum beam is transmitted, and sum and difference beams are formed by the receiver. When normalized by the sum signal (Σ), which takes out the effects of variations in target RCS, the amplitude and phase of the difference (Δ) return give the magnitude and direction of the error. Thus, the monopulse error voltage ϵ is given by

$$\epsilon = k \ \frac{\Delta}{\Sigma} \qquad\qquad\qquad (2\text{-}7)$$

where k is simply a proportionality constant.

Phase comparison monopulse is less often used than amplitude monopulse, because it is less efficient for a given aperture size. It is basically an interferometric technique which compares the phases of the signals at the two antenna feeds to calculate target direction. Beam squint is not required for phase monopulse, so it is often used in planar antenna configurations to reduce the complexity of the antenna design. In any event, monopulse requires either a coherent multichannel receiver or multiplexing of the sum and difference signals, and thus is more expensive to implement than conical scan. However, the relative immunity of monopulse to simple jamming techniques has made it the preferred modern tracking technique [20].

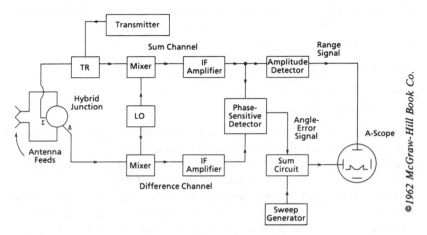

Figure 2-8 Block diagram of an amplitude-comparison-monopulse radar (reprinted from Skolnik [1])

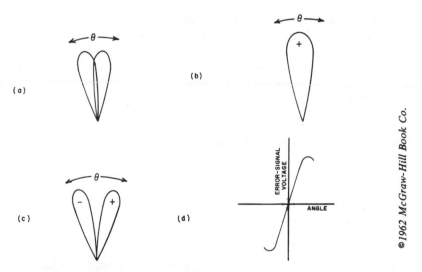

Figure 2-9 Amplitude monopulse antenna patterns and error signal:
    (a) Squinted antenna patterns;
    (b) Sum pattern;
    (c) Difference pattern;
    (d) Error signal. (adapted from Skolnik [1])

2.6 THE RADAR EQUATION

The radar range equation provides the most useful mathematical relationship available to the engineer in assessing both the need for, and the resulting effectiveness of, efforts to reduce radar target cross section. In its complete form [21, 22], the radar equation accounts for:

- Radar system parameters
- Target parameters
- Background effects (clutter and noise)
- Propagation effects (refraction and diffraction)
- Propagation medium (absorption and scatter)

When fully implemented, the radar equation can be used to estimate radar system performance, and the bottom line for any RCS control task is its effect on radar performance. Thus, a thorough knowledge of the radar equation and its implications are vitally necessary in the area of RCSR.

Assume that a radar transmitter has a power output of P_t watts, which is delivered to an antenna with an isotropic pattern. At a distance R from the antenna, the power density is simply the power transmitted divided by the area of the sphere over which it has evenly spread.

$$\text{Power density} = P_t/4\pi R^2 \text{ (watts/m}^2) \tag{2-8}$$

If a directional antenna is used, instead of an isotropic one, then the power density at a point in space is concentrated by the power gain of the antenna in that direction, $G_t(\theta, \phi)$, where θ and ϕ define the principal plane angles from the main beam of the antenna. We will only consider the simplest case, that for the target in the main beam of the radar, and thus the power density at the target, located at a distance R from the radar, is simply

$$\text{Power density} = P_t\, G_t/4\pi R^2 \text{ (watts/m}^2) \tag{2-9}$$

where G_t is the peak gain of the antenna.

The concept of RCS is discussed in significant detail in Ch. 6, so it will suffice here to simply define the RCS, σ, as the projected area which would be required to intercept and radiate isotropically the same power as the target radiates toward the radar receiver. Thus, we can treat the problem as though the target intercepts the power

$$\text{Power intercepted} = P_t\, G_t\, \sigma/4\pi R^2 \text{ (watts)} \tag{2-10}$$

and radiates it isotropically, so that the power density at the radar receiving antenna (which for simplicity is assumed collocated with the transmitting antenna) is

$$\text{Power density} = P_t\, G_t\, \sigma/(4\pi)^2\, R^4 \text{ (watts/m}^2) \tag{2-11}$$

The power received by the radar antenna is simply the power density at the antenna, multiplied by the effective capture area of the antenna, but it is usually more convenient to work with antenna gain, where the gain and capture area are related by

$$A_c = G_r \lambda^2/4\pi \text{ (m}^2) \qquad (2\text{-}12)$$

Finally, if we assume that the same antenna is used for transmission and reception, so that $G_t = G_r = G$, then the received power is

$$P_r = P_t \ G^2 \ \lambda^2 \ \sigma/(4\pi)^3 \ R^4 \text{ (watts)} \qquad (2\text{-}13)$$

This is the simplest form of the radar equation and ignores a number of effects which can be critical in detailed radar performance analysis. Nevertheless, it is invaluable for rough performance calculations, and it is particularly handy for assessing expected changes in radar performance for a given change in RCS.

For detection range estimates it is convenient to cast the radar equation in a slightly different form. In the simple case of detection of a target in receiver noise, a required minimum signal-to-noise ratio can be defined based on required detection probability, target statistics, and radar characteristics. Because receiver noise can be considered to be a constant, the minimum signal-to-noise ratio defines the maximum detection range by defining a minimum level of received signal, P_{min}, which can be tolerated. Thus, the maximum detection range is given by

$$R_{max} = [P_t \ G^2 \ \lambda^2 \ \sigma/(4\pi)^3 \ P_{min}]^{1/4} \text{ (m)} \qquad (2\text{-}14)$$

The sobering thought for the RCS specialist, gleaned from Eq. (2-14), is that the maximum detection range (in free space) varies only as the fourth root of the RCS. Thus, a 12 dB reduction in RCS will be required to halve the maximum detection range (i.e., reduce the maximum detection range by 3 dB). For detection in clutter the relationship between RCS and maximum detection range becomes more complicated, because clutter will typically exhibit a received power *versus* range dependency varying from R^{-3} to R^{-7} [23].

2.7 RADAR DETECTION

The entire premise that RCSR is beneficial is predicated on the fact that a radar must detect a target in a background of other signals. If there were no signals competing with the target return, additional amplification could be added in the radar receiver to provide a detectable output, no matter how small the input. However, there are always other background signals in which the target must be detected. These include cosmic and atmospheric noise (which may usually be neglected at L band and above), terrain backscatter (land and sea clutter), atmospheric clutter (backscatter from hydrometeors), and ECM (chaff, jammers, *et cetera*). Models exist which describe the effects of all of the above types of background interference [21, 22, 23, 24]. However, those effects obviously complicate the detection process, and it is beyond the scope of this book to delve into the subtleties of radar detection in jamming and clutter environments. Therefore, this section will concentrate on the simplest case, that of detection of a target in receiver noise.

2.7.1 The Decision Process

The radar detection problem is normally approached through the concepts of statistical decision theory, because it involves the detection of a random time-varying signal in a randomly varying background. There are a number of excellent texts on detection theory, both general [25, 26] and dealing specifically with radar detection [27, 28]. The radar detection decision is typically posed as a hypothesis test, where the two hypotheses are

H_0: no target is present (noise only)

H_1: a target is present (target return plus noise)

and the goal is to formulate a test which maximizes the choice of the correct hypothesis, based on chosen constraints concerning errors in the decision.

The error constraint normally used in radar is to limit the number of times H_1 is chosen when in fact no target is present (i.e., a false alarm) to a certain level. This essentially involves correct placement of a threshold (see Fig. 2-10) so that noise alone will cross, on the average, at an acceptable rate. Threshold placement obviously requires a knowledge of the statistics of the receiver noise.

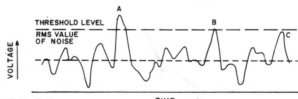

Figure 2-10 Typical radar receiver output envelope as a function of time (reprinted from Skolnik [1])

2.7.2 Noise Statistics

For a typical radar receiver, thermal noise power generated by the random thermal motion of conduction electrons in the input stages limits the signal which can be detected. The available thermal noise power is a function of the temperature T and the bandwidth B_n of the receiver, and is given by

$$P_n = kTB_n \text{ (watts)} \tag{2-15}$$

where k is Boltzmann's constant $= 1.38 \cdot 10^{-25} \text{ J/}^\circ\text{K}$. At room temperature (290° K), $P_n = -114 \text{ dBm}$ (dB relative to a milliwatt) for a receiver with a 1

MHz bandwidth. The noise bandwidth of the receiver B_n is a function of the IF filter response shape, but the three-dB IF bandwidth is normally used without appreciable error [1].

The ideal receiver would add no noise to the signal to be amplified, and the input and output signal-to-noise ratio would be the same. Actual receivers add some noise of their own, and the noise figure, F, defined for a linear system as

$$F = \frac{S_{in}/N_{in}}{S_{out}/N_{out}} \tag{2-16}$$

is the measure of how much the receiver degrades the input signal-to-noise ratio.

Additional losses (scanning, beamshape, collapsing, intergration, *et cetera*) in the radar system can be defined which further degrade the received signal power. If those losses are lumped, and are designated by L, then the radar equation can be expressed in terms of signal-to-noise ratio as

$$SNR = \frac{P_t G^2 \lambda^2 \sigma}{(4\pi)^3 R^4 k T B_n \, FL} \tag{2-17}$$

The thermal noise voltage in the IF stages before the second detector can be described statistically as a bivariate Gaussian process (in-phase and quadrature signals), with each component having a zero mean and a variance equal to half the noise power, as shown in Fig. 2-11. The envelope of the noise voltage after passage through a linear detector is Rayleigh distributed (also Fig. 2-11), and it is to this detected noise signal that a threshold is normally applied.

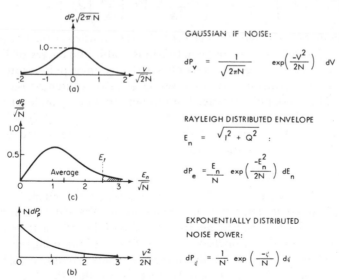

Figure 2-11 Thermal noise distributions in the radar receiver (adapted from Brookner [4])

2.7.3 Single Pulse Probability of Detection for a Non-Fluctuating Target

The Neyman-Pearson decision criterion, one which is often applied in radar, maximizes detection probability for a fixed false alarm probability. Thus, the calculation of detection probability first requires that the position of the threshold be determined, and, then, based on that threshold, probability of detection can be calculated from the target-plus-noise density function.

The threshold level for a given false alarm rate can be set based on the noise distribution, where the probability of false alarm is given by

$$P_f = \int_{V_t}^{\infty} p_n(v)\, dv = \int_{V_t}^{\infty} \frac{v}{N}\, \exp\left(\frac{-v^2}{2N}\right) dv$$

$$(2\text{-}18)$$

$$P_f = \exp(-V_t^2/2N)$$

where V_t is the threshold voltage and N is the noise power.

For a non-fluctuating target with a voltage amplitude A, Rice[29] gives the combined target plus noise probability density function at the output of the detector as

$$p(v)\, dv = \frac{v}{N}\, \exp\left(\frac{-v^2 + A^2}{2N}\right) I_0\left(\frac{vA}{N}\right) dv \qquad (2\text{-}19)$$

where $I_0(\ )$ is the modified Bessel function of order zero. The probability of detection is given by

$$P_d = \int_{V_t}^{\infty} p(v)\, dv = \int_{V_t}^{\infty} \frac{V}{N}\, \exp\left(\frac{-v^2 + A^2}{2N}\right) I_0\left(\frac{vA}{N}\right) dv \qquad (2\text{-}20)$$

which cannot be evaluated in closed form. However, when evaluated numerically and expressed in terms of signal-to-noise ratio, the curves of Fig. 2-12 result. These curves represent detection based on a single pulse. Note that for reasonable false alarm rates (10^{-6} or lower) large values of signal-to-noise ratio are required for a significant detection probability. For example, at a 1 MHz bandwidth a false alarm rate of less than $1 \cdot 10^{-6}$ is required, if at least one second between false alarms is desired. Even at that rate, a signal-to-noise ratio of almost 12 dB is required for a $P_d = 0.5$.

2.7.4 Pulse Integration

In general, many pulses are received from a target on a single scan by the radar search antenna. Thus, the single pulse probability of detection calculated in Eq. (2-20) is not realistic because the sum of the returns from multiple pulses is used in the detection decision. The process of summing the available pulses to enhance detection is called integration, and it relies on the fact that the noise signal decorrelates from pulse-to-pulse to allow the signal return to be enhanced relative to the noise.

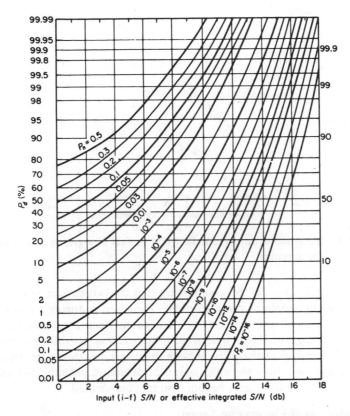

Figure 2-12 Probability of detection as a function of the signal-to-noise ratio (reprinted from Brookner [4])

Integration can be accomplished at IF or at video in the receiver. If the integration is carried out at IF with the phase preserved (coherent intergration), an increase in single pulse signal-to-noise ratio by a maximum factor of the number of pulses integrated can be achieved. Thus, for 20 pulses integrated, the required single pulse signal-to-noise ratio for a given detection probability might be reduced by 13 dB because of the integration gain.

If post-detection integration is used, the phase information is discarded, and a less efficient integrator is obtained. One common method of post-detection integration is on the CRT screen using phosphor persistence and the integration ability of the operator's eye. Figure 2-13 illustrates the perfect coherent integrator *versus* the operator-CRT combination. For a few pulses integrated, the difference between the two methods is not great. However, for a large number of pulses integrated, the difference in integration gain between noncoherent and coherent integrators will be significant.

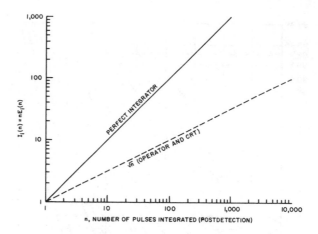

Figure 2-13 Integration improvement for a perfect integrator and an operator/CRT combination

Figure 2-12 is based on the assumption that the target does not fluctuate and the noise decorrelates pulse-to-pulse. However, if during the scan time across the target neither the target nor background fluctuates (as might be the case for some types of clutter), then pulse integration provides no improvement in the detection probability. If both the target and the clutter fluctuate, as is often the case, then the calculation of detection probabilities and the assessment of integration gain becomes more complicated.

2.7.5 Detection of Fluctuating Targets

The basic work on detection of fluctuating targets was done by Peter Swerling [30], and in that work he defined four cases of fluctuating targets based on two RCS distributions and two fluctuation rates. They are:

Amplitude Distribution	*Slow Fluctuation*	*Fast Fluctuation*
$dP = \dfrac{1}{<\sigma>} \exp \dfrac{-\sigma}{<\sigma>} \, d\sigma$	Case 1	Case 2
$dP = \dfrac{4\sigma}{<\sigma>^2} \exp -2\sigma/<\sigma> \, d\sigma$	Case 3	Case 4

where $<\sigma>$ is the average RCS.

The first amplitude distribution can be recognized as Rayleigh, while the second is the square root of a chi-squared distribution with four degrees of freedom. Physically, a Rayleigh distribution is given by a large number of nearly equal amplitude scatterers adding together with random phases. The chi-squared distribution represents a body with one dominant plus many small scatterers. The slow fluctuation case assumes constant RCS during a scan, with scan-to-scan variations. The fast fluctuation cases represents targets whose return is independent pulse-to-pulse.

Figure 2-14 illustrates the additional loss due to target fluctuation for the slowly varying Swerling cases 1 and 3. Note that loss is greatest at the higher probabilities of detection, and that the curve is not particularly sensitive to false alarm rate. For the rapidly varying targets the fluctuation loss is related to that for the slowly fluctuating target and to the number of pulses integrated, n.

$$L_{F2} \simeq [L_{F1}]^{1/n} \quad \text{(Case 2 target)}$$

$$(2\text{-}21)$$

$$L_{F4} \simeq [L_{F3}]^{1/n} \quad \text{(Case 4 target)}$$

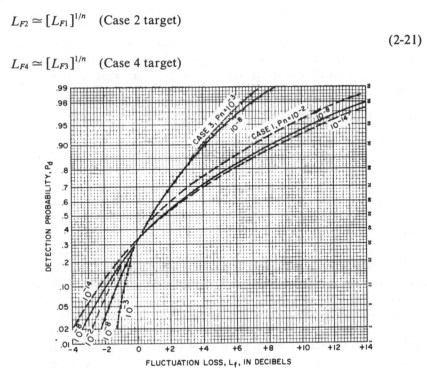

Figure 2-14 Target fluctuation loss for Swerling case 1 and 3 targets as a function of probability of detection, with false alarm rate as a parameter (reprinted from Brookner [4])

2.7.6 Radar Performance Prediction

Based on detection analysis, the signal-to-noise ratio required for a given probability of detection can be established. The description above has concentrated on the cases of a non-fluctuating target and some standard fluctuating cases in Gaussian noise. "Cookbook" procedures exist [4] which lead through the calculation of the radar loss factors which we have simply lumped and called L in Eq. (2-17). These procedures can also account for other target fluctuation statistics than those in the Swerling models (e.g., lognormal, Weibull, *et cetera*). In many cases the interfering signal will be clutter, rather than receiver noise, and the clutter envelope may not be Rayleigh distributed. Much recent work has been done in calculating performance and postulating optimum detection strategies for various clutter and target characteristics. Reference [22] contains an excellent compilation of papers in the detection area.

After calculating the required SNR for a given detection situation, a maximum detection range can be calculated based on the radar equation. Again, a large number of factors such as refraction, diffraction, and multipath were not discussed in this presentation. Where those must be considered, the references [21 or 22] provide the necessary information allowing a more detailed analysis.

2.8 ELECTRONIC COUNTERMEASURES

This final topic in radar fundamentals is included because of its importance to the engineer concerned with RCS prediction and control. RCSR is in itself a passive countermeasure, and sometimes sufficiently low RCS can be obtained to allow the treated vehicle to be hidden in clutter and to escape detection altogether. However, if detection occurs, then a reduced RCS is beneficial, because ECM techniques rely on distraction of the radar from the target by a larger signal.

ECM may rely on a number of techniques, such as creating saturation of the radar screen to hide the desired target by using a stand-off jammer, creating false targets with chaff, or using a deception jammer to break radar track on the target. In the case of chaff, the idea is to force the tracking radar off the target (and preferably onto the chaff cloud). However, particularly when used from aircraft, the rate at which the chaff slows and moves out of the Doppler filter pass band of MTI or Pulse Doppler systems is extremely rapid. Thus, chaff is typically employed in the terminal phase of a missile engagement, and it is only required to break track for a sufficient period that the missile is unable to recover. Obviously, a significant amplitude advantage in chaff RCS over target RCS is desired, both to pull the track centroid as far off the target as possible and to maintain the chaff RCS advantage in the Doppler pass band as long as possible.

Deception jammers are generally carried on the jamming vehicle. Thus, spatial separation of the jammer and target cannot be used to break track, as can be done with chaff. In addition, most modern missiles have a home-on-jam (HOJ) mode, and thus simple barrage jamming will also be unsuccessful. Therefore, deception jammers must produce a signal which appears to the radar to come fro.n somewhere other than the target. One successful deception jamming technique, mentioned previously in connection with conical scan radars, is to produce a jamming signal amplitude modulated at the conical scan frequency. If sufficiently strong, such a signal will mask the signal from the target and produce a false error signal likely to cause a loss of track.

Monopulse systems are immune to amplitude modulation jamming because they produce an error signal based on each pulse. Thus, other techniques such as range gate pull off (RGPO), or velocity gate pull off (VGPO), must be used to break track. In RGPO, the jamming pulses are given increasing delays with time so that the range gate of the tracking radar will walk off the target with the jamming pulse. VGPO performs a similar function by slowly ramping the frequency to move a velocity gate off the target.

All of these jamming methods require that the jamming signal overcome the skin return from the target. The basic measure of the return from the jammer *versus* the normal radar return is called the *J*-over-*S* ratio (J/S). For a jammer with an output power P_J and an antenna gain G_J, the power received by a radar with antenna gain G is

$$J = \frac{P_J G_J G \lambda^2}{(4\pi)^2 R^2} \text{ (watts)} \tag{2-22}$$

The skin return is simply given by the radar equation, (2-13), and so

$$J/S = \frac{4\pi P_J G_J R^2}{P_t G \sigma} \tag{2-23}$$

Several things are noteworthy about this equation. First, J/S is proportional to R^2. Thus, J/S decreases as range closes, and at some range the minimum value of J/S for successful jamming will be reached. This range is normally called the "burn-through range." Second, J/S varies inversely with σ, as expected, so that the lower the RCS, the more effective the jamming. In addition, because J/S varies only as R^2, a 6 dB reduction in RCS will halve the burn-through range, rather than the 12 dB required in the non-jamming situation to halve the detection range.

RCSR is only one method in a "bag of tricks" for reducing vulnerability. Intelligence, tactics, and countermeasures can also play a large part in increasing mission effectiveness. Therefore, the requirements for and cost of RCSR should be evaluated for the integrated system, and not separated from the total engagement problem.

2.9 SUMMARY

The purpose of this chapter has been to provide a brief overview of radar fundamentals. It has purposely been long on verbiage and short on equations, and the reader is encouraged to consult the references for details on the areas discussed. The intent has been to give a flavor for some of the types of radars normally deployed, present the radar equation, tie the radar equation to the concept of detection theory, and, finally, show how ECM alters the parameters of the radar-target engagement.

One key point to remember from this chapter is that RCSR rarely provides a one-to-one trade-off on detection range. In free space, 12 dB of RCSR is required to halve the detection range. The situation may be better in other environments than receiver noise, but clutter is typically not homogenous, and the RCSR designer should not rely too heavily on the belief that clutter will hide his vehicle. In an environment where ECM is called for, jamming effectiveness varies with the square of the range, rather than the inverse fourth power dependence seen in the radar equation, so the situation is more favorable than for detection.

RCSR always exists as a systems' trade-off. An adequate evaluation of RCSR needs and effectiveness requires knowledge of the threat radars, the environment, available countermeasures, and measures of mission effectiveness. For practical systems development, RCS design cannot exist in a vacuum. It must be considered as an integrated part of a platform's defensive weapons suite.

REFERENCES

1.   M.I. Skolnik, *Introduction to Radar Systems*, New York, McGraw-Hill, 1962.
2.   D.K. Barton, *Radar System Analysis*, Dedham, MA, Artech House, 1976.
3.   F.E. Nathanson, *Radar Design Principles*, New York, McGraw-Hill, 1969.
4.   E. Brookner, *Radar Technology*, Dedham, MA, Artech House, 1978.
5.   M.I. Skolnik, *Radar Handbook*, New York, McGraw-Hill, 1969.
6.   MIT Radiation Laboratory Series, New York, McGraw-Hill, 1947.
7.   J.L. Eaves, *Principles of Modern Radar*, Georgia Institute of Technology, Atlanta, GA, 1981.
8.   G.W. Ewell, *Radar Transmitters*, New York, McGraw-Hill. 1981.
9.   J.D. Kraus, *Antennas*, New York, McGraw-Hill, 1950.
10. W.L. Weeks, *Antenna Engineering*, New York, McGraw-Hill, 1968.
11. C.A. Balanis, *Antenna Theory*, New York, Harper and Row, 1982.
12. R.C. Hansen, ed., *Microwave Scanning Antennas*, New York, Academic Press, 1964.
13. R.C. Johnson, and H. Jasik, *Antenna Engineering Handbook*, New York, McGraw-Hill, 1984.
14. G.H. Knittel, "Phased Array Antennas—An Overview," Ch. 21, *Radar Technology*, E. Brookner, ed., Dedham, MA, Artech House, 1978.

15. S. Silver, *Microwave Antenna Theory and Design*, MIT Radiation Laboratory Series, Vol. 12, New York, McGraw-Hill, 1949.

16. N.T. Alexander, "Antennas," *Principles of Modern Radar*, Georgia Institute of Technology, Atlanta, GA, 1981.

17. J.M Smith, and R.H. Logan, "AN/APS-116 Periscope—Detecting Radar," *IEEE Trans. AES*, Jan. 1980, pp. 66-73.

18. C.E. Cook, and M. Bernfeld, *Radar Signals—An Introduction to Theory and Applications*, New York, Academic Press, 1967.

19. D.K. Barton, ed., *C.W. and Doppler Radars*, *Radars*, Vol. 7, Dedham, MA, Artech House, 1978.

20. D.K. Barton, ed., *Monopulse Radar*, *Radars*, Vol. 1, Dedham, MA, Artech House, 1974.

21. L.V. Blake, "A Guide to Basic Pulse-Radar Maximum-Range Calculations," NRL Report 6930, December 1969 (AD701321); see also Ch. 2 of *Radar Handbook*, M. Skolnik, ed., New York, McGraw-Hill, 1970.

22. D.K. Barton, ed., *The Radar Equation*, *Radars*, Vol. 2, Dedham, MA, Artech House, 1974.

23. M.W. Long, *Radar Reflectivity of Land and Sea*, Dedham, MA, Artech House, 1983.

24. J.A. Boyd, *et al.*, *Electronic Countermeasures*, Los Altos, CA, 1978.

25. J.M. Wozencraft, and I.M. Jacobs, *Principles of Communication Engineering*, New York, John Wiley and Sons, 1968.

26. H.L. Van Trees, *Detection, Estimation, and Modulation Theory*, New York, John Wiley and Sons, 1965.

27. J.V. DiFranco, and W.L. Rubin, *Radar Detection*, Dedham, MA, Artech House, 1980.

28. D.P. Meyer, and H.A. Mayer, *Radar Target Detection*, New York, Academic Press, 1973.

29. S.O. Rice, "Mathematical Analysis of Random Noise," *B.S.T.J.* 23, No. 3, July 1944, and *B.S.T.J.* 24, No. 1, January 1945; reprinted, N. Wax, Dover Publications, 1954.

30. P. Swerling, "Probability of Detection for Fluctuating Targets," *IRE Trans. Info. Theory*, Vol IT-6, No. 2, April 1960.

CHAPTER 3

PHYSICS AND OVERVIEW OF ELECTROMAGNETIC SCATTERING

J. F. Shaeffer

3.1 INTRODUCTION

The objective of this chapter is to introduce the concepts of electromagnetic scattering in an overview fashion as a prelude to the material presented in succeeding chapters. The topics to be presented are:

- The definition of radar cross section
- The three electromagnetic scattering regimes, which depend on the ratio of wavelength to body size, and the predictive tools used for each scattering regime
- The physics of the scattering process involving induced currents and the resultant radiation
- A review of Maxwell's equations of electromagnetic theory, including the wave equation, properties of EM waves, and the Stratton-Chu formulation for currents and charges as sources of scattered fields

Chapter 4 will cover numerical prediction techniques useful for the low and resonant frequency regimes in greater detail. Chapter 5 will cover the high frequency topics of geometric optics, physical optics, diffraction, and traveling wave phenomena.

3.2 RADAR CROSS SECTION

Radar cross section is a measure of the power that is returned or scattered in a given direction, normalized with respect to the power density of the incident field. This scattered power is further normalized so that the decay due to spherical spreading of the scattered wave is not a factor in computing the RCS, σ. The purpose of this normalization is to remove the effect of the range, R, and thereby arrive at a signature description that is independent of the distance between the target and radar. Formally, the radar cross section is

$$\sigma = 4\pi \lim_{R \to \infty} R^2 \frac{|\vec{E}^s|^2}{|\vec{E}^i|^2} = 4\pi \lim_{R \to \infty} R^2 \frac{|\vec{H}^s|^2}{|\vec{H}^i|^2} \tag{3-1}$$

where $\vec{E}^s$, $\vec{H}^s$ are the scattered electric and magnetic fields, respectively, and $\vec{E}^i$, $\vec{H}^i$ are the incident fields. The scattered field is due to the presence of the target, so that the total field is the incident plus scattered fields

$$\vec{E}^T = \vec{E}^i + \vec{E}^s \tag{3-2}$$

The unit of cross section is area, usually given in square meters, but sometimes in square wavelengths.

This definition is made more recognizable by examination of the basic radar equation for power received, P_r, in terms of transmitted, scattered, and returned power:

$$P_r = \underbrace{\left(\frac{P_t G_t}{4\pi R^2} \right)}_{\#1} \underbrace{\left(\frac{\sigma}{4\pi R^2} \right)}_{\#2} \underbrace{\left(A_r \right)}_{\#3} \tag{3-3}$$

The first term is the power density at the target due to radiation from the transmitter. This term has the units of watts per square meter. The second term represents the amount of the incident power reflected or scattered back to the receiver. The third term represents the amount of the returned power which is captured by the receiving antenna as a result of its effective area.

The radar cross section is also sometimes defined in the context of antenna terminology because σ is the area which would intercept sufficient power from the incident wave to produce a given echo, if the reflection were isotropic.

Radar cross section is a function of:
- Target configuration
- Frequency
- Incident polarization
- Receiver polarization
- Angular orientation of the target with respect to the incident field

Hence, in general, σ could be specified as

$$\sigma_{ij} (\theta, \phi) \tag{3-4}$$

where i and j refer to the incident and received polarizations, such as horizontal and vertical, and where (θ,ϕ) are the polar spherical angles of view.

The radar cross section of a body is also a function of the pulse width τ of the incident radiation. When τ is large enough, $\tau > 2L/c$, *where L* is the body size and c the speed of light, the entire target is illuminated at once. This is loosely equivalent to the target being illuminated by a continuous wave at some frequency f. This is known as "long pulse" illumination and is the usual RCS measurement case. However, when very short pulses are used to illuminate a target ($\tau < 2L/c$), each scatterer on the target may contribute independently to the return. Thus, the total target return is a collection of individual scattering returns separated in time. Short pulse radars are used for diagnostic purposes to identify these scattering centers on a complex target.

3.2.1 Scattering Matrix

Radar cross section, as a scalar number, is a function of the polarization of the incident and received waves. A more complete description of the interaction of the incident wave and the target is given by the polarization scattering matrix $\bar{\bar{S}}$, which relates $\vec{E}^s$ to $\vec{E}^i$, component-by-component. In matrix notation, this is

$$\vec{E}^s = \bar{\bar{S}} \cdot \vec{E}^i \tag{3-5}$$

Since $\vec{E}$ can be decomposed into two independent directions or polarizations, S is the four-element matrix

$$\begin{bmatrix} E_1^s \\ E_2^s \end{bmatrix} = \begin{bmatrix} S_{11} & S_{12} \\ S_{21} & S_{22} \end{bmatrix} \begin{bmatrix} E_1^i \\ E_2^i \end{bmatrix} \tag{3-6}$$

The components of S are related to target cross section by

$$S_{ij} = (4\pi r^2)^{-1/2} \sqrt{\sigma_{ij}} \tag{3-7}$$

where we recognize $\sqrt{\sigma}$ as a complex number that contains phase as well as amplitude information. The received voltage, V_r energy still depends on the polarization of the receiver, $\hat{n}_r$, by

$$V_r \propto \hat{n}_r \cdot \vec{E}^s \tag{3-8}$$

and on the polarization of the transmitted wave by

$$\vec{E}^i = \alpha \vec{E}_1^i + \beta \vec{E}_2^i \tag{3-9}$$

where α and β are the transmitted components of each polarization along $\vec{E}_1^i$ and $\vec{E}_2^i$, respectively.

Eight quantities (four amplitudes and phase angles or, alternatively, four complex numbers) specify the scattering matrix. One phase angle may be used as a reference for the other three. If the radar is monostatic, then $S_{12} = S_{21}$, and $\bar{S}$ can then be specified by five quantities. If we had a coherent radar which transmitted and received two orthogonal polarizations, then the scattering matrix could be determined for a given aspect (θ, ϕ) and frequency f. For a specific target and orientation with respect to the radar, we can extract no more signal information than that contained in the scattering matrix. The

wrong - rather, if body is symmetrical

scattering matrix is further discussed by Huynen[1], who considers the eigenvalues and eigenvectors of the scattering matrix as functions of:
- Target size
- Target orientation
- Target non-symmetry
- Double bounce depolarization
- Target characteristic angle

Such information can be useful for target identification purposes.

Radar cross section is usually given in square meters and is often expressed in logarithmic form as dB relative to a square meter,

$$\sigma_{dBsm} = 10 \log_{10} \sigma \tag{3-10}$$

where the reference is one square meter. The returns from many targets may have a dynamic range of 80 dB. In terms of radar cross-section reduction (RCSR), a reduction of x dB means a percentage reduction of $(1 - 10^{-x/10})$ × 100% of the returned scattered power. Thus, a 10 dB reduction corresponds to 90%, 20 dB to 99%, 30 dB to 99.9%, and so on. Put another way, 10 dB reduction means 1/10 of the original power remaining, 20 dB means 1/100 power remaining, and 30 dB means 1/1000 power remaining.

Radar cross-section reduction requirements can be estimated from the radar range equation in terms of detection range reduction in a noise background. If a target of cross section σ is detected at range R_1, then with the same radar, a target of cross section σ_2 is detectable at range R_2,

$$\frac{R_1}{R_2} = \left(\frac{\sigma_1}{\sigma_2}\right)^{1/4} \tag{3-11}$$

Thus, a 10 dB decrease in σ, $\sigma_1/\sigma_2 = 0.1$ corresponds to a range reduction of 56%, a 20 dB decrease ($\sigma_1/\sigma_2 = 0.01$) a 32% range reduction, and so on. Because the received power varies inversely with the fourth power of range, but directly with the RCS, we must reduce the RCS by four orders of magnitude (40 dB) for a reduction in detection range of only one order of magnitude. Radar detection in noise is a statistical process and, hence, these numbers represent only a conceptual comparison. Multipath considerations change this value of r to the fourth dependence.

The detection process, which is a central problem of radar, must deal with target echo fluctuations due to to changes in aspect angle as well as from:
- Changes in multipath from target motion
- Changes in beam-pointing direction due to antenna scanning
- Variations in transmitter frequency
- Clutter
- Multipath

Thus, the detection of a radar target is a statistical process with separate statistics to describe the target, the noise, and clutter.

Detection of complex targets is described in terms of probability functions for the power returned. A complex target with many scattering centers, none of which is dominant, is described by the Rayleigh probability density function. A complex target with one dominant scatterer and many lesser scatterers is characterized by a Ricean probability density function. Figures 3-1 and 3-2 show the distribution function $f(\sigma)$ for each of these distributions, where $f(\sigma)d\sigma$ is the probability that the value of σ lies between σ and $\sigma + d\sigma$. The area under each curve is unity.

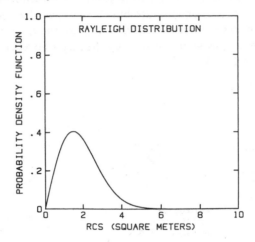

Figure 3-1 Rayleigh distribution for many scatterers, none dominant

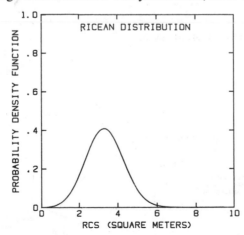

Figure 3-2 Ricean distribution for one dominant scatterer and many lesser scatterers

3.3 SCATTERING REGIMES

When an electromagnetic wave impinges upon an object, electric and magnetic currents flow in and upon the object in accordance with Maxwell's equations and the corresponding boundary conditions. These induced currents, in turn, generate an electromagnetic field of their own. This field is the "scattered field" from the object, and generally propagates in all directions with various amplitude and phase, as suggested by Fig. 3-3. When the scattered field of interest is back toward the radar transmitter, the field is called the "monostatic" signature. This is the usual case when the transmitter and receiver use the same antenna, or separate but closely spaced antennas. When the scattered field in some other direction is desired, the field is called the "bistatic" signature, and is characterized by the angle between the transmitter and receiver, as in Fig. 3-4.

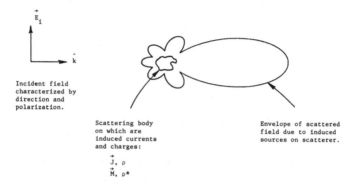

Figure 3-3 Basic scattering process

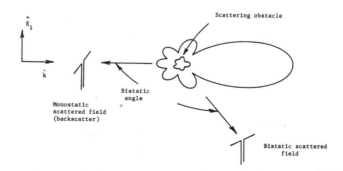

Figure 3-4 Monostatic and bistatic scattering.

Three RCS regimes characterize the relationship between the wavelength λ and body size L. The radar cross section for an object is characteristic of the ratio of $L/\lambda \sim kL$, where $k = 2\pi/\lambda = 2\pi f/c$ is the wave number, as shown in Fig. 3-5 for σ *versus* ka for a sphere. (Note that in this figure, σ is normalized with respect to the geometric cross section πa^2.) At low frequencies, for which the wavelength is much greater than the sphere circumference, the radar cross section is proportional to $(ka)^4$. Thus, at long wavelengths, σ is small, but increases as the fourth power of frequency. When the wavelength is between 0.1 and 10 circumferences $(1.0 \leq ka \leq 10)$, the cross section displays strong oscillatory behavior. This region is known as the resonant region. When the wavelength is less than 0.1 circumferences, $ka > 10$; that is, when the wavelength is small, the oscillatory behavior dies out and approaches a constant value of πa^2, the projected area of the sphere. This region is known as the optics region.

The RCS behavior of the sphere is unusual because it is independent of aspect angle. However, the three general frequency regimes are characteristic of all targets.

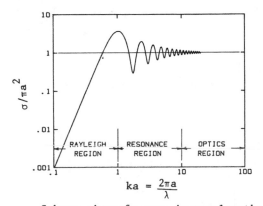

Sphere circumference in wavelengths

Figure 3-5 The radar cross section of a metallic sphere illustrates the three scattering regimes

3.3.1 Low-Frequency Scattering

When the incident wavelength is much greater than the body size, the scattering is called Rayleigh scattering, after Lord Rayleigh and his analysis of why the sky is blue in terms of scattering by particles much smaller than a wavelength. For the low-frequency case, we can assume that there is essentially no phase variation of the incident wave over the scattering body; each part of the scatterer "sees" the same incident field at each instant of time. This situation is equivalent to a static field problem, except that now the incident field is slowly modulated in time. For linear polarization, the vector direction of the incident field does not change with time, as shown in Fig. 3-6. For circular incident polarization, the situation can be understood by decomposing the incident polarization into two orthogonal linear polarizations, one shifted in phase by 90 degrees with respect to the other. This quasi-static field builds up opposite charges at the ends of the body; in effect, a dipole moment is induced by the incident field. The strength of this induced dipole is a function of the size and orientation of the body relative to the vector direction of the incident field. For example, when the applied field is perpendicular to a long body, the induced dipole moment is less than the moment induced when the applied field is parallel to the body axis. This is suggested in Fig. 3-7.

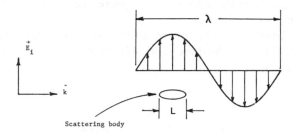

Figure 3-6 In the low frequency region, there is little variation in either the amplitude or phase of the incident field over the body length

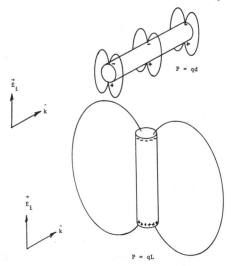

Figure 3-7 Low frequency induced dipole moments. *P* is the induced dipole moment, *L* is the body length, and *d* is the body diameter

The salient characteristic of Rayleigh scattering is that the cross section is proportional to the fourth power of the frequency or the wave number.

$$\sigma \propto \omega^4 \text{ or } k^4 \tag{3-12}$$

This behavior can be explained qualitatively by reference to the expression for scattered field, Eq. (3-79) where $E^s \sim \omega J$. There, J is a current density related to the charge by $J \propto dq/dt = i\omega q$. Hence, $E^s \propto \omega^2 q$ and the radar cross section, which is proportional to $(E^s)^2$, therefore depends on ω^4.

Since Rayleigh scattering is essentially a static field problem, all the analytical procedures for electrostatics can be invoked. These include the integral equation approach (the solution to Poisson's equation), and the dipole and multipole expansions. A scalar approach, rather than a vector approach, is possible because induced charge density is the chief physical mechanism. For low-frequency scattering, the entire body participates in the scattering process. Details of the shape are not important and, therefore, only a basic or crude geometric description is required because the volume of the scatterer is important.

For most applied problems, the wavelength is usually small compared to body size so that Rayleigh scattering is of little importance. The low-frequency approach can be used until there is appreciable phase change of the incident wave over the length of the scatterer, for example, $\pi/8$ radians (22.5°) or $ka \leq \pi/8$.

3.3.2 Resonant Scattering

When the incident wavelength is on the order of the body size, the phase of the incident field changes significantly over the length of the target, as shown in Fig. 3-8. This is the resonance region, for which, as in statics, every part of the scatterer affects every other part. The field at any point on the body is the sum of the incident field and a scattered field due to every part of the body. This collective interaction determines the resultant current density. Thus, the overall geometry is important, even if very small scale details are not.

In this scattering regime, the exact Stratton-Chu integral equations must be solved to obtain the induced currents and, hence, the scattered field. Techniques for solving the integral equations have received great attention in the last twenty years. The availability of large, very high speed computers has enabled the implementation of integral equation matrix methods. To effect a solution, we first invoke the appropriate boundary conditions on the total fields at the surface of the body. The unknown quantities to be determined are the electric or magnetic current densities on the body. The unknown source current is expressed in terms of a finite series of basis functions with complex coefficients to be determined. This is much like expressing a function in terms of a Fourier series, and the formula has the form:

$$\vec{J} = \sum_{n=1}^{N} [\hat{u}_\theta f_n (t) a_n + \hat{u}_\phi f_n (t) b_n] \qquad (3\text{-}13)$$

where $\hat{u}_\theta$ and $\hat{u}_\phi$ are typically orthogonal surface unit vectors, $f_n(t)$ are the basis functions, and a_n and b_n are the unknown coefficients to be determined. The solution then proceeds by the enforcement of the integral equation around the body profile, using appropriate testing or weighting functions on each part of the body to obtain a system of linear equations for the unknown coefficients. The result is a matrix equation, and when the applied excitation is specified (an incident field of given polarization and direction), the unknown coefficients of the current distribution may be determined. After the body currents have been found by matrix inversion, then the scattered field in the desired direction is obtained from a radiation integral which relates the scattered field to the induced surface currents. These procedures are called the *method of moments* (MOM or MM).

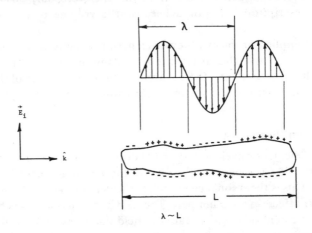

Figure 3-8 In resonance region scattering, the phase of the incident field changes several times along the body length

The general characteristics are:
*   The matrix is a function of geometry and frequency such that it requires computation and inversion only once at each frequency of interest
*   The current distribution is obtained for each excitation of interest by multiplying the excitation or voltage vector by the matrix inverse
*   The scattered field is found from the computed surface current distribution.

Specific geometries which have received much attention in recent years are:
- Wires on which the current flow is axial
- Bodies of Revolution (BOR) for which there is circumferential symmetry
- Bodies of Translation (BOT) for which there is translational symmetry
- Surface patch models

Computer codes are generally available for many of these geometries from the authors or government agencies.

Integral equation methods are limited to bodies with small surface areas measured in terms of square wavelengths. Typically the surface must be sampled five to ten times per wavelength, which can quickly lead to very large matrices when $L \gg \lambda$. Hence, the method is limited to bodies not much greater than 10λ in size. For higher frequencies, other approximations, such as physical or geometrical optics, must be used. In fact, the higher the frequency, the less useful the integral equation approach becomes, because the method accounts for the effect of each part of a body on every other part. In the low and resonant frequency regions this is very necessary, but at high frequencies, the interaction between body regions becomes very small, and the scattering becomes mostly (with few exceptions) a local, rather than collective, phenomenon. In addition to being impractical in the high-frequency region, detailed matrix interaction methods are not an important part of the physics of the problem.

3.3.3 High-Frequency Scattering

When the incident wavelength is much smaller than the length of the scattering body, $\lambda \ll L$, a completely different approach to computing the scattered field is necessary. In the high-frequency region, collective interactions are very small, so that a body can be treated as a collection of independent scattering centers. Detailed geometries now become important in the scattering process. High frequency techniques include:
- Geometrical optics (GO), which is the high frequency limit of zero wavelength in which the scattering phenomenon is treated by classical ray tracing
- Physical optics (PO), which is similar to the integral equations description since it is based on source currents
- Geometric theory of diffraction (GTD), which extends the usefulness of GO for regions where diffracted fields are important, such as in shadow areas
- Physical theory of diffraction (PTD), which explicitly removes the physical-optics surface effects from GTD
- Method of equivalent currents (MEC), which represents edge-diffracted fields in terms of fictious filamentary source currents

Physical optics uses the integral equation representation, along with the physically reasonable high frequency assumption that the scattered field from one point on a body to any other point is insignificant compared to the incident field strength. For a given point on the scattering body, the field H^s scattered by other parts of the body is small compared to the incident field H^0. Therefore, the total field at each surface point is $\vec{H}^T = \vec{H}^0 + \vec{H}^s \simeq \vec{H}^0$ where $\vec{H}^0$ is the incident or applied field. This now results in a tremendous simplification of the expression for the scattered field,

$$\vec{H}^s(\vec{r}) = 2 \int_s (\hat{n} \times \vec{H}^0) \times \nabla \psi \, ds; \quad \psi = \frac{e^{ikR}}{4\pi R} \tag{3-14}$$

because $\vec{H}^s$ no longer appears under the integral sign.

The physical optics surface current density is $\vec{J}_{po} = 2\,(\hat{n} \times \vec{H}^0)$ where $\hat{n}$ is the surface unit normal, and the factor of 2 is required to satisfy the boundary conditions for a perfectly conducting surface. The tangential surface fields, having been specified in terms of the incident field, can be thought of as the source current for the scattered field, where

$$\vec{J}_{PO} = \begin{cases} 2\,\hat{n} \,\times\, \vec{H}^0 & \text{illuminated part of surface} \\ 0 & \text{shadow part of surface} \end{cases} \tag{3-15}$$

The integral of Eq. (3-14), can be interpreted as the sum of Huygen wavelets from the surface where $\nabla \psi$, the gradient of the Greens' function, can be loosely interpreted as the Huygen wavelet. The application of *PO* requires specification of a geometry and the incident field.

Thus, *PO* has the following features:

- The *PO* integral is taken over a non-closed illuminated surface
- The tangent plane approximation for the total fields at the surface is in terms of the incident field, where the radius of curvature of the scattering surface is much greater than λ
- The fields on the shadowed side of the object are zero

When an incident wave impinges on a surface and induces currents $\vec{J}_{PO} = 2(\hat{n} \times \vec{H}^0)$ over the illuminated surface, only those currents near the specular point contribute to the scattered field. These are the currents in the region of the ray optics reflection point, as shown in Fig. 3-9. Thus, when *PO* is applied to surfaces much greater than λ, the scattering arises only from regions near the specular point. This phenomenon is the basis for the analytical technique called stationary phase. It permits us to expand the phase term under the integral about the specular point (neglecting terms of higher order than R^2) and to extend the surface limits of integration to infinity because only regions near the specular point contribute to the far scattered field.

The physical optics formalism retains some concepts of low-frequency scattering techniques in the sense that the "exact" Stratton-Chu formalism is modified for the high frequency region. Thus, the concept of surface currents is retained. The theory of physical optics is limited, however, because it does not account for edge or surface wave scattering, whose interaction fields were assumed to be negligible.

The high frequency approaches of geometric optics and the geometrical theory of diffraction depart entirely from the "exact" low-frequency approach and start instead with ray tracing, in which the very high frequency limit of Maxwell's equations is expressed in optics theorems. When the wavelength becomes small enough that energy flow is along the ray paths (not necessarily straight lines) and $\lambda \ll L$, then optic principles govern the behavior of the scattered field. In GO, the main task becomes determining how, and by what paths, energy may reach the observation point from the source point. In general, we speak of direct, reflected, and diffracted rays, as in Fig. 3-10. GO normally deals with the first two types of rays, and GTD is the extension of GO to include the propagation of energy into shadow regions, namely, diffracted rays.

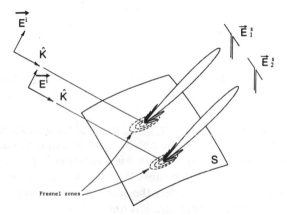

Figure 3-9 Currents are induced over the entire scattering surface S, but those in the vicinity of the specular points (the Fresnel zones) are the dominant contributors to the scattered field

The principal ingredients of GO are:
- Ray paths
- Ray spreading
- Reflection coefficients

The GTD extension then adds the following to the list:
- diffracted ray paths, including surface rays
- diffraction coefficients

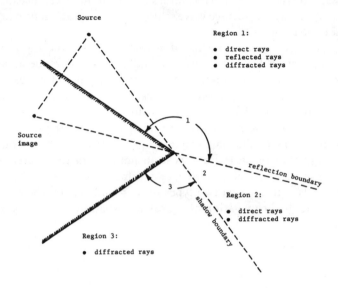

Figure 3-10 Depending on the location of the observer, energy may be received via several ray mechanisms. This example illustrates diffraction by a wedge

Ray paths are detrmined from Fermat's principle, which states that the ray trajectory is the one for which the optical path length is the minimum path distance. When the index of refraction n is a function of position, the path is not a straight line. However, for most RCS applications, we take air as the medium with $n \cong 1$, and we neglect the small variations in n compared with unity. Minimum paths are then straight lines.

When a ray is reflected, the specular reflection point is located where the total path length is a minimum and where the angle of incidence is equal to the angle of reflection. When the scattering surface is planar, the specular point is the intersection of the line from the source image to observation point with the surface (Fig. 3-11). At the specular point, the angle of incidence is equal to the angle of reflection. As will be shown in Ch. 5, the field scattered in GO is governed by the geometric surface curvature at the specular point.

GO fails when we must consider fields scattered from edges, tips, corners, tangent points, or shadow regions. This is because the electric and magnetic fields are no longer transverse to the direction of propagation in the vicinity of discontinuities, as assumed in the optical theory of rays. GTD is the procedure developed by Keller to account for these effects. Examples of diffracted rays are shown in Fig. 3-12 for edge, wedge, corner tip, multiple diffracted, and curvature. The diffracted field is determined solely by the local geometry of the scattering edge or tip, the direction of illumination, the position of the observer, and the polarization of the incident field.

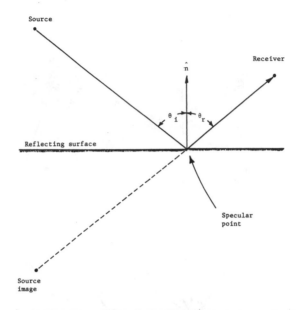

Figure 3-11 Specular point may be found for a flat surface by connecting the receiver and source image with a straight line

The general form used to implement GTD is to express the scattered field as a product of

$$\vec{E}^s = \Gamma A_d(s) e^{iks} \overline{\overline{D}} \cdot \vec{E}^i \tag{3-16}$$

where Γ is the reflection coefficient, $A_d(s)$ is the ray spreading factor, e^{iks} is the phase variation relative to the specular point, and $\overline{\overline{D}}$ is a diffraction matrix which relates the incident to scattered electric field at the diffraction point.

$$\overline{\overline{D}} = \begin{bmatrix} D_{\parallel} & 0 \\ 0 & D_{\perp} \end{bmatrix} \tag{3-17}$$

where $D_{\parallel}$ and $D_{\perp}$ are scalar diffraction coefficients. The D are based on the exact solution of the canonical problem of diffraction such as a wedge, and for practical use are treated as "cookbook" formulas.

The application of GO and GTD with general purpose computer programs for arbitrary scattering geometry becomes very difficult due to the complexity of tracing rays, multiple bounces, shadowing, *et cetera*. Thus, most computer applications are fairly geometrically specific in that they apply to narrow classes of shapes.

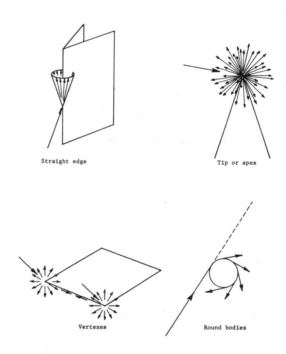

Figure 3-12 Some diffraction mechanisms

3.4 ELECTROMAGNETIC THEORY

Radar cross section analyses require a knowledge of the electric field $\vec{E}$ or the magnetic field $\vec{H}$ of the wave as the field interacts with a scattering body. The purpose of this section is to specify the laws governing electromagnetic phenomena (Maxwell's equations), to develop the wave equation, to discuss the solution to the wave equation, and to discuss waves at boundaries. In order to more fully understand the physics of the *EM* wave, we must start with a description of Maxwell's equations, which are the fundamental laws governing all electromagentic behavior. The four basic electromagnetic vector field quantities are:

$\vec{E}$ = electric field intensity, volts/meter;
$\vec{D}$ = displacement flux, coulombs/meter^2;
$\vec{B}$ = magnetic induction flux, Tesla or Webers/meter^2;
$\vec{H}$ = magnetic field intensity, amperes/meter;

and their associated source densities:

ρ = electric charge density, coulombs/meter^3;
$\vec{J}$ = electric current density, amperes/meter^2;
ρ^* = magnetic charge density, (fictitious);
$\vec{M}$ = magnetic current density, (fictitious).

All electromagnetic behavior from static fields to optical frequencies is governed by a set of four equations known as Maxwell's equations, which relate the above field and flux variables among themselves and to sources. These equations totally summarize electromagnetic behavior and are usually expressed in differential form. However, they can also be expressed in integral form. Maxwell is associated with these laws because he completed the set by recognizing the need to add a displacement term which predicts the propagation of electromagnetic waves, later shown experimentally by Hertz. Three of the four physical relationships are also known under the separate names of Gauss, Faraday, and Ampere. Collectively they are called Maxwell's equations. These equations specify both the divergence and curl of the field components, which as shown in Appendix A, is sufficient to completely characterize the field.

Gauss' law is a statement relating electrical displacement flux to its source, the electrical charge. This law is a restatement of Coulomb's law:

$$\nabla \cdot \vec{D} = \rho \qquad \text{(differential form)}$$

$$\oint \vec{E} \cdot \vec{dS} = q/\epsilon \qquad \text{(integral form)} \qquad \text{(3-18)}$$

ALT. STATEMENT $\int_S \mathbf{D} \cdot da = \int_{vol} \nabla \cdot D \, dV$

where $\vec{D} = \epsilon \vec{E}$. It states that an electric field is associated with electric charge.

Faraday's law is a statement saying that an electric field can also be induced or caused by the time rate of change of a magnetic field:

$$\nabla \times \vec{E} = -\partial \vec{B}/\partial t \qquad \text{(differential form)}$$

$$\oint \vec{E} \cdot \vec{dl} = -\frac{\partial}{\partial t} \int \vec{B} \cdot d\vec{S} = -\frac{\partial \phi}{\partial t} \qquad \text{(integral form)} \qquad \text{(3-19)}$$

where ϕ is the magnetic flux through surface S.

from STOKES THEOREM: (i.e. ...

$\int_S (\nabla \times A) \cdot da = \oint A \cdot ds$; $\nabla \times E = -\partial B/\partial t$

$\int_S \nabla \times E \cdot da = \oint E \cdot dl = -\int \partial B/\partial t \cdot ds = -\partial \phi/\partial t$

Ampere's law states that a magnetic field can be caused by an electric current $\vec{J}$ and a time changing displacement current $\partial \vec{D}/\partial t$. This latter term was Maxwell's contribution, leading immediately to the notion of wave propagation:

$$\nabla \times \vec{H} = \vec{J} + \partial \vec{D}/\partial t \qquad \text{(differential form)}$$

$$(3\text{-}20)$$

$$\oint \vec{H} \cdot d\vec{l} = \text{total current enclosed} \qquad \text{(integral form)}$$

The statics form of this law is a restatement of the Biot-Savart law.

Lastly, there is the statement that there are no free magnetic poles or charges:

$$\nabla \cdot \vec{B} = 0 \qquad \text{(differential form)}$$

$$(3\text{-}21)$$

$$\oint \vec{B} \cdot d\vec{S} = 0 \qquad \text{(integral form)}$$

These four physical laws, Eqs. (3-19) through (3-21), summarized in Fig. 3-13 through 3-16 and Table 3-1, along with boundary conditions on the fields at interfaces, form the mathematical basis for all electromagnetic phenomena.

Table 3-1
Summary of Maxwell's Equations
(in Vacuum)

I. Sources for Displacement $\vec{D}$ and Electric field $\vec{E}$ are:
    (1) Electric charge ρ
    (2) Time changing magnetic field $\vec{B}$

II. Sources for Magnetic flux $\vec{B}$ and Intensity $\vec{H}$ are:
    (1) Electric current $\vec{J}$
    (2) Time changing $\vec{D}$ $(= \epsilon_o \vec{E})$

In the presence of material bodies, the electric field $\vec{E}$ and displacement flux $\vec{D}$ are related by the polarization $\vec{P}$,

$$\vec{D} = \epsilon_0 \vec{E} + \vec{P} \qquad (3\text{-}22)$$

where ϵ_0 is the permittivity of free space. Usually the material polarization $\vec{P}$ is assumed as a linear function of electric field, $\vec{P} = \chi \vec{E}$ where the proportionality constant χ is called the electric susceptibility. Then

$$\vec{D} = \left(1 + \frac{\chi}{\epsilon_0}\right)\epsilon_0 \vec{E} = \epsilon_m \epsilon_0 \vec{E} \qquad (3\text{-}23)$$

where ϵ_m is the relative dielectric constant, ϵ_0 has units of farads/meter.

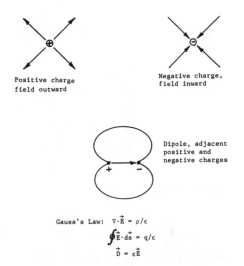

Positive charge
field outward

Negative charge,
field inward

Dipole, adjacent
positive and
negative charges

Gauss's Law: $\vec{\nabla} \cdot \vec{E} = \rho/\epsilon$

$$\oint \vec{E} \cdot \vec{ds} = q/\epsilon$$

$$\vec{D} = \epsilon \vec{E}$$

Relates source charge density to field strength E. Charge is
related to scalar potential and the conservative component of
the electric field.

Figure 3-13 Gauss' Law: Electric charge as the source of electric fields

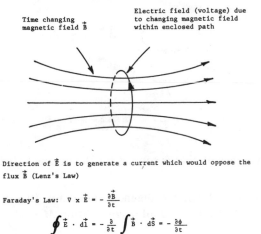

Time changing
magnetic field $\vec{B}$

Electric field (voltage) due
to changing magnetic field
within enclosed path

Direction of $\vec{E}$ is to generate a current which would oppose the
flux $\vec{B}$ (Lenz's Law)

Faraday's Law: $\nabla \times \vec{E} = -\dfrac{\partial \vec{B}}{\partial t}$

$$\oint \vec{E} \cdot \vec{dl} = -\frac{\partial}{\partial t} \int \vec{B} \cdot \vec{dS} = -\frac{\partial \phi}{\partial t}$$

Time changing $\vec{B}$ is related to the vector potential and the solenoidal
(non-conservative) component of the electric field.

Figure 3-14 Faraday's Law: Time changing magnetic field as the source of
electric fields

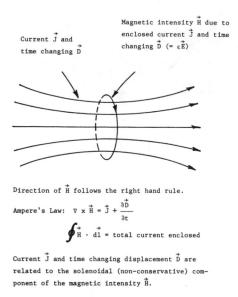

Current $\vec{J}$ and time changing displacement $\vec{D}$ are related to the solenoidal (non-conservative) component of the magnetic intensity $\vec{H}$.

Figure 3-15 Ampere's Law: Current density $\vec{J}$ and time changing displacement $\vec{D}$ as sources of the magnetic intensity $\vec{H}$

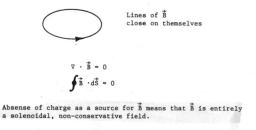

Absense of charge as a source for $\vec{B}$ means that $\vec{B}$ is entirely a solenoidal, non-conservative field.

Figure 3-16 Absence of charge sources for the magnetic field B

For the magnetic field, the magnetic induction flux is the sum of the magnetic intensity $\vec{H}$ and material magnetization $\vec{M}$:

$$\frac{1}{\mu_0} \, \vec{B} \, = \, \vec{H} \, + \, \vec{M} \tag{3-24}$$

where μ_0 is the permeability of free space. In general, $\vec{M}$ is a function of $\vec{B}$ or $\vec{H}$ as given by a conventional *B-H* or *M-H* curve for magnetic materials and, as such, exhibits the effect known as hysteresis. Many classes of materials, however, may be characterized as isotropic as well as linear, so that

$$\vec{M} = \chi_m \vec{H} \tag{3-25}$$

where χ_m is the magnetic susceptibility. Then

$$\vec{B} = (1 + \chi_m)\ \mu_0\ \vec{H} = \mu_m \mu_0 \vec{H} \tag{3-26}$$

where $\mu_m = 1 + \chi_m$ and is called the relative magnetic permeability. μ_0 has units of henrys per meter. It may take on very large values in magnetic materials such as iron and nickel. We add a word of caution, however: the concept of μ_m requires very careful justification when working with magnetic materials.

The connection between the electric field $\vec{E}$ and current density $\vec{J}$ is given by Ohm's law:

$$\vec{J} = \sigma \vec{E} \tag{3-27}$$

where the constant of proportionality σ is the conductivity of the medium and has units of Siemens (mhos/m).

The above relationships connecting $\vec{D}$ to $\vec{E}$, $\vec{B}$ to $\vec{H}$, and $\vec{J}$ to $\vec{E}$ are called the constitutive relationships.

3.4.1 Electromagnetic Vector and Scalar Potentials

As was pointed out in the appendix on vector analysis, a vector field must be described both by its divergence and its curl. For the magnetic field, Maxwell's equations require that $\nabla \cdot \vec{B} = 0$. Hence,

$$\vec{B} = \nabla \times \vec{A} \tag{3-28}$$

where $\vec{A}$ is a vector potential. Using Eq. (3-28) in Faraday's law Eq. (3-19), we obtain

$$\nabla \times \left(\vec{E} + \frac{\partial \vec{A}}{\partial t} \right) = 0 \tag{3-29}$$

which means that the quantity in parentheses is the gradient of a scalar function ϕ, or

$$\vec{E} = -\nabla \phi - \frac{\partial \vec{A}}{\partial t} \tag{3-30}$$

which again states that an electric field is due to both a charge density and a time changing magnetic field.

Further development of the required expressions for the potentials ϕ and $\vec{A}$ leads to a set of inhomogeneous wave equations. The decoupling of these equations requires the choice of a specific relationship between ϕ and $\vec{A}$ known as the "gauge." The usual choice is to let

$$\nabla \cdot \vec{A} = \frac{\partial \phi}{\partial t} \tag{3-31}$$

Then Maxwell's equations reduce to two equations for each of the potentials:

$$\nabla^2 \phi - \frac{\partial^2 \phi}{\partial t^2} = -\rho$$

or $\nabla^2 \phi - k^2 \phi = -\rho$

and   (3-32)

$$\nabla^2 \vec{A} - \frac{\partial^2 \vec{A}}{\partial t^2} = -\vec{J}$$

or $\nabla^2 \vec{A} - k^2 \vec{A} = -\vec{J}$

whose time harmonic solutions are

$$\phi = \frac{1}{\epsilon} \int \frac{\rho\, e^{ikR}}{4\pi R}\, dv$$

                                                                                             (3-33)

$$\vec{A} = \mu \int \frac{\vec{J}\, e^{ikR}}{4\pi R}\, dv$$

Thus, the scalar potential is due to an electric charge density ρ and the vector potential is due to an electrical current $\vec{J}$.

If the current and charge sources are known, then the potentials ϕ and $\vec{A}$ can be acquired and the fields $\vec{E}$ and $\vec{B}$ obtained. However, we seldom know the sources and, hence, development of this line of thought will not yield the required insight into wave phenomena. Further, modern numerical solutions solve the integral equations directly rather than solving for these potentials from which the fields are obtained via the indicated differentiations.

Wave Equation

Maxwell's equations can be manipulated in such away that under certain conditions, a second-order partial differential equation can be devleoped involving only the electric field or only the magnetic field. The solution of this equation reveals the nature of *EM* waves. Let us specify time harmonic fields by setting $\vec{E}(\vec{r}, t) = \vec{E}(\vec{r})\, e^{-i\omega t}$, remembering that the physical field is obtained by taking the real part. The choice of $(-i\omega t)$ follows the physics convention and, if we had followed the normal engineering convention, the choice would have been $(+j\omega t)$. The real part of $\vec{E}$ is $\cos(\omega t + \phi)$ where ϕ is the phase of $\vec{E}(\vec{r})$.

Let us begin by writing Maxwell's equations with $e^{-i\omega t}$ time dependence and using the constitutive relations to obtain

$$\nabla \cdot \vec{B} = 0 \qquad\qquad\qquad \nabla \times \vec{E} = i\omega\mu\vec{H}$$

                                                                                             (3-34)

$$\nabla \cdot \vec{E} = \rho/\epsilon \qquad\qquad \nabla \times \vec{H} = (\sigma - i\omega\epsilon)\, \vec{E}$$

To develop the wave equation, we take the curl of Faraday's Law to obtain

$$\nabla \times (\nabla \times \vec{E}) = i\omega\mu \; (\nabla \times \vec{H}) = i\omega\mu \; (\sigma - i\omega\epsilon) \; \vec{E}$$
$$= i\omega\mu\sigma\vec{E} + \omega^2 \mu\epsilon\vec{E} \qquad (3\text{-}35)$$

The next step is to expand $\nabla \times (\nabla \times \vec{E})$ for a charge-free region where $\nabla \cdot \vec{E} = 0$. This yields

$$\nabla \times (\nabla \times \vec{E}) = \nabla(\nabla \cdot \vec{E}) - \nabla^2 \vec{E} = -\nabla^2 \vec{E} \qquad (3\text{-}36)$$

We note that, to obtain the wave equation, we take the curl of the curl of a vector field. Because the curl is a measure of rotation, we compute the "rotation of the rotation," a kind of second derivative or acceleration term. We shall see that this "acceleration" of the field (when $\sigma = 0$) is proportional to the negative of itself and, hence, an oscillatory solution is obtained. This occurs in mechanics when the force is opposite to the direction of motion. When σ is not zero, the acceleration has another term $i\omega\mu\sigma\vec{E}$, which leads to a damping of the wave.

Using Eq. (3-36) in Eq. (3-35), we obtain the wave equation for the electric field:

$$\nabla^2 \vec{E} = -i\omega\mu\sigma\vec{E} - \omega^2 \mu\epsilon\vec{E} \qquad (3\text{-}37)$$

If we further specify the case of a dielectric or free-space medium, for which $\sigma = 0$, and write $k^2 = \omega^2 \mu\epsilon = (2\pi/\lambda)^2$, we obtain

$$\nabla^2 \vec{E} + k^2 \vec{E} = 0 \qquad (3\text{-}38)$$

which is a standard expression for the wave equation. To better understand the wave number k, we note that the velocity of propagation of the wave is $v = 1/\sqrt{\mu\epsilon}$, which for $\epsilon_m = \mu_m = 1$ is denoted by $c = 3 \times 10^8$ meters per sec. Then, noting that $k = \omega/c$ and $f\lambda = c$, we finally obtain $k = 2\pi/\lambda$ where λ is the wavelength. The wavenumber is sometimes denoted as a vector aligned in the direction of wave propagation.

The index of refraction is a measure of how much slower the wave travels in a dielectric medium than in free space, and is given by

$$n = \frac{c}{v} = \sqrt{\epsilon_m \mu_m} \qquad (3\text{-}39)$$

Because μ_m is seldom greater than unity for nonmagnetic materials, the index of refraction for that case is equal to the square root of the dielectric constant.

The solution to the wave equation can take several forms depending on the coordinate system chosen. In a rectangular system, one solution is

$$\vec{E}(\vec{r}, t) = \vec{E}_0 \; e^{i(\vec{k} \cdot \vec{r} - \omega t)} \qquad (3\text{-}40)$$

and the corresponding magnetic field is obtained by using Eq. (3-40) in Faraday's law to obtain

$$\vec{H}(\vec{r}, t) = \frac{k}{\omega \mu} \quad \hat{k} \times \vec{E}_0 \tag{3-41}$$

If we denote

$$k/\omega\mu = \frac{1}{c\mu} = \frac{\sqrt{\epsilon\mu}}{\mu} = \sqrt{\frac{\epsilon}{\mu}} = \frac{1}{Z} \, ,$$

where Z is the intrinsic impedance of the medium, then the ratio of absolute magnitude of E to H field is the impedance of the wave:

$$\frac{|\vec{E}|}{|\vec{H}|} = Z = \sqrt{\frac{\mu}{\epsilon}} \tag{3-42}$$

which is equal to 377 ohms for free space. With this definition,

$$Z \vec{H} = \hat{k} \times \vec{E} \tag{3-43}$$

where $\hat{k}$ is a unit vector along the direction of propagation of the wave. The most important result here is that the wave is transverse with $\vec{E}$ and $\vec{H}$ perpendicular to each other and to the direction of propagation $\hat{k}$. A summary of the wave nature of *EM* fields is shown in Table 3-2.

Table 3-2
Wave Equation Summary
(in Vacuum)

Maxwell says:	$B (= \mu_o H)$ is related to time changing $D (= \epsilon_o E)$
	$D (= \epsilon_o E)$ is related to time changing $B (= \mu_o H)$

The *wave equation* is derived from *Maxwell's equations* and shows that an electromagnetic wave has the following characteristics:

- $\vec{E}$ and $\vec{H}$ are wave-like, oscillating in time and space.
- An *EM* wave has both $\vec{E}$ and $\vec{H}$ components.
- $\vec{E}$ is perpendicular to $\vec{H}$.
- $\vec{E}$ and $\vec{H}$ are perpendicular to the direction of propagation $\vec{k}$.
- Velocity of propagation $= \dfrac{1}{\sqrt{\epsilon\mu}} = \dfrac{1}{\sqrt{\epsilon_o\mu_o}} = c$, the speed of light in free space.
- Ratio of E to H is the wave impedance $\sqrt{\dfrac{\mu}{\epsilon}} = 377\Omega$ for free space.
- In a plane perpendicular to $\vec{k}$, $\vec{E}$ is arbitrary, hence the concept of polarization.

The constant vector $\vec{E}_0$ of the solution is complex, having amplitude and phase. Because the only requirement of the solution is that $\vec{E}_0$ be perpendicular to direction of propagation $\hat{k}$, $\vec{E}_0$ has two possible orientations, called polarizations. These are usually referenced to the local vertical and horizontal, or left and right circular, as shown in Fig. 3-17.

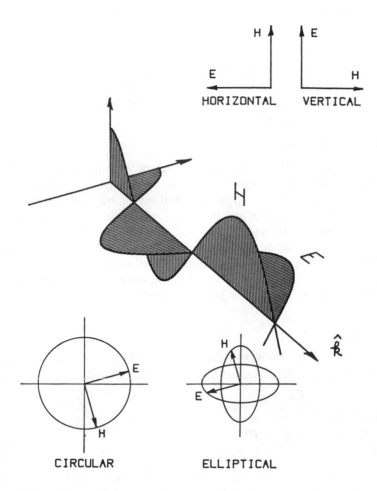

Figure 3-17 Remote from any source, E and H are perpendicular to each other and the direction of propagation $\hat{k}$. The polarization is referred to the local orientation of E. In the four polarization examples here, the direction of propagation is into the plane of the page

Plane waves are convenient for analytical purposes, but waves in the real world are typically spherical. Their magnitudes fall off inversely with distance:

$$\frac{e^{i\vec{k} \cdot \vec{r}}}{r} \tag{3-44}$$

so that energy flux, E^2, falls off as $1/r^2$. For two-dimensional coordinate systems cylindrical waves are used and have the form:

$$\frac{e^{i\vec{k} \cdot \vec{r}}}{\sqrt{r}} \tag{3-45}$$

with energy flux which goes as $1/r$. The plane wave, however, is a very useful concept because a spherical wave at very large distances does indeed seem locally to be a plane wave. Thus, a spherical wave from a distant radar incident on a scatterer can be treated as planar.

3.4.2 Waves at Boundaries

Radar cross section phenomenology is inherently a study of what happens when a wave strikes a boundary. This study involves boundary conditions and reflection coefficients. Based on the physical reasoning that charge can be neither created nor destroyed, and that electromagnetic fields are created by charge and current distributions, the fields must satisfy certain boundary conditions at the interface between two different media. The conditions can be derived from the integral form of Maxwell's equation applied to an interface. The use of a small "pillbox" volume and the two divergence equations requires that the normal components of $\vec{B}$ be continuous,

$$B_{n1} = B_{n2}$$
$$\text{or} \quad \hat{n} \cdot (\vec{B}_1 - \vec{B}_2) = 0 \tag{3-46}$$

where $\hat{n}$ is a unit vector normal to the surface. The normal component of $\vec{D}$ is discontinuous by the amount of free surface charge density,

$$D_{n1} - D_{n2} = \rho_s$$

This statement is equivalent to

$$\hat{n} \cdot (\vec{D}_1 - \vec{D}_2) = \rho_s$$
$$\text{or} \quad \hat{n} \cdot (\epsilon_1 \vec{E}_1 - \epsilon_2 \vec{E}_2) = \rho_s \tag{3-47}$$

Placing a small rectangular loop at an interface and using the two curl equations leads to the requirement that the tangential electric field be continuous across the boundary:

$$E_{t1} = E_{t2} \tag{3-48}$$

or

$$\hat{n} \times (\vec{E}_1 - \vec{E}_2) = 0 \tag{3-49}$$

Turning to the magnetic field, the tangential component of $\vec{H}$ is discontinuous by the amount of surface current density $\vec{J}_s$

$$H_{t1} - H_{t2} = J_s$$

or $\quad \hat{n} \times (\vec{H}_1 - \vec{H}_2) = \vec{J}_s$ (3-50)

The surface current can have a non-zero value when the integration loop is reduced to zero only when J_s is infinite. This requires that the conductivity σ be infinite, meaning that the surface must be perfectly conducting. For finite conductivity, the tangential magnetic field is continuous across the boundary,

$$\hat{n} \times (\vec{H}_1 - \vec{H}_2) = 0$$ (3-51)

A summary of boundary conditions is given in Fig. 3-18.

Boundary Conditions

σ	E_t	D_n	H_t	B_n
$\sigma_1 = \sigma_2 = 0$	$E_{1t} = E_{2t}$	$D_{1n} = D_{2n}$	$H_{1t} = H_{2t}$	$B_{1n} = B_{2n}$
$\sigma_2 = \infty$	$E_{2t} = 0$ $E_{1t} = 0$	$D_{2n} = 0$ $D_{1n} = \rho_s$	$H_{2t} = 0$ $H_{1t} = J_s$	$B_{2n} = 0$ $B_{1n} = 0$
σ_1, σ_2 arb. $\neq \infty$	$E_{1t} = E_{2t}$	$(\epsilon_1 + \frac{\sigma_1}{\omega}) E_{1n}$ $= (\epsilon_2 + \frac{\sigma_2}{\omega}) E_{2n}$ $= \epsilon E$	$H_{1t} = H_{2t}$	$B_{1n} = B_{2n}$

where $E = \epsilon' + \epsilon'' = \epsilon + i\frac{\sigma}{\omega}$

Figure 3-18 Summary of boundary conditions

When a wave impinges on an interface, part of the wave is reflected and part is transmitted, as shown in Fig. 3-19. The incident, reflected, and transmitted waves must have the same phase, independent of position, at the interface,

$$\vec{k}_i \cdot \vec{r} = \vec{k}_r \cdot \vec{r} = \vec{k}_t \cdot \vec{r}$$ (3-52)

This requires that the angle of reflection be equal to the angle of incidence, $\theta_r = \theta_i$, which is known as Snell's Law.

The requirement that the angle of incidence be equal to the angle of reflection is the central concept for high frequency scattering and is called the specular concept of scattering. Equation (3-52) also allows us to calculate the angle of the transmitted ray:

$$k_1 \sin \theta_1 = k_2 \sin \theta_2 \tag{3-53}$$

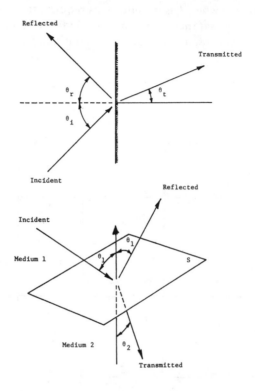

Figure 3-19 Reflection and transmission at an interface

Because

$$k = \frac{\omega}{c/n} = \left(\frac{\omega}{c} \right) n \qquad \cdots \text{because } n = \frac{c}{v} \text{ or } v = \frac{c}{n}$$
$$\text{where } v = \text{velocity in free w/l vice fra space}$$

we may express Snell's law in terms of the indices of refraction,

$$n_1 \sin \theta_1 = n_2 \sin \theta_2 \tag{3-54}$$

where n_1 and n_2 are the indices of refraction of the two media separated by the interfaces. When a wave passes from air ($n_1 = 1$) to a more dense medium ($n_2 > 1$), the ray bends toward the normal by the angle θ_2, where

$$\sin \theta_2 = \frac{n_1}{n_2} \sin \theta_1 \tag{3-55}$$

The bending of rays in passing from one medium to another is known as refraction. Because the bending is governed only by the electromagnetic properties of the media on either side of the interface, the refractive indices are sufficient to characterize the effect.

3.4.3 Reflection and Transmission Coefficients

The strengths of the electric and magnetic fields reflected at a plane interface can be expressed in terms of the reflection and transmission coefficients associated with the interface. In general, if E_i, E_r, and E_t are the incident, reflected, and transmitted fields, respectively, then the reflection and transmission coefficients are defined as

$$R = \frac{E_r}{E_i} \quad \text{and} \quad T = \frac{E_t}{E_i} \tag{3-56}$$

These coefficients are functions of the electromagnetic material parameters $\epsilon_1 \mu_1$ and $\epsilon_2 \mu_2$, the angle of incidence, and the polarization of the incident wave. When the electric field is polarized perpendicular to the plane of incidence, the boundary conditions require that [2]:

$$R_\perp = \frac{Z_2/Z_1 \cos \theta_1 - [1 - (k_1/k_2)^2 \sin^2 \theta_1]^{1/2}}{Z_2/Z_1 \cos \theta_1 + [1 - (k_1/k_2)^2 \sin^2 \theta_1]^{1/2}} \tag{3-57}$$

When the electric field E lies in the plane of incidence:

$$R_\| = -\frac{Z_1/Z_2 \cos \theta_1 - [1 - (k_1/k_2)^2 \sin^2 \theta_1]^{1/2}}{Z_1/Z_2 \cos \theta_1 + [1 - (k_1/k_2)^2 \sin^2 \theta_1]^{1/2}} \tag{3-58}$$

where Z_1 and Z_2 are the intrinsic impedances of the two media, respectively, and similarly for the wavenumbers k_1 and k_2.

The Z's are the intrinsic impedances of each medium, as in Eq.(3-42), and in general μ and ϵ are both complex numbers,

$$\mu = \mu' + i\mu'' \tag{3-59}$$

$$\epsilon = \epsilon' + i\epsilon'' \tag{3-60}$$

The ratio k_1/k_2 is $\sqrt{(\mu_1 \epsilon_1)/(\mu_2 \epsilon_2)}$, which for $\mu_1 = \mu_2$ is the ratio of the indices of refraction, $k_1/k_2 = n_1/n_2$.

When the angle of incidence is normal to the interface, $\theta_1 = 0°$, the two reflection coefficients reduce to the same value,

$$R = R_\perp = R_\| = \frac{Z_2 - Z_1}{Z_2 + Z_1} \tag{3-61}$$

If medium 2 is a very good conductor (metal or sea water at 10 GHz) so that its wave impedance is very low, then $Z_2 \to 0$ and $R \to -1$, meaning that the wave is entirely reflected and suffers a phase change of 180 degrees. If medium 2 has a very high impedance ($Z \to \infty$), then $R \to 1$, so that all the incident energy is reflected, but without a phase change. The analogy between transmission line theory and wave propagation is given in Fig. 3-20.

	Transmission Line	E M Wave
Basic Equations	$\dfrac{\partial V}{\partial \ell} = -ZI$ $\dfrac{\partial I}{\partial \ell} = -YV$	$\nabla \times \vec{E} = -i\omega\mu\vec{H}$ $\nabla \times \vec{H} = (\sigma - i\omega\varepsilon)\vec{E}$
Wave Equations	$\dfrac{d^2 V}{d\ell^2} - \gamma^2 V = 0$	$\nabla^2 \vec{E} + k^2 \vec{E} = 0$
Propagation Constant	$\gamma = \sqrt{ZY} = \alpha + i\beta$ $Z = R + i\omega L$ $Y = G + i\omega C$ $\beta = \omega\sqrt{LC}$ ‡	$\dfrac{d^2 E}{dx^2} + k^2 E = 0$ $k = \dfrac{\omega}{c} = \omega\sqrt{\varepsilon\mu}$ $\varepsilon = \varepsilon'(1 + i\dfrac{\sigma}{\omega\varepsilon})^{1/2}$
Wave Impedance	$Z_o = \dfrac{V_o^+}{I_o^+} = \sqrt{\dfrac{Z}{Y}} = \sqrt{\dfrac{L}{C}}$ ‡	$Z_o = \dfrac{|\vec{E}|}{|\vec{H}|} = \sqrt{\dfrac{\mu}{\varepsilon}}$
Solution	$V = V_o^+ e^{-\gamma x} + V_o^- e^{\gamma x}$	$E = E_o^+ e^{ikx} + E_o^- e^{-ikx}$
Reflection Coefficient	$\rho = \dfrac{Z_L - Z_o}{Z_L + Z_o}$	$\rho = \dfrac{Z_L - Z_o}{Z_L + Z_o}$ Normal Incidence
Standing Wave Ratio	$SWR = \dfrac{1 + |\rho|}{1 - |\rho|}$	--------
Power	$P = VI$ ‡	$\vec{S} = \vec{E} \times \vec{H}^*$

‡ Lossless Case

Figure 3-20 Transmission line/EM wave analogy

The reflection and transmission coefficients have been defined for field quantities when no absorption takes place,

$$|R|^2 + |T|^2 = 1 \tag{3-62}$$

Following Ruck's development [2], the relationship between T and R is

$$1 + R = T \tag{3-63}$$

because the tangential components of E_1 and E_2 must be continuous, and $E_1 = E_i + E_r$ and $E_2 = E_t$. Thus, for normal incidence, the transmission coefficient is

$$T = \frac{2 Z_2}{Z_1 + Z_2} \tag{3-64}$$

The reader is directed to the literature for the case when absorption also needs to be included, e.g., when medium two has non-zero conductivity or a loss tangent.

The dielectric constant ϵ can take on complex values if the medium has non-zero conductivity. In this case, the wave equation is

$$\nabla^2 \vec{E} + (i\omega\mu\sigma + \omega^2\mu\epsilon) \vec{E} = 0 \tag{3-65}$$

Now $k^2 = \omega^2\mu\epsilon + i\omega\mu\sigma$ is a complex number, and may be written

$$k^2 = \omega^2 \mu\epsilon \left(1 + \frac{i\sigma}{\omega\epsilon}\right) \tag{3-66}$$

Thus, we can identify ϵ as

$$\epsilon = \epsilon' + i\epsilon'' = \epsilon (1 + i\sigma/\omega\epsilon) = \epsilon + \frac{i\sigma}{\omega} \tag{3-67}$$

The imaginary part of k leads directly to absorption of the wave:

$$k = \omega \sqrt{\mu\epsilon} \left(1 + i \frac{\sigma}{\omega\epsilon}\right)^{1/2} \tag{3-68}$$

For a good conductor, $\sigma/\omega\epsilon \gg 1$ the wavenumber k is

$$k = \frac{(1 + i)}{\delta} \tag{3-69}$$

where the skin depth δ is equal to $(\pi f \mu\sigma)^{-1/2}$. Thus, the wave is attenuated by 63% in a distance δ. These wave reflection formulas show that if small reflections are desired, such as for radar cross-section reduction (RCSR), then the wave should never see large changes in impedance. Rather, gradual impedance changes are desired.

In summary, we see that the reflection of waves from plane boundaries depends on polarization, angle of incidence, and electromagnetic material parameters (μ, ϵ), which, in general, are complex and frequency dependent.

3.4.4 Wave Reflection from Surface Current Point of View

The development of reflection coefficients in terms of intrinsic material impedance and boundary conditions represents but one point of view. Another recognizes that the scattering processes of reflection and transmission occur because the incident wave induces currents at a material interface. These currents, in turn, produce a scattered wave which re-radiates energy in various directions. In this context, the total field $\vec{H}^T$ is the sum of an incident part, $\vec{H}^0$, plus a scattered part, $\vec{H}^s$.

$$\vec{H}^T = \vec{H}^0 + \vec{H}^s \tag{3-70}$$

The boundary conditions that must be satisfied by the fields may be represented as equivalent source currents and charges, which become the sources of the scattered field. The surface electric and magnetic currents are then interpreted in terms of the total tangential fields at the surface. The tangential magnetic field is expressed as an electric current:

$$\hat{n} \times \vec{H}^T = \vec{J}_s \tag{3-71}$$

while the tangential electric field is expressed as a fictitious magnetic current:

$$\hat{n} \times \vec{E}^T = -\vec{M}_s \tag{3-72}$$

The corresponding electric and magnetic charge densities are related to their respective currents through the notion of conservation of charge,

$$\rho = \frac{1}{i\omega\epsilon} \nabla \cdot \vec{J}$$

$$\rho^* = \frac{1}{i\omega\mu} \nabla \cdot \vec{M} \tag{3-73}$$

For the special case of a perfect conductor, at whose surface the total tangential electric field must be zero, the magnetic current and charge are also zero,

$$\hat{n} \times \vec{E}^T = -\vec{M} = 0$$

$$\rho^* = 0 \tag{3-74}$$

In the general case of scattering from electric or magnetic bodies, $\vec{E}_t^T$ and $\vec{H}_t^T$ are finite on an interface; hence, both $\vec{J}$ and $\vec{M}$ must be considered. As pointed out by Stratton, magnetic currents and charges have never been observed in nature, yet they form a useful mathematical artifice when one enforces arbitrary boundary conditions in terms of induced source currents.

3.4.5 Stratton-Chu Equations for the Scattered Field

When several regions of space are involved, we can use Maxwell's equation in conjunction with the vector Green's theorem to arrive at a set of field equations for the scattered field. This has been done by Stratton. Consider the geometry of Fig. 3-21 where Region I is separated from Region II by a surface S. Assume that there are magnetic and electric source currents and charges in each region. The field anywhere in Region I is given by the Stratton-Chu equations as the sum of a volume integral over the sources in Region I and a surface integral over the fields on surface S caused by the sources in Region II.

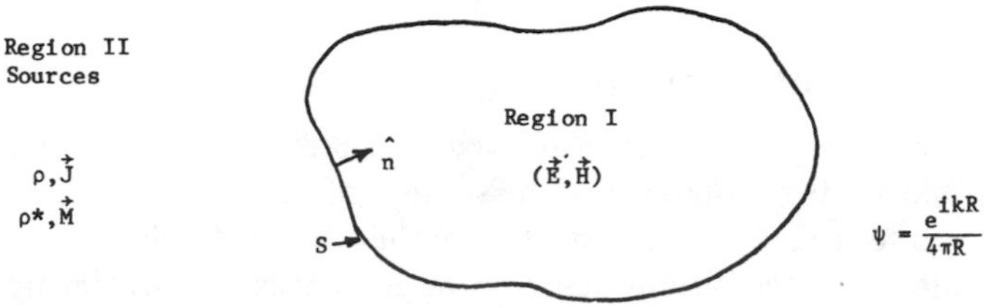

Region II
Sources

$\rho, \vec{J}$
$\rho*, \vec{M}$

Region I

$\hat{n}$

$(\vec{E}, \vec{H})$

S

$\psi = \dfrac{e^{ikR}}{4\pi R}$

$$\vec{E}(\vec{r}_I) = \int_v (i\omega\mu\vec{J}\psi - \vec{M} \times \nabla\psi + \frac{1}{\epsilon} \rho\nabla\psi)dv$$

$$+ \oint_s [i\omega\mu(\hat{n} \times \vec{H})\psi + (\hat{n} \times \vec{E}) \times \nabla\psi + (\hat{n} \cdot \vec{E})\nabla\psi]ds$$

$$\vec{H}(\vec{r}_I) = \int_v [i\omega\mu\vec{M}\psi + \vec{J} \times \nabla\psi + \frac{1}{\mu} \rho^*\nabla\psi]dv$$

$$- \oint_s [i\omega\epsilon(\hat{n} \times \vec{E})\psi - (\hat{n} \times \vec{H}) \times \nabla\psi - (\hat{n} \cdot \vec{H})\nabla\psi]ds$$

Figure 3-21 Stratton-Chu equations

$$\vec{E}^s = \int_V (i\omega\mu\vec{J}\psi - \vec{M} \times \nabla\psi + \frac{1}{\epsilon} \rho \nabla\psi) dV$$

$$+ \oint_S [i\omega\mu(\hat{n} \times \vec{H})\psi + (\hat{n} \times \vec{E}) \cdot \nabla\psi + (\hat{n} \cdot \vec{E})\nabla\psi] dS$$

$$\tag{3-75}$$

$$\vec{H}^s = \int_V (i\omega\epsilon\vec{M}\psi + \vec{J} \times \nabla\psi + \frac{1}{\mu} \rho^*\nabla\psi) dV$$

$$- \oint_S [i\omega\epsilon(\hat{n} \times \vec{E})\psi - (\hat{n} \times \vec{H}) \times \nabla\psi - (\hat{n} \cdot \vec{H})\nabla\psi] dS$$

$$\tag{3-76}$$

where the integral over the surface sums the contribution from each part of the scatterer, and where $\psi = e^{ikR}/4\pi R$ is the free space Green's function with $R = |\vec{r} - \vec{r}_s|$. These equations hold for any frequency and their solution requires integral equation techinques. Great simplification, however, occurs in the low and high frequency regions.

The interpretation that the fields at the surface are sources in the form of currents is apparent from

$$\vec{J} = \hat{n} \times \vec{H} \qquad\qquad \vec{M} = -\hat{n} \times \vec{E}$$

$$\rho/\epsilon = \hat{n} \cdot \vec{E} \qquad\qquad \rho^*/\mu = \hat{n} \cdot \vec{H} \qquad\qquad (3\text{-}77)$$

The interpretation of the tangential fields in terms of surface currents is a useful formalism for representing the sources of the field. If we can control these surface sources, we can then control the scattered fields.

The Stratton-Chu equations describe the general case of scattering from an arbitrary body. For the case of a perfectly conducting body, the total tangential electric field must be zero ($\hat{n} \times \vec{E} = 0$), so that magnetic current $\vec{M}$ and the magnetic charge ρ^* are zero. Then, using the equation of continuity, the electric charge is

$$\rho = \frac{1}{i\omega\epsilon} \nabla \cdot \vec{J}$$

Since $\omega = ck$, where $c = (\epsilon\mu)^{-1/2}$ we obtain

$$\vec{E}^s = i\omega\mu \oint_S \left\{ \vec{J}\psi + \frac{1}{k^2} (\nabla \cdot \vec{J})\nabla\psi \right\} dS \qquad \text{(E-field equation)}$$

$$\hspace{9cm} (3\text{-}78)$$

$$\vec{H}^s = \oint_S \vec{J} \times \nabla\psi \, dS \qquad\qquad \text{(H-field equation)}$$

These equations for a perfectly conducting body have the following features:

1. In general, $\vec{J} = \hat{n} \times \vec{H}^T = \hat{n} \times (\vec{H}^0 + \vec{H}^s)$, so that the equations are integral equations; the unknown field $\vec{H}^s$ appears on both sides of the equation.

2. In the far field $\vec{E}^s$ and $\vec{H}^s$ are related by $|E|/|H| = 377$ ohms. The only time the E field and H field equations need simultaneous solution is for frequencies near interior body resonances of the scatterer.

3. The equation for the magnetic field is of the form of the magnetostatics Biot-Savart law for the magnetic field from a current distribution. For statics, $k = 0$, hence $\psi = [1/(4\pi R)]$ and $\nabla\psi = [(1/4\pi)(\hat{R}/R^2)]$.

4. In the far field, the fields are transverse and decay inversely with increasing distance as $1/R$. Hence, the electric field is given by the transverse components of

$$\vec{E}^s_{\text{far field}} \simeq i\omega\mu \int_S \vec{J}\psi \, dS \qquad \text{(transverse components)} \qquad (3\text{-}79)$$

5. The far magnetic field is

$$\vec{H}^s_{\text{far field}} = (ik/4\pi) \int_S [(\vec{J} \times \hat{R})/R] \, e^{-ikr} \, dS \qquad (3\text{-}80)$$

because the first term of $\nabla \psi$ is small compared to ikR.

3.5 SUMMARY

This chapter has been an overview of electromagnetic scattering. We have seen that RCS is a measure of power returned from the incident wave, and that it is a function of the angular orientation and shape of the scattering body, the frequency, and the polarization of the transmitter and receiver. The scattered wave, of which RCS is a measure, is caused by re-radiation of currents induced on the scattering body by the incident wave. The scattering process breaks into three natural regimes: the low frequency or Rayleigh region, where the wavelength is much longer than the scattering body size and where the scattering process is due to induced dipole moments when only gross size and shape of the body are of importance; the resonant region, where the wavelength is of the same order as the body size and the scattering process is collective such that every part of the body electrically affects every other part of the body and hence the scattered energy; and the high frequency region, where the wavelength is much smaller than the scattering body and the scattering process is principally a summation of the return from isolated, non-interacting scattering centers.

Maxwell's equations tell us that *EM* waves are a combination of electric and magnetic fields which are perpendicular to each other and to the direction of propagation. When an *EM* wave is incident on a body, the boundary conditions on the fields require that surface currents flow. These currents, in turn, re-radiate a scattered *EM* wave. The strengths of the reflected and transmitted waves for planar surfaces are given by the Fresnel coefficients, which are functions of the incident polarization and material properties. The formal expressions relating the surface electric and magnetic currents and charges to the scattered fields are the Stratton-Chu equations. These expressions are integrals which perform the phasor and vector sums over the scattering body surface.

References and Select Bibliography

It is impossible to give a complete list of references for electromagnetic scattering. A good place to start would be the references listed in [2] and below. The list here is mostly from [2].

1. J.H. Huynen, "Phenomenological Theory of Radar Targets," *Electromagnetic Scattering*, L.E. Uslenghi, ed., New York, Academic Press, 1978.

2.   G.T. Ruck, ed., *Radar Cross Section Handbook*, Vols. 1 and 2, New York, Plenum Press, 1970.

3.   J.A. Stratton, *Electromagnetic Theory*, New York, McGraw-Hill, 1941.

4.   M. Born and E. Wolf, *Principles of Optics*, New York, Pergamon Press, 1959.

5.   M. Kline and I.W. Kay, *Electromagnetic Theory and Geometrical Optics*, New York, John Wiley and Sons, 1965.

6.   R.G. Kouyoumjian, "An Introduction to Geometrical Optics and the Geometrical Theory of Diffraction," *Recent Advances in Antenna and Scattering Theory*, short course notes, Ohio State University, 1965.

7.   J.B. Keller, "A Geometrical Theory of Diffraction," *Symposium on the Calculus of Variations and Its Application*, New York, McGraw-Hill, 1958, p. 27, and a discussion at the end of Chapter IX.

8.   J.B. Keller, "Geometrical Theory of Diffraction," *J. Opt. Soc. Am.*, 52:116, 1962.

9.   J.B. Keller, "Diffraction by an Aperture," *J. Appl. Phys.*, 28:426, 1957.

10.   R.E. Kleinman, "The Rayleigh Region," *Proc. IEEE*, 53:848, 1965.

11.   R.F. Harrington, L.F. Chang, A.T. Adams, *et al.*, "Matrix Methods for Solving Field Problems," Syracuse University (August 1966), AD 639744.

12.   L. Peters, "End-Fire Echo Area of Long Thin Bodies," *Trans. IRE*, AP-6:133, 1958.

13.   J.W. Crispin, Jr. and K.M. Siegel, editors, *Methods of Radar Cross Section Analysis*, New York, Academic Press, 1968.

14.   J.J. Bowman, T.B.A. Senior, and P.L.E. Usleughi, *Electromagnetic and Acoustic Scattering from Simple Shapes*, Amsterdam, North-Holland, 1969.

15.   R. Mittra, ed., *Computer Techniques for Electromagnetics*, London, Pergamon Press, 1973.

16.   R.F. Harrington, *Field Computation by Moment Methods*, New York, MacMillan, 1968.

17.   J.F. Shaeffer, "EM Scattering from Bodies of Revolution with Attached Wires," *IEEE APS*, Vol. AP-30, No. 3, May 1982.

18.   L.E. Uslenghi, ed., *Electromagnetic Scattering*, New York, Academic Press, 1978.

19.   O.D. Jefimenko, *Electricity and Magnetism*, New York, Meredith Publishing, 1966.

20.   J.R. Reitz and F.J. Milford, *Foundations of Electromagnetic Theory*, London, Addison-Wesley, 1967.

21.   R.F. Harrington, *Time Harmonic Electromagnetic Fields*, New York, McGraw-Hill, 1961.

22. J.A. Stratton, *Electromagnetic Theory*, New York, McGraw-Hill, 1941.
23. S.Ramo, J.R. Whinnery, and T. Van Duzer, *Fields and Waves in Communication Electronics*, New York, John Wiley and Sons, 1965.
24. W.H. Panofsky and M. Phillips, *Classical Electricity and Magnetism*, London, Addison-Wesley, 1955.
25. J.D. Jackson, *Classical Electrodynamics*, New York, John Wiley and Sons, 1962.

CHAPTER 4

EXACT PREDICTION TECHNIQUES

J.F. Shaeffer

4.1 INTRODUCTION

The objective of this chapter is to briefly review the classical exact solutions for the scattering of simple objects and then to examine the very powerful modern techniques used to solve Maxwell's equations as expressed by the Stratton-Chu integral equation formulation. Examples of numerical solutions then will be presented.

Exact solutions for practical geometries for scattering are rarely found. This is because the wave equation is solvable by historical analytical methods when the scattering geometry coincides with one of the few separable coordinate systems for which exact series solutions are available. Unfortunately, few practical geometries match the solutions available.

The advent of computer solutions in the last twenty years has lead to techniques for solving the exact integral equations of electromagnetics. Until the computer age, these formulations were not considered useful. Education in electromagnetics, which centered on classical solutions to Maxwell's equations applied to separable coordinate systems and special series solutions, is now giving way to the modern numerical methods.

4.2 CLASSICAL SOLUTIONS

We saw in Ch. 3 that Maxwell's equations for time harmonic waves in source-free regions can be combined to form the wave equation:

$$\nabla^2 \vec{F} + k^2 \vec{F} = 0 \tag{4-1}$$

where $\vec{F}$ represents either the electric or magnetic field. This is a second-order partial differential equation whose solution can yield the fields scattered by simple bodies [1]. The solution is limited to simple bodies because a coordinate system must be found for which the body surface coincides with one of the coordinates. An example is the infinite cylinder, whose surface coincides with radial coordinates $\rho = a$, where a is the cylinder radius. Another example is the sphere, whose surface coincides with the radial coordinate $r = a$, which is the radius of the sphere. There are not many coordinate systems for which this requirement can be satisfied.

Equation (4-1) must be satisfied by each of the three rectangular components of the vector field $\vec{F}$. If we represent any one of those vector components by a function V, then V is a solution of the scalar wave equation:

$$\nabla^2 V + k^2 V = 0 \tag{4-2}$$

In order to solve the scalar wave equation, we must be able to represent the function V in terms of three other functions, each of which depends on only one coordinate:

$$V(u_1, u_2, u_3) = V_1(u_1) V_2(u_2) V_3(u_3) \tag{4-3}$$

where u_1, u_2, u_3 represent the three coordinates. These could be the x, y, z of the rectangular coordinate system, or the r, θ, ϕ of the spherical coordinate system for example. The following are eleven coordinate systems in which the scalar wave equation is separable:

    rectangular
    circular cylindrical
    elliptic cylindrical
    parabolic cylindrical
    conical
    parabolic
    spherical
    prolate spheroidal
    oblate spheroidal
    paraboloidal
    ellipsoidal

In obtaining three ordinary differential equations from the partial differential equation implied by Eq. (4-1), a pair of separation constants must be introduced. The allowed values of the separation constants may be discrete numbers, or they may represent a continuous spectrum, but they must be determined by invoking the boundary conditions that the electromagnetic fields must satisfy at the surface of the obstacle. It is possible to determine the separation constants if the obstacle is perfectly "soft" or perfectly "hard," terms that arise in acoustical scattering. These boundary conditions are

$$V = 0 \qquad \text{(perfectly soft)} \tag{4-4}$$

$$\frac{\partial V}{\partial n} = 0 \quad \text{(perfectly hard)} \tag{4-5}$$

where n is the coordinate normal to the surface over which Eq. (4-4) or (4-5) is imposed. If the body is penetrable, or has a surface impedance that is neither zero nor infinity, then the scalar wave equation is completely separable only for the rectangular, spherical, and circularly cylindrical coordinate systems. Thus, complete solutions are possible in all eleven systems for perfectly conducting bodies, but in only three for imperfectly conducting bodies.

The solutions of the vector wave equation always involve polynomials or infinite series which are not necessarily easy to generate. In much the same way that an infinite series of sine and cosine functions (the Fourier series) can be combined to describe any periodic waveform, an infinite collection of Bessel functions, for example, can be combined to duplicate the total fields around the surface of a circular cylinder illuminated by a single plane wave. In the case of a sphere, we find the solution in terms of the associated Legendre polynomials. Whatever the case, we always truncate the infinite series at a point where additional terms have no significant impact on the terms already summed. However, as the obstacle becomes progressively larger, more terms are required to reach that point, and for bodies much larger than 10 to 20 wavelengths, the exact solutions are no longer attractive nor even interesting.

When separable, the wave equation yields the exact solution for the total field everywhere in space, and typically the incident field is expanded in terms of elemental waves in the coordinate system used. The scattering in any direction is obtainable this way, and it will be found that the scattered field decays with increasing distance from the obstacle. When the scattered field is normalized with respect to the incident field, squared, and then multiplied by the area of a sphere whose radius is the distance to the observation point, we obtain the radar cross section. See page 48 The same kind of procedure is used for two-dimensional problems, except that after normalizing with respect to the incident field strength and squaring, we multiply by the circumference of a cylinder whose radius is the distance to the far field observation point. In this case, we obtain a scattering width (not an area), which may be interpreted as the radar cross section per unit length of the two-dimensional structure.

In any two-dimensional coordinate system, we can distinguish two separate cases, depending on whether the incident electric or magnetic field is parallel to the axis (generally the z-coordinate) of the obstacle. The result for the perfectly conducting circular cylinder is perhaps the easiest of all the two-dimensional problems to obtain, and the scattered fields for the two polarization cases are

$$E^s = - \sum_{n=0}^{\infty} \epsilon_n (-i)^n \frac{J_n(ka)}{H_n^{(1)}(ka)} H_n^{(1)}(k\rho) \cos n\phi \qquad (4\text{-}6)$$

$$H^s = - \sum_{n=0}^{\infty} \epsilon_n (-i)^n \frac{J_n'(ka)}{H_n^{(1)\prime}(ka)} H_n^{(1)}(k\rho) \cos n\phi \qquad (4\text{-}7)$$

where the primes indicate the derivative with respect to the argument, and

$$\epsilon_n = \begin{cases} 1, \ n = 0 \\ 2, \ n \neq 0 \end{cases} \qquad (4\text{-}8)$$

The Hankel function is a linear combination of the Bessel functions of the first and second kinds,

$$H_n^{(1)}(ka) = J_n(ka) + i Y_n(ka) \qquad (4\text{-}9)$$

In these equations, a is the radius of the cylinder, ρ is the distance from the cylinder axis to the point of observation of the scattered field, and ϕ is the bistatic angle subtended at the cylinder axis between the directions of incidence and scattering. The Hankel function in the numerator represents an outward traveling wave, and for large arguments its intensity decays as $(k\rho)^{-1/2}$.

The scattering width of the cylinder (the radar cross section per unit length) for the two cases is

$$\sigma_e = \frac{2\lambda}{\pi} \left| \sum_{n=0}^{\infty} \epsilon_n (-1)^n \frac{J_n(ka)}{H_n^{(1)}(ka)} \cos n\phi \right|^2 \qquad (4\text{-}10)$$

$$\sigma_h = \frac{2\lambda}{\pi} \left| \sum_{n=0}^{\infty} \epsilon_n (-1)^n \frac{J_n'(ka)}{H_n^{(1)\prime}(ka)} \cos n\phi \right|^2 \qquad (4\text{-}11)$$

A bistatic scattering pattern calculated using Eq. (4-10) may be found in Fig. 6-1 for a cylinder 10 wavelengths in circumference. The backscattering width can be obtained from these formulas simply by allowing ϕ to be zero, and Fig. 5-6 illustrates how the backscattering depends on the size of the cylinder.

In the case of the sphere, the scattered field must be represented in terms of θ and ϕ components, where θ is the bistatic angle subtended in the directions of incidence and scattering at the center of the sphere, and ϕ is the angle between the plane of scattering (formed by the directions of incidence and scattering) and the plane containing the incident electric field and direction of incidence (see Fig. 4-1 for details of the scattering geometry). The components of the scattered field are

$$E_\theta^s = \frac{-i e^{ika} \cos \phi}{kr} \sum_{n=1}^{\infty} (-1)^n \frac{2n+1}{n(n+1)} \left[b_n \frac{\partial P_n^1(\cos \theta)}{\partial \theta} - a_n \frac{P_n^1(\cos \theta)}{\sin \theta} \right]$$

$$(4\text{-}12)$$

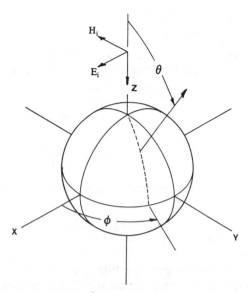

Figure 4-1 Spherical polar scattering geometry

$$E_\phi^s = \frac{i\,e^{ika}\,\sin\phi}{kr} \sum_{n=1}^{\infty} (-1)^n \frac{2n+1}{n(n=1)} \left[b_n \frac{\partial P_n^1(\cos\theta)}{\sin\theta} - a_n \frac{\partial P_n^1(\cos\theta)}{\partial\theta} \right]$$

$$(4\text{-}13)$$

where r is the distance to the point of observation, a is the radius of the sphere, and $P_n{}^1(\cos\theta)$ is the associated Legendre function of order n and degree 1. The general function is defined as

$$P_n^m(x) = \frac{(1-x^2)^{m/2}}{2^n n!} \cdot \frac{d^{n+m}(x^2-1)^n}{dx^{n+m}}$$

$$(4\text{-}14)$$

hence, the first three orders of $P_n^1(\cos\theta)$ have the values:

$$P_0^1(\cos\theta) = 0$$

$$P_1^1(\cos\theta) = \sin\theta$$

$$(4\text{-}15)$$

$$P_2^1(\cos\theta) = \frac{3}{2}\sin 2\theta$$

The coefficients a_n and b_n are

$$a_n = \frac{j_n(ka)}{h_n^{(1)}(ka)}$$

$$(4\text{-}16)$$

and

$$b_n = \frac{ka\, j_{n-1}(ka) - n j_n(ka)}{ka\, h_{n-1}^{(1)}(ka) - n h_n^{(1)}(ka)} \tag{4-17}$$

where $h_n^{(1)}(x) = j_n(x) + i y_y(x)$, in which $j_n(x)$ and $y_n(x)$ are the spherical Bessel functions of the first and second kinds, respectively. These are ordinary Bessel functions of half-order, and the first pairs are

$$j_0(x) = \frac{\sin x}{x} \qquad\qquad y_0(x) = -\frac{\cos x}{x}$$

$$j_1(x) = \frac{1}{x}\left(\frac{\sin x}{x} - \cos x\right) \qquad y_1(x) = -\frac{1}{x}\left(\frac{\cos x}{x} + \sin x\right) \tag{4-18}$$

In the backscattering direction, Eq. (4-13) becomes zero and the radar cross section is

$$\sigma = \frac{\lambda^2}{\pi}\left|\sum_{n=1}^{\infty} (-1)^n (n+1/2)(b_n - a_n)\right|^2 \tag{4-19}$$

This is the expression used to generate Figs. 3-5 and 6-4.

As recently as 1960, the computation of these functions was often performed by groups of computers, people operating desk calculators, and working with published tables of functions, such as those found in the NBS *Handbook of Mathematical Functions* [2]. The desk calculators were mechanical and had little capability for internal storage of intermediate results, and the computers had to print column after column of numbers to be used in subsequent calculations. Anyone who has not had to generate literally reams of numbers cannot appreciate the value of even the simplest hand-held electronic calculator. Even the large, high-speed electronic computer, however, does not solve all our problems. For example, the ascending series representation of $J_n(x)$, the Bessel function of the first kind of order n, becomes unstable as the argument approaches 20 or 30, and we must generate the functions using a technique known as backward recursion. $Y_n(x)$, the Bessel function of the second kind of order n, can be computed using forward recursion. However, even these techniques also become troublesome for large orders because the recursion generates very large numbers before the numbers are normalized (i.e., arithmetic overflow).

4.3 INTEGRAL EQUATION SOLUTIONS

The objective of this presentation of the integral equation formulation will not be to instruct the reader on how to work out the equations, but rather to present an overview of the method. As a starting point, we will present the basic electric field and magnetic field integral equations. From these we will obtain the equations which must be solved after the *EM* boundary conditions have been applied to the surface of the scattering body. The procedure for solving for the unknown surface currents will then be presented, including a discussion of operator notation, the expansion of the induced currents in a finite series of basis functions, and the development of the interaction matrix. Finally, once the surface currents have been found, the scattered field will be determined. Several examples of the scattering from simple bodies will be given showing the solution for induced currents and the resultant scattered field. The physical meaning of the results will be interpreted in terms of the Green's function and its gradient (Huygen's wavelets), and the matrix elements will be related to the field at one point on a scattering body produced by the currents flowing on another part of the body.

Although the integral equation formulation is exact, some analysts may regard the solutions as numerical methods, which inherently have less than exact characteristics. However, the power and utility of these methods lie in their application to arbitrary geometries for arbitrary electromagnetic excitations. The integral equation methods are limited to the low frequency and resonant frequency ranges. Although the methods apply in principle to any frequency range, they suffer two practical limitations in the high frequency region. First, the matrix for large body problems (relative to a wavelength) becomes very large, precluding solution even by today's large, high speed, main-frame computers. Second, the need for matrix element interaction in the high frequency region decreases significantly because the scattering becomes more a localized phenomenon than a collective phenomenon.

The term "method of moments," and the associated acronym MOM or MM, has been applied to the matrix solution of the *EM* integral equations. The problems which have been solved by MOM include wires, two-dimensional and three-dimensional surfaces using surface patches, bodies of revolution (BOR) and bodies of translation (BOT). The solution goal, whether viewed as the end itself or as a necessary step for obtaining the scattered fields, is to determine the currents induced on the scattering body. These currents are functions of the incident polarization, the direction of arrival of the incident wave, and of how various parts of the scattering body interact.

4.3.1. EM Integral Equations

The electromagnetic integral equations were obtained by Stratton and Chu using the vector Green's theorem in conjunction with Maxwell's equations. These equations give the prescription for scattered fields in terms of surface current sources. Surface $\bar{E}$ and $\bar{H}$ vectors are decomposed into tangential and perpendicular components.

The total electric and magnetic fields are written as the sum of the incident and scattered fields,

$$\vec{E}^T = \vec{E}^i + \vec{E}^s$$
$$\vec{H}^T = \vec{H}^i + \vec{H}^s \tag{4-20}$$

The scattered $\vec{E}$ and $\vec{H}$ fields are given by the Stratton-Chu integrals [3]:

$$\vec{E}^s = \oint_s \left[i\omega\mu\,(\hat{n} \times \vec{H})\psi + (\hat{n} \times \vec{E}) \times \nabla\psi + (\hat{n} \cdot \vec{E})\,\nabla\psi \right] dS$$

$$\vec{H}^s = -\oint_s \left[i\omega\epsilon\,(\hat{n} \times \vec{E})\psi - (\hat{n} \times \vec{H}) \times \nabla\psi - (\hat{n} \cdot \vec{H})\,\nabla\psi \right] dS \tag{4-21}$$

where ψ is the free-space Green's function, ω is the radian frequency, μ and ϵ are the permeability and permittivity, and $\hat{n}$ is the outward unit normal on surface S. The scattered electric field has as its surface sources the tangential components of the total magnetic and electric fields (electric and magnetic currents) and the perpendicular component of ~~magnetic field~~ (~~magnetic~~ *Electric field (electric charge)* charge).

The tangential and perpendicular components of the surface field are interpreted as currents and charges:

$$\begin{aligned}
\vec{J} &= \hat{n} \times \vec{H}^T & &\text{electric current} \\
\vec{M} &= -\hat{n} \times \vec{E}^T & &\text{magnetic current} \\
\rho &= \epsilon\,\hat{n} \cdot \vec{E}^T & &\text{electric charge} \\
\rho^* &= \mu\,\hat{n} \cdot \vec{H}^T & &\text{magnetic charge}
\end{aligned} \tag{4-22}$$

where $\hat{n}$ is the unit normal to the surface.

The Green's function ψ and its gradient $\nabla\psi$ are the mathematical equivalents of Huygen's wavelets. That is, each elemental surface current or charge is related to the scattered field by means of the Huygen wavelet and the total field is simply the sum (integral) over all such surface current elements. Huygen's wavelets are graphically shown in Fig. 4-2, in which the scattering from a flat surface due to an incident plane wave is shown by the heavy arrow. Each wavelet has a radius corresponding to the time history of excitation by the incident wave. The scattered waves generated by the summation of wavelets have two components, the first being the specularly reflected wave whose angle of reflection is equal to the angle of incidence. The second is a forward scattered wave propagating in the same direction as the incident wave, but with the opposite phase. As a result, when the forward scattered wave is added to the incident wave, it creates a shadow or null total field on the back side of the flat surface.

Mathematically, the Green's function relates an elemental source current or charge to the field at the observation point. The three-dimensional Green's function is an outward scalar spherical wave whose intensity falls off as inverse distance,

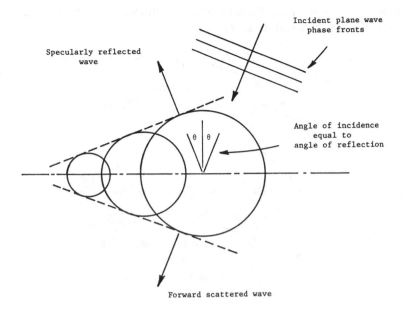

Figure 4-2 Huygen's wavelet wave-front construction

$$\psi = \frac{e^{ikR}}{4\pi R} \tag{4-23}$$

where an $e^{-i\omega t}$ time dependence has been assumed and where R is the distance from the elemental source (prime coordinates) to the observer (unprimed coordinates),

$$R = [(x - x')^2 + (y - y')^2 + (z - z')^2]^{1/2} \tag{4-24}$$

Recalling that the gradient points in the direction of the maximum rate of change of the function, the gradient of the Green's function is an outward vector spherical wave,

$$\nabla \psi = (1 + ikR) \frac{e^{ikR}}{4\pi R^2} \hat{R} \tag{4-25}$$

$$= (1 + ikR) \psi \hat{R} / R$$

whose vector direction $\hat{R}$ is radially outward from each elemental source. The far zone E and H fields must be perpendicular to the radial vector $\nabla \psi$, and this requirement is inherently contained within the integral formulation, as we shall shortly see. The decay of this wave in the near field, $kR \ll 1$, is inversely as the square of the distance, while in the far field, $kR \gg 1$, it is inversely as the first power of the distance. Figure 4-3 further illustrates the physical nature of the Green's function and its gradient.

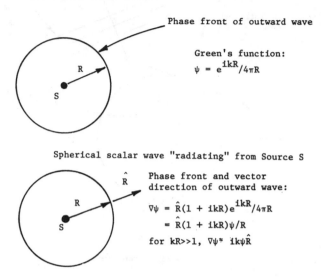

Phase front of outward wave

Green's function:
$\psi = e^{ikR}/4\pi R$

Spherical scalar wave "radiating" from Source S

$\hat{R}$ Phase front and vector direction of outward wave:

$\nabla\psi = \hat{R}(1 + ikR)e^{ikR}/4\pi R$
$= \hat{R}(1 + ikR)\psi/R$
for $kR \gg 1$, $\nabla\psi \approx ik\psi\hat{R}$

Spherical vector wave "radiating" from Source S

Figure 4-3 Green's function and its gradient as a Huygen wavelet

This definition of the Green's function is not valid when the source and field points coincide, since $R = 0$ and ψ and $\nabla\psi$ are infinite. In these cases, we must recognize that a Green's function can still be defined in the sense that it represents a "self" term. Self terms for currents and charge sheets are derivable from Maxwell's equations using the integral form of the curl and divergence equations with elemental loops (lines) and pillboxes (volumes). The results are

$$(\hat{n} \times \vec{H})_{self} = \frac{1}{2}\vec{J}, \qquad (\hat{n} \times \vec{E})_{self} = \frac{1}{2}\vec{M},$$

$$(\hat{n} \cdot \vec{E})_{self} = \frac{\rho}{2\epsilon} \qquad\qquad (\hat{n} \cdot \vec{H})_{self} = \frac{\rho^*}{2\mu}$$

(4-26)

where E and H are the total fields on the surface due to currents J and M and charge densities ρ and ρ^*. When the Stratton-Chu equations include spatial locations where source and field points coincide, we must write out the self term explicitly and specify that the surface integrals exclude the troublesome self region. The symbol for such an integral is the integral sign with a bar through it and is called the principal value integral.

When we consider scattering from imperfect conductors, such as dielectric and magnetic bodies, we must include magnetic as well as electric currents and charges. For the discussions which follow, we will specialize the scattering formulation to that from a perfect conductor, so that the total tangential electric field is zero at the surface. Then we need not consider magnetic currents or charges as sources of the scattered E and H fields. Our specialization still enables us to examine the nature of the solutions for a practical case without undue complication. For perfect conductors, with the magnetic sources set to zero, the integral forms for the E and H fields become

$$\vec{E}^s = \int_S [i\omega\mu(\hat{n} \times \vec{H})\psi + (\hat{n} \cdot \vec{E})\nabla\psi]dS = \int_S [i\omega\mu\vec{J}\psi + \frac{1}{\epsilon}\rho\nabla\psi]dS$$

$$(4\text{-}27)$$

$$\vec{H}^s = \int_S (\hat{n} \times \vec{H}) \times \nabla\psi\, dS = \int_S \vec{J} \times \nabla\psi\, dS$$

These equations are known as the electric and magnetic field integral equations, or EFIE and MFIE, respectively. These forms show that the scattered E field is caused by electric currents and charges, while the scattered H field is caused only by electric currents. The next step is to apply the boundary conditions.

4.3.2 Boundary Conditions

The EFIE and MFIE are not yet in the form required to effect a solution. To do so, we must apply Eq. (4-27) to the surface of a perfectly conducting scattering body, for which we know that the tangential components of the fields are zero.

First, the surface charge density is rewritten invoking the conservation of charge using the continuity equation,

(see page 72)

$$(\hat{n} \cdot \vec{E}) = \frac{1}{\epsilon}\rho = \frac{i}{\omega\epsilon} (\nabla \cdot \vec{J}) \qquad (4\text{-}28)$$

This states that charge can be neither created nor destroyed, hence the charge density must change as current flows into or out of a given location.

When the observation point is on the surface, where the field values are known from the boundary conditions, the resulting forms for the EFIE and MFIE may be obtained as follows [3]:

$$\hat{n} \times \vec{E}^T = n \times (\vec{E}^i + \vec{E}^s) = 0 \qquad (4\text{-}29)$$

and

$$\hat{n} \times \vec{H}^T = \hat{n} \times (\vec{H}^i + \vec{H}^s) = \vec{J}$$

which leads to

[handwritten annotations:]

$\nabla \times H = (6 - i\omega\epsilon) E$ *(p.72)*

$\nabla \times H = - i\omega\epsilon E$ *(for free space or dielectric)*

one: since $\nabla \times H = J$ *then*

$J = - i\omega\epsilon E$

$\nabla \cdot J = \nabla D (-i\omega)$

on $(n \cdot E) = \frac{i}{\omega\epsilon}(\nabla \cdot \vec{J}) = \frac{\delta}{\omega\epsilon} \cdot (-i\omega DD)$

or $(n \cdot E) = D/\epsilon$

56.

$$\hat{n} \times \vec{E}^i = - \hat{n} \times \vec{E}^s = \hat{n} \times \int [i\omega\mu\vec{J}\,\psi + \frac{i}{\omega\epsilon}\nabla \cdot \vec{J}\,\nabla\psi]dS \qquad (4\text{-}30)$$

and

$$\hat{n} \times \vec{H}^i = \frac{1}{2}\vec{J} - \hat{n} \times \oint \vec{J} \times \nabla\psi\,ds \qquad (4\text{-}31)$$

Equations (4-30) and(4-31) are thus the starting points for obtaining the unknown surface current density. Except for frequencies corresponding to interior body resonances, either form may be used. When the surface becomes very thin, such as for wires and thin cylinders, the EFIE must be used because of the difficulty in adequately representing $\nabla\psi$ in the MFIE for these cases. For closed smooth conductors, the MFIE is often applied.

The EFIE is an integral equation of the first kind because the unknowns occur only within the integral. The MFIE is an integral equation of the second kind because the unknown current J occurs by itself and within the integral. The MFIE is particularly noteworthy because the integral contribution to the current density can be of secondary importance compared to the incident field. In fact, as will be shown in the next chapter, the physical optics assumption for the high frequency region explicitly ignores the field expressed by the integral, so that the current density is given by twice the tangential component of the incident magnetic field without the need to solve an integral equation.

The next step is to solve for the unknown current density.

4.3.3 Solution Procedures

The procedures required to find the unknown current density involve:
- Expressing the unknown in terms of a set of basis functions with unknown coefficients
- Defining weighting or testing functions
- Explicitly defining interaction matrix elements
- Inverting the matrix
- Specifying the polarization and direction of the incident field and computing the resultant current density
- Computing the bistatic scattered field radiated by these induced currents

Monostatic scattering patterns require more computation than bistatic patterns because the induced currents must be computed for each angle of incidence. However, scattering in only one direction needs to be computed, that being back toward the source of illumination.

Because our goal is an overall appreciation of the technique, we will specialize our solution to the slightly simpler MFIE for a two-dimensional closed surface. The solution of integral equations is mathematically associated with the theory of linear vector spaces.

The unknown current density is a surface vector function. Thus, the series expansion must have unknown coefficients for two orthogonal directions on any surface patch. For wires, however, current is usually assumed to flow axially along the wire so that only one vector direction is needed for each segment. The key assumption in breaking up the surface into patches is that the phase of the current is constant over each patch or segment. Because the actual phase angle varies spatially with distance, the constant phase requirement over each surface patch requires five to ten samples per wavelength to adequately sample the actual variation. This requirement, in turn, dictates the matrix size for a given problem. Because the number of unknowns increases as the square of the number of surface patches, MOM is typically used only for resonant region scattering.

The unknown surface currents are typically expanded as

$$\vec{J} = \sum_{i=1}^{N} b_{x,i} f(t) \, \hat{u}_x + b_{y,i} f(t) \, \hat{u}_y \qquad \text{(for 3-D surfaces)}$$

$$\text{(4-32)}$$

$$= \sum_{i=1}^{N} b_i \vec{f}_i(t) \qquad \text{(for wires or 2-D surfaces)}$$

where $\hat{u}_x$ and $\hat{u}_y$ are the orthogonal unit surface vectors, $f(t)$ is an expansion function, and b is the complex (magnitude and phase) unknown current coefficient.

Before going further, we shall introduce the vector operator formalism used to considerably shorten the mathematical notation. It is traditional to compactly write the EFIE and MFIE integrals with linear operators, which are "short-hand" mathematical notations, and in no way does the notation change the physics or numerical procedures. The electric field operator is defined as

$$L_E(\vec{J}) = \hat{n} \times \int [i\omega\mu\vec{J}\psi + \frac{1}{i\omega\epsilon} (\nabla \cdot \vec{J}) \nabla\psi] dS \qquad \text{(4-33)}$$

and the magnetic field operator is defined as

$$L_H(\vec{J}) = \frac{1}{2} \, \vec{J} - \hat{n} \times \oint \vec{J} \times \nabla\psi \, dS \qquad \text{(4-34)}$$

The physical interpretation of these operators is that they give the tangential scattered field on the surface due to a surface current J.

With the aid of this operator notation, the solution is obtained by inserting the series expansion for the unknown currents into the MFIE and remembering that the unknown coefficients are constants which may be brought out from under the integral operators:

$$L_H(\vec{J}) = \sum_{i=1}^{N} b_i L_H(\vec{f}_i) = \hat{n} \times \vec{H}^i \qquad (4\text{-}35)$$

The next step is to multiply each side of Eq. (4-35) by a vector weighting function $\vec{W}_j$ and to integrate the result over each surface patch. The mathematical notation for this step is that of a generalized inner product and is given the following symbols:

$$< \vec{W}, L_H(\vec{J}) > = \int \vec{W} \cdot L_H(\vec{J}) dS \qquad (4\text{-}36)$$

This step can be physically interpreted as the way in which continuous boundary conditions over a surface region are expressed as single-average values at one localized point on the surface. The inner product is applied to each of the N surface current patch samples to arrive at a set of N equations in N unknowns.

$$\sum_{i=1}^{N} b_i < \vec{W}_j, L_H(\vec{f}_i) > = < \vec{W}_j, \hat{n} \times \vec{H}^i > \qquad (4\text{-}37)$$

for $j = 1$ to N.

This set of linear equations for the unknown current coefficients b_i can be compactly expressed in matrix notation as

$$\bar{\bar{Z}}\vec{b} = \vec{H}^i \qquad (4\text{-}38)$$

where the matrix elements are given by

$$Z_{ij} = < \vec{W}_j, L_H(\vec{f}_i) > \qquad (4\text{-}39)$$

and the unknown current coefficients are expressed as a generalized column vector,

$$\vec{b} = \begin{bmatrix} b_1 \\ \cdot \\ \cdot \\ \cdot \\ b_N \end{bmatrix} \qquad (4\text{-}40)$$

The known incident fields, which represent the forcing function for the solution, are also expressed as a generalized column vector:

$$\vec{H}_j^i = < \vec{W}_j, \hat{n} \times \vec{H}^i > \qquad (4\text{-}41)$$

The physical meaning of each term is clear. The matrix elements express the electrical interaction of each part of the scattering surface with every other part. The *ijth* matrix element is a measure of the fields produced at the *ith* surface patch created by a unit current located at the *jth* surface patch. For general 3-D problems, each matrix element is a four-fold integral which results when the surface effect of a source segment (two-fold integral) on the sample surface (another two-fold integral) is computed. Because the matrix elements are for unit currents, the matrix and its elements are independent of the actual electrical excitation of the body. The matrix is simply a measure of the electrical interaction of the body with itself, and as such is a function of body geometry and frequency.

The unknown current coefficients are constants for each small surface patch and may be removed from within the integrals to form the unknown surface current column vector. The excitation vector on the right-hand side is the forcing function for the solution. Its elements are the integral of the weighting function W with the incident field. Physically, this integral is again a finite measure of how the incident field varies over each surface patch. The excitation vector is sometimes called the voltage vector, particularly when the EFIE is used, in which case the applied excitation is an incident electric field.

The physical meaning of the matrix equation can be made clearer if we examine the *j*th row of Eq. (4-37) which represents the fields on the *j*th segment from the incident field plus that from all the other surface segments. The term on the right-hand side represents the excitation field, while the terms on the left-hand side represent the field at the *j*th segment from each of the other segments, each with current b_i. The *j*th term is the self term and represents the fields at the *j*th location due to *j*th location currents.

The solution for the surface currents is formally given by

$$[b] = [Z]^{-1} [H] \tag{4-42}$$

The solution procedure is to compute the matrix elements, specify the polarization and incident direction of the illuminating field, compute the excitation vector, invert the impedance matrix, and then solve for the currents.

4.3.4 Scattered Fields

Once the currents are known for a given excitation, the scattered fields due to these currents may be computed from the EFIE or MFIE expressions, Eq. (4-27). Usually, only the far field values are of interest and, therefore, the far field Green's function gradient is used in the EFIE and MFIE to obtain [4]:

$$\bar{E}^s = \frac{i\omega\mu}{4\pi R} e^{ikR} \int_S [\bar{J} - (\bar{J} \cdot \hat{R})] e^{-ikr} dS \tag{4-43}$$

$$\bar{H}^s = \frac{-i\omega\epsilon}{4\pi R} e^{ikR} \int_S \sqrt{\frac{\mu}{\epsilon}} (\bar{J} \times \bar{R}) e^{-ikr} dS \tag{4-44}$$

where R is the distance from a local origin to the observation point and r is a local surface coordinate.

The choice between the far field EFIE or MFIE is not significant because in the far field E and H are related by Maxwell's equation. For bistatic computations, either Eq. (4-43) or (4-44) is evaluated for the angles of interest. For monostatic computations, only the backscattered field is computed, and then the entire process must be repeated for determining a new set of surface currents for the next illumination angle. Once the scattered fields are known, the RCS may be computed using Eq. (3-1).

Further insight into the far zone fields as a function of current sources can be gained from examining the above expressions. As we have already seen, elemental sources are related to observed fields via a Green's function which is spherical in nature. However, the far field patterns produced by elemental current sources are shaped like doughnuts with nulls along the vector direction of the current element (see Fig. 4-4). From the far zone expression for magnetic field, we see that $\vec{H}^s$ is a cross product of the current element $\vec{J}$ and the radial vector of $\nabla \psi$. This creates a doughnut shape with an $\vec{H}$ vector transverse to the radial direction, as is required for an outward wave. From the expression for the far zone electric field, we see that $\vec{E}^s$ is the product of the scalar Green's function and the transverse components of the current element $\vec{J}$, that is, the radial component of $\vec{J}$ is vectorially subtracted and does not contribute to $\vec{E}^s$ in the far field. When the far field point lies along the $\vec{J}$ direction, $\vec{J}$ is entirely radial, hence there is no contribution to the far field in this direction. This again creates the doughnut pattern. The vector direction of $\vec{E}^s$ is, of course, transverse to the radial direction because only the transverse components of $\vec{J}$ contribute to the far field. The integration process forms the phasor and vector sum of all such current elements to arrive at the net field due to all elemental sources $\vec{J}$.

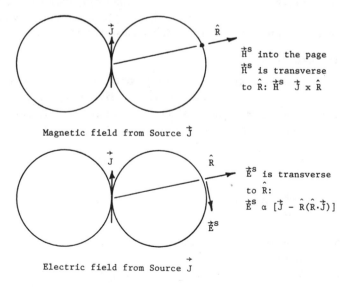

Magnetic field from Source $\vec{J}$

$\vec{H}^s$ into the page
$\vec{H}^s$ is transverse
to $\hat{R}$: $\vec{H}^s$ $\vec{J} \times \hat{R}$

Electric field from Source $\vec{J}$

$\vec{E}^s$ is transverse
to $\hat{R}$:
$\vec{E}^s \propto [\vec{J} - \hat{R}(\hat{R} \cdot \vec{J})]$

Figure 4-4 Far-zone fields from an elemental current

When we actually compute the far zone $\vec{E}^s$ field from Eq. (4-43), we subtract the radial component of $\vec{J}$ prior to integration, as indicated. An alternative is to compute $\vec{E}^s$ from

$$\vec{E}^s = \frac{i\omega\mu}{4\pi R} e^{ikR} \int \vec{J} e^{-ikr} dS \qquad (4\text{-}45)$$

without the radial subtraction, and then to perform a dot product of the resultant $\vec{E}^s$ with two orthogonal unit vectors to find the transverse $\vec{E}$ field components.

$$
\begin{aligned}
E_\theta &= \hat{n}_\theta \cdot \vec{E}^s \\
E_\phi &= \hat{n}_\phi \cdot \vec{E}^s
\end{aligned}
\qquad (4\text{-}46)
$$

where $\hat{n}_\theta$ and $\hat{n}_\phi$ are orthogonal unit vectors transverse to the radial direction $\hat{R}$.

4.3.5 Example Solutions

The scattering process can be understood by examining the solution for the induced currents and resultant scattered field for a collection of simple body geometries. For this purpose, we have chosen some rather elementary shapes: a wire, the infinite circular cylinder, a rectangular cylinder, and a body of revolution with attached wires. Each configuration reveals something different about the induced currents and the far field that results from those currents.

The first example of scattering predictions is for a wire 2λ in length illuminated by a wave whose incident E field has a component along the wire. The results presented here are from Harrington [5], who used a MOM wire code to compute the induced currents from which the bistatic scattering patterns were computed. The results for six angles of illumination are presented in Fig. 4-5. The most interesting characteristic is that, for even a short wire, the specular reflectivity nature of high frequency scattering is obvious. We can see that the angle of the major scattering lobe is equal to the incident angle. This behavior becomes more pronounced as the wire becomes longer, for which the major lobes become narrower in width and more intense. The wire is not an effective scatterer when viewed near end-on incidence because the component of the incident E field along the wire direction is small.

Note the presence of the traveling wave lobe, which appears in the case for an incident angle of 30 degrees. This lobe is aligned back in the general direction of the illumination and peaks near 35 degrees. It is caused by a reflection of the induced forward current on the wire, giving rise to a backward current which radiates the traveling wave lobe. The angular location of this lobe is discussed in Ch. 5. The formula given there predicts a location of 34.6 degrees, which agrees quite well with the numerical results.

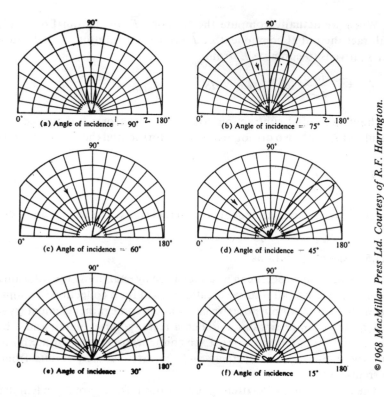

Figure 4-5 Bistatic RCS patterns of a straight wire 2λ in length (from Harrington [5])

Before we examine the scattered patterns for two-dimensional structures, the reader should bear in mind that only the scattered field is presented. The total field is the sum of the incident and scattered fields.

$$\vec{H}^T = \vec{H}^i + \vec{H}^s$$

and in the shadow region behind the scatterer the total field is small. To create the shadow, the forward scattered field must cancel the incident field. Thus, it must have nearly the same magnitude, but the opposite phase, as the illuminating field, so that phasor addition (subtraction in this case) of the two creates a small value. In the scattering patterns below, the shadow-creating forward scatter lobe is the strongest lobe.

The next examples are for two-dimensional scattering results for the case where there is a component of the incident E field along the surface (H field transverse to plane of Fig. 4-6). The numerical method used is that given by Harrington [5] for two-dimensional MFIE MOM computations. The reader

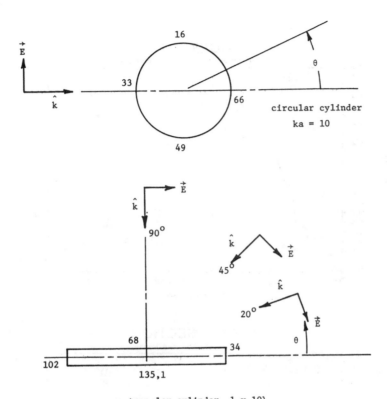

rectangular cylinder, 1 x 10λ

Figure 4-6 Two-dimensional MOM geometries. Numerical values signify segment numbers

is urged to examine the computed current distributions of the following examples. Because these were obtained using numerical solution techniques for the exact *EM* formulation, we may take these results as "exact" currents. It is instructive to compare them with the assumed physical optics currents, as discussed in Ch. 5, in which currents on the illuminated surfaces are taken as twice the tangential magnetic field strength, while the currents on the shadow surfaces are taken as zero.

The currents induced on a circular cylinder and the resultant bistatic scattering patterns are shown in Fig. 4-7 for *ka* = 10. The amplitude of the computed current density is normalized to the tangential component of the incident magnetic field and is plotted as a function of the circle segment number defined in Fig. 4-6. The angle of illumination was 180 degrees. We see that the illuminated side currents have peak values of twice the incident magnetic field,

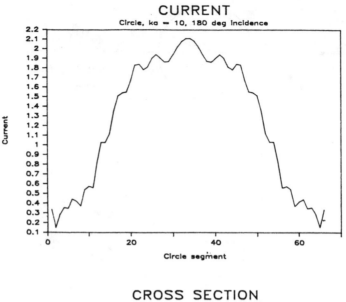

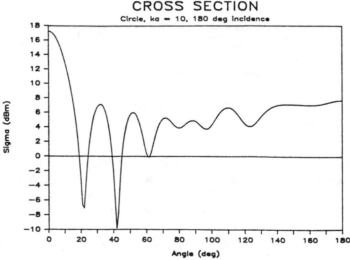

Figure 4-7 Circular cylinder current distribution and bistatic scattering pattern. Illumination is from $\theta = 180°$

while those on the shadow side decrease toward zero. The RCS in the backward direction (at 180 degrees) is fairly constant, agreeing very well with the high frequency 2-D geometric optics result of $\sigma = \pi a = 5m = 7\,dBm$. This scattering pattern should be compared with that presented in Fig. 6-1 computed using an infinite series solution.

The results for a rectangular cylinder with a transverse cross section 1λ by 10λ are shown for illumination angles of $90°$, $45°$, and $20°$ in Figs. 4-8 through 4-10, respectively. Again the currents are plotted *versus* the segment index number defined in Fig. 4-6. The scattering patterns show the specular nature of the scattering process; the major scattering lobe is aligned along the direction where the angle of reflection is equal to the angle of incidence.

The current distribution on the rectangular cylinder for incidence normal to the broadside agrees well with the *PO* approximation. Near the corners, however, we see decaying ripples which suggest the presence of a surface current wave due to edge diffraction. These ripples indicate that the corners are secondary scattering centers. When the angle of incidence approaches the normal to the narrow side, the currents are not as well approximated by the *PO* values. As has been observed experimentally, this suggests that scattered fields predicted by *PO* can be expected to agree well for incidence angles in the broadside region, but not near end-on.

A final example of an exact MOM solution involves a solution for wire elements attached to a body of revolution. MOM solutions for BORs have been obtained by Mautz and Harrington [6] using current expansion functions which take into account azimuthal symmetry, resulting in azimuthal modal expansion for the currents. The result is that the BOR impedance matrix can be reduced to a set of independent submatrices for each circumferential mode. The MOM solutions for wires in the presence of BORs [7] involve not only the wire solutions, but also the electrical interaction between the wires and the BOR.

The attachment of the wires to the BOR requires the introduction of a special current basis function which mathematically represents the physical nature of the currents in the vicinity of the junction [8,9]. The scattering geometry illustrated is for a hemispherical capped cylinder (the BOR) with wire loops attached (Fig. 4-12). The system matrix (Fig. 4-11) illustrates six types of component electrical interaction elements:

- BOR-BOR in block diagonal modal form
- Wire-wire
- Junction-junction
- BOR-wire
- BOR-junction
- Wire-junction

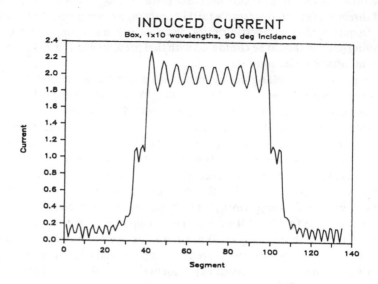

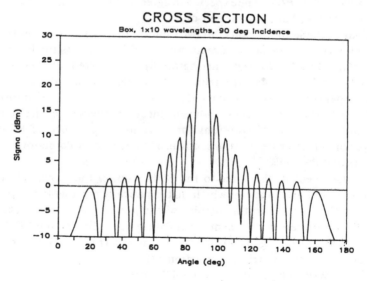

Figure 4-8 Rectangular 1 x 10λ cylinder illuminated at θ = 90°

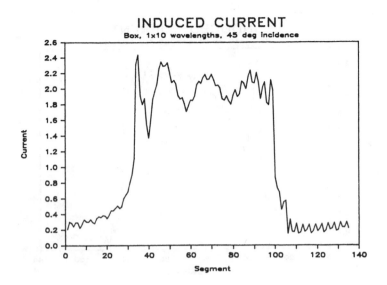

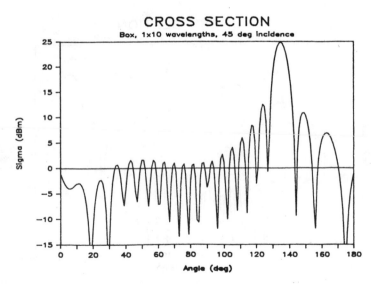

Figure 4-9 Rectangular 1 x 10λ cylinder illuminated at θ = 45°

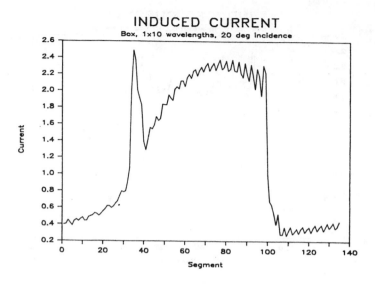

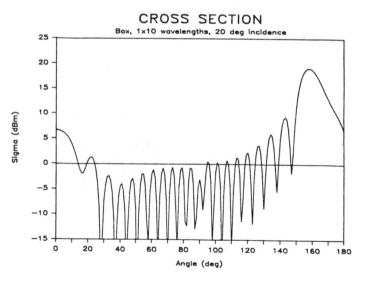

Figure 4-10 Rectangular 1 x 10λ cylinder illuminated at $\theta = 20°$

These submatrices express the electrical interaction of each part of the body with itself and with the other parts of the body. The current expansion coefficients form the unknown current vector and the right-hand side is the known voltgage vector, which is different for each excitation (incident wave). An example of the measured and computed scattering for a BOR with attached wires is shown in Fig. 4 - 12.

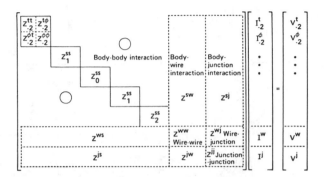

Figure 4-11 Composite BOR-wire-junction system matrix

4.4 COMPARISONS WITH HIGH FREQUENCY SOLUTIONS

The solution techinques for the high frequency region are presented in the next chapter. This section examines some of the results of geometric optics, physical optics, and stationary phase with examples just presented. The goal is to relate the high frequency solutions to the exact solutions.

Geometric Optics

The geometric optics prescription tells us that electromagnetic scattering is confined to rays whose angle of reflection is equal to the angle of incidence. The results in sec. 4.3 show that the scattering is indeed specular. However, the scattered or reflected energy is not confined exactly to the specular direction, but is distributed in a scattering lobe of finite magnitude and width whose maximum is aligned in the specular direction. As the incident wavelength becomes shorter (toward the optic limit of zero), the main scattering lobe increases in magnitude and becomes narrower in width, hence more of the reflected energy is concentrated in exactly the specular direction. The two-dimensional cylinder result shows that the reflected energy is uniformly scattered in the backward half-plane. In terms of GO, this means that the requirement for equal incident and reflected angles is satisfied over this entire range. The magnitude of the GO reflection is given in terms of surface curvature and is constant over the scattering region, as is the exact result.

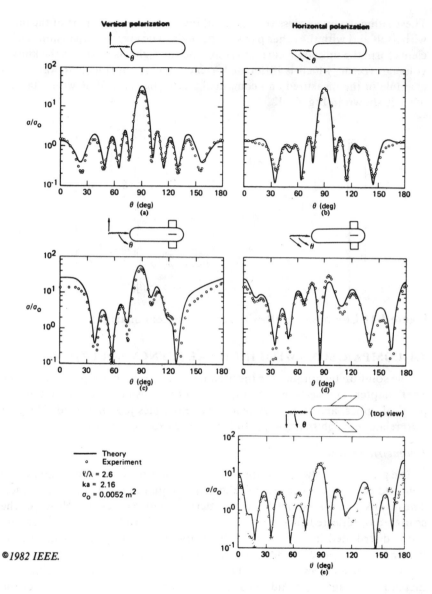

Figure 4-12 Scattering from cylinder with hemispherical endcaps, $l/\lambda = 2.6$, $ka = 2.16$. (a) Clean configuration, vertical polarization; (b) Clean configuration, horizontal polarization; (c) Wire-loop fins, vertical polarization; (d) Wire-loop fins, horizontal polarization; (e) Wire-loop wings, horizontal polarization (from Shaeffer [8]).

Physical Optics

Physical optics assigns explicit values to the induced currents on the surface of the scatterer. The electric current is fixed at twice the tangential component of the incident magnetic field over the illuminated region and at zero over the shadow region. From the exact current distributions given in Figs. 4-7 through 4-10, we find that the *PO* approximation is a good representation of the actual current density. In addition, the MFIE as expressed by Eq. 4-31 indicates that if the principal value integral is ignored, the currents are indeed $2\hat{n} \times \vec{H}^i$. The integral represents interactions due to other parts of the scatterer, and for high frequencies they can be ignored. The fact that the induced currents are taken as localized phenomena is in many cases quite acceptable.

Stationary Phase

The principle of stationary phase states that the scattered field is due primarily to the currents in the Fresnel zone surrounding the specular point. This is not a part of the formalism of the exact solution as expressed in the far zone scattered fields given by Eqs. (4-43) and (4-44). In principle, however, the stationary phase argument also applies to these expressions. The integrations indicated are carried out over the entire surface, but only the Fresnel zone currents contribute significantly to the scattered field. The application of stationary phase is not contained within the exact solutions, and expressions (4-43) and (4-44) are evaluated over the entire surface. To do otherwise would require *a priori* knowledge of the location of the specular point and the size of the Fresnel zone, and this information is seldom incorporated into exact solutions. In addition, performing numerical integration over the entire surface is not nearly as complex a job as the computation of the matrix and its inverse.

4.5 SUMMARY

The classical exact scattering solutions are restricted to bodies whose surfaces coincide with the coordinates of systems in which the wave equation is solvable by the separation of variables technique. The most important of these are the infinite series solutions for the sphere and the infinite two-dimensional circular cylinder. While these early solutions may have given needed insight into the scattering process, they were seldom useful for practical geometries.

The advent of the modern computer has led to numerical solutions for *EM* scattering as formulated by the Stratton-Chu integrals for arbitrary geometries. The range of application has comprised arbitrary wire geometries, bodies of revolution and translation, and surface patch models. MOM is the favored approach for scattering bodies in or below the resonant region because the usual high frequency methods do not apply in this region. MOM, of course, applies to actual targets when we are interested in the scattering in the

UHF region, which is typical for search or acquisition radars. However, in the high frequency region, although MOM is applicable in principle, it is seldom applied because the matrix grows prohibitively large for even today's computers. Moreover, high frequency scattering is mostly due to non-interacting scattering centers, for which the detailed matrix interaction methods entail much useless computation.

If we make the effort, the MOM solutions can yield great insight into the scattering process. This is because the induced current density, which is the source for the scattered radiation, is computed. While MOM solutions probably will never be applied to a target as complicated as a tank in the microwave region, the method is still of great use in understanding scattering phenomenology. MOM can be used to explore surface and traveling wave phenomena in the high frequency region, which is not inherently within the framework of high frequency methodology.

Future trends in MOM applications are toward combining MOM with high frequency methods to obtain hybrid approaches, thereby combining the best features of each method.

REFERENCES

1.   J.J. Bowman, T.B.A. Senior, and P.L. Uslenghi, editors, *Electromagnetic and Acoustic Scattering by Simple Shapes*, Amsterdam, North-Holland, 1969.

2.   Milton Abramowitz and Irene Stegun, *Handbook of Mathematical Functions,* National Bureau of Standards, published by the U.S. Government Printing Office, Washington, D.C., ninth printing, June 1964.

3.   R. Mittra, ed., *Computer Techniques for Electromagnetics*, Oxford, Pergamon, 1973.

4.   S. Silver, *Microwave Antenna Theory and Design*, volume 12 of the MIT Radiation Lab series; McGraw Hill, 1949; Dover, 1965.

5.   R.F. Harrington, *Field Computation by Moment Methods,* New York, MacMillan, 1968.

6.   J.R. Mautz and R.F. Harrington, "Radiation and Scattering from Bodies of Revolution," *Appl. Sci. Res.*, vol. 20, June 1969, pp. 405-435.

7.   J.R. Mautz, "Scattering from Loaded Wire Objects near a Loaded Surface of Revolution," Syracuse Univ. Res. Corp., SURC TN 74-030, January 1974.

8.   J.F. Shaeffer, "EM Scattering from Bodies of Revolution with Attached Wires," *IEEE Trans. Antennas Propag.*, vol. AP-30, May 1982, pp. 426-431.

9.   J.F. Shaeffer and L.N. Medgyesi-Mitschang, "Radiation from Wire Antennas attached to Bodies of Revolution: The Junction Problem," *IEEE Trans. Antennas Propag.*, vol. AP-29, May 1981, pp. 479-487.

CHAPTER 5

HIGH FREQUENCY RCS PREDICTION TECHNIQUES

E. F. Knott

5.1 OVERVIEW

We have seen in Chapters 3 and 4 how scattering solutions may be effected in the Rayleigh and resonant regions, and in this chapter we will discuss high frequency RCS prediction techniques. It must be emphasized at the outset that the term "high frequency" refers not so much to the actual frequency of the incident wave as it does to the size of the target when compared to the incident wavelength. In the high frequency range, the scattering obstacle should be at least five wavelengths in size, although reasonably accurate results may be obtained for some bodies even smaller than this.

The high frequency region is of great practical importance, as may be seen from a consideration of typical threat radars and targets. Threat radars range from low frequency systems used for surveillance to the higher frequencies used for fire control and command-guided surface-to-air missile systems. Long range surveillance radars use frequencies down to the VHF region or lower because the signals propagate well beyond the visible horizon. The wavelengths can be as long as about 10m, hence airborne targets are not likely to be more than two or three wavelengths in size. This is the edge of the high frequency RCS region, and high frequency prediction techniques may not be useful for such targets. However, the wavelengths of radars operating in the UHF bands are on the order of a meter or shorter, hence most airborne targets will be at least 10 wavelengths long. At fire control radar wavelengths, the targets may be hundreds of wavelengths long, and high frequency prediction methods are very useful. Thus, although the methods may not be applicable to the very low, over-the-horizon radars, most targets of practical interest are assuredly electrically large for radar frequencies of 1 GHz and above.

The simplicity of high frequency prediction methods is due to the fact that each part of the body scatters energy essentially independently of all other parts. Therefore, the fields induced on a portion of the target are only due to the incident wave and not the energy scattered by other parts. This makes it relatively simple to estimate the induced fields, and to integrate them over the body surface to obtain the far scattered field and, therefore, the RCS. There are a few exceptions to this general assumption, however, such as re-entrant structures for which some internal features may be illuminated by specular reflections from other internal features. Examples are engine intakes and corner reflectors for which other procedures must be used [1].

The electrical size requirement actually applies to individual scattering features, and not the overall target length. In practice, high frquency methods can be applied only to relatively simple shapes that are easily described in mathematical terms. Examples are elementary surfaces and simple curves representing edges. If surfaces or edges cannot be described by simple mathematical expressions, they must be approximated or replaced by those that can. Thus, we are ultimately forced to approximate the actual target by a collection of simple shapes. As such, the high frequency size requirement (that the body be at least several wavelengths in size) applies to these simple shapes, and not necessarily to the overall target. Even so, most target features are still within the high frequency scattering region, and the methods described below can be used.

Probably the oldest and simplest is the method of geometric optics (GO), developed many years ago in early studies of light. The radar cross section, even in bistatic directions, is given by a very simple formula that involves only the local radii of curvature at the specular point. However, this simple prescription fails when one or both radii of curvature becomes infinite, as in the case of a cylinder or flat plate, and we might then turn to the method of physical optics (PO). The local radii of curvature of the surface may be infinite, yet physical optics gives the correct result if the surface is not too small and if the scattering direction does not swing too far from the specular direction. However, physical optics fails at wide angles from the specular direction.

The reason for the failure of physical optics is that the contributions from edges are ignored, and at this point we can invoke Keller's geometrical theory of edge diffraction (GTD) [2]. Noted for its "cookbook" simplicity, GTD is based on the canonical solution for diffraction by a wedge and gives remarkably good answers for a wide variety of scattering problems. However, it is a wide-angle theory, and it fails in the transition regions of the shadow and reflection boundaries, which will be explained below. Uniform asymptotic solutions have been devised that overcome these difficulties [7], but GTD suffers yet another shortcoming, giving infinite results at caustics, among which is the important case of scattering along the axis of a ring discontinuity.

The method of equivalent currents (MEC) was developed to overcome the caustic difficulties of GTD [3,4,5], but the method does not address the singularities in the diffraction coefficients. In an effort parallel to that of Keller, Ufimtsev developed his physical theory of diffraction (PTD) for treating edges [6]. Like Keller, he turned to the solution of the canonical wedge problem, but he sought to approximate the edge contribution by subtracting the incident field and the physical optics fields from the exact solution for the total field. As a result, his diffraction coefficients can be expressed as the difference between Keller's diffraction coefficients and a set of physical optics coefficients. It turns out that each set is singular along the reflection or shadow boundaries, but the difference remains finite. Despite this serendipity, both GTD and PTD apply only to scattering directions lying on the Keller cone, which will be explained below.

In an attempt to overcome this restriction, Mitzner [8] devised his incremental length diffraction coefficient (ILDC), thus extending Ufimtsev's theory to arbitrary directions, but this is precisely what the method of equivalent currents did for Keller's theory. Thus, there is an exact parallel between Mitzner's extension of PTD and Michaeli's extension of GTD. Unfortunately, none of the four theories adequately treats the surface traveling wave contributions from long structures illuminated at grazing incidence.

Surface traveling waves are induced on long surfaces, even wires, when there is a component of the incident electric field tangent to the surface and in the direction of propagation of the surface wave. This kind of current wave is responsible for the familiar end-fire radiation characteristics of long-wire antennas. The forward traveling wave does not contribute significantly to the backscattered energy, but if the far end of the surface is bounded by a discontinuity, as is usually the case, the current wave is reflected. The backward traveling current wave also radiates in an end-fire mode, but this time in the direction of the radar. Thus, the surface traveling wave can give rise to large contributions to the radar cross section of long, smooth structures. None of the high frequency theories predicts the effect in and of itself, although the repeated application of GTD to account for multiply diffracted rays seems to work for relatively simple structures [9].

5.2 GEOMETRIC OPTICS

The theory of geometric optics was used for many years by astronomers and lens makers in designing and building optical systems. It accounts not only for the way light rays are reflected from smooth surfaces, but for the change in the angle of the transmitted ray when light passes from one medium to another. This was of considerable importance in the design of lenses, because the bending of the transmitted ray depends not only on the wavelength of the light, but on the refractive index of the lens materials, which itself varies with the wavelength. When used to predict the scattering of radar waves from objects of practical interest, the body surfaces are usually assumed to be perfectly conducting, although this is not necessarily a restriction.

Geometric optics is a ray-tracing procedure in which the wavelength is allowed to become infinitesimally small. This being the case, energy propagates along slender tubes according to the formula.

$$u = Pe^{ikS} \tag{5-1}$$

where the amplitude P represents the intensity of either the magnetic or electric field intensity transverse to the direction of propagation. The amplitude P and the phase factor S are both functions of position in space and may be complex numbers. Propagation is in the direction given by ΔS, hence surfaces of constant S are surfaces of constant phase. Equation (5-1) is a solution of the wave equation in the limit of vanishing wavelength, and because the field components are transverse to the direction of propagation, the solution is not valid near discontinuities such as edges. (The actual fields near edges have radial as well as transverse components.)

When a ray strikes a smooth flat surface separating two media of different refractive index, part of the energy is reflected and part is transmitted across the boundary into the second medium. When the electromagnetic boundary conditions are invoked, we find that the transmitted ray propagates in a direction different from that of the incident ray (refraction), and the angle of the reflected ray, as measured from the surface normal, is equal to the angle of the incident ray. The effect is known as Snell's law, and the amplitude and phase of both the reflected and transmitted rays can be calculated. When normalized with respect to the amplitude of the incident ray, the complex amplitudes of the reflected and transmitted rays are none other than the classic Fresnel reflection and transmission coefficients discussed in Ch. 3.

The reflection coefficient for a perfectly conducting surface is −1, implying a 180 degree phase shift and no reduction of the intensity of the reflected wave. This is true only at the point of reflection (the specular point), and if the ray is due to a point source of energy some finite distance from the surface, the reflected ray decays in intensity as it travels away from the specular point. The decay in intensity is due, of course, to the spreading of energy, and the effect can be accentuated if the reflecting surface is curved. It is also possible for the energy in a ray bundle to increase, which is precisely the effect desired of focussed mirrors. In this case, a caustic may be formed when an infinity of rays converge at a point or a line.

The decay or increase in energy can be calculated by invoking the principle of conservation of energy along a ray tube, such as shown in Fig. 5-1. By demanding that all the energy entering the tube at one end be transmitted to the other, we will find that the ratio of the power density at the output to that at the input is

$$\frac{|A(s)|^2}{|A(o)|^2} = \frac{\rho_1 \rho_2}{(s + \rho_1)(s + \rho_2)} \tag{5-2}$$

where $A(o)$ and $A(s)$ are the field intensities at the input and output, respectively, s is the distance along the tube between the two ends, and ρ_1 and ρ_2 are the principal radii of curvature of the wavefront at the output of the tube.

If the body is illuminated by a spherical wave due to a point source located a finite distance away, as in Fig. 5-1, we can find the image of the source in the surface by extending the reflected ray tube backward until the sides of the tube intersect. Because of the differences in the surface radii of curvature, the two sides of the tube will intersect along a line that does not necessarily coincide with the intersection of the top and bottom of the tube, hence the image of a point source in a curved surface does not generally yield another point. The effect is called astigmatism, and the result is a blurred image.

The radii of curvature of the reflected wavefront at the specular point can be related to the radii of curvature of the incident wavefront and the radii of curvature of the body there. The relationship is not a simple one, and it involves not only the angle of arrival of the incident ray, but also the angle by which the principal planes of the body curvature are rotated out of the plane of incidence. (The reader may find the complete formula in [7].) This relationship may be inserted in Eq. (5-2) and the distance s may be measured from the specular point to the point of observation. If the direction of observation is now taken to be back toward the source, s becomes the distance R in Equation (3-1). If R is forced to infinity, as required by the formula, the angular dependence on the local angle of arrival of the incident ray disappears. In addition, the angular rotation of the principal planes of the body radii of curvature out of the plane of incidence also drops out of the expression. The results of the calculation are simply

$$\sigma = \pi a_1 a_2 \tag{5-3}$$

where a_1 and a_2 are the principal radii of curvature of the body at the specular point.

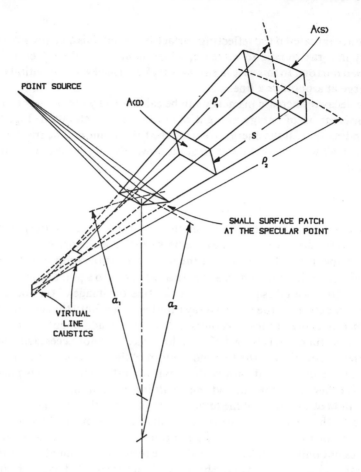

Figure 5-1 Geometrical optics reflection from a doubly curved surface. The curvature of the reflected wavefront is given by ρ_1 and ρ_2 while that of the reflecting surface is a_1 and a_2. In general, the planes containing a_1 and a_2 are neither parallel nor perpendicular to the plane of incidence. The radii ρ_1 and ρ_2 are measured at the caustics (here below the actual surface) formed by extending the reflected ray tube backward until the rays meets

Note that this very simple expression exhibits absolutely no dependence on the frequency of the incident wave. Moreover, even though we have not demonstrated it, this is also the result for bistatic directions. Thus, when applied to the very simple case of scattering by a sphere, the radar cross section is simply the projected area of the sphere, no matter what the bistatic angle. For general spheroids, because the specular point shifts over the surface of the body as the scattering direction moves, and the body curvature changes over the surface, the bistatic scattering will change with the bistatic angle. The formula can be used as it stands for perfectly conducting bodies, but other terms must be included for non-conducting penetrable bodies such as raindrops. The radii of curvature of the body should be large compared to the wavelength, but reasonably accurate results can be obtained for bodies as small as two or three wavelengths in diameter. The theory of geometric optics can even be applied to the scattering of soap bubbles by estimating the reflection coefficient of a thin membrane [10].

Note that the radii of curvature used in Eq. (5-3) must be those at the specular point. Thus, one of the problems that must be addressed in practical computation schemes is the identification of the specular point. For arbitrary bodies, this is not always an easy task and, even for simple structures such as spheroids, the analyst may generate more solutions than needed. Unless care is taken in the solution, the analyst may generate a fourth-order (quartic) equation to be solved, for which there are four solutions (roots), only one of which is the desired one. Nevertheless, all four must be found before the proper one is selected. The multiplicity of roots usually stems from an all too casual definition of the problem, such as the failure to distinguish negative angles from positive ones. In any event, identification of the specular point is a necessary step in the application of the theory of geometric optics.

A serious shortcoming of the theory can be noted immediately from the form of the result. Since the RCS depends on the radii of curvature of the body at the specular point, the formula predicts infinite results for flat or singly curved surfaces, for which one or both radii of curvature are infinite. Fortunately, there is a way around the dilemma.

5.3 PHYSICAL OPTICS

The theory of physical optics overcomes the catastrophe of the infinities of flat and singly curved surfaces by approximating the induced surface fields and integrating them to obtain the scattered field. Because the induced fields remain finite, the scattered fields are finite as well. The beginning point is the Stratton-Chu integral equations, presented in Ch. 3. These expressions hold for a closed scattering surface, and Stratton has demonstrated [11] that if the surface is not closed, additional terms must be added (line integrals around the edge bounding the open surface) to account for the edge discontinuity. There are two simplifications which can be made immediately in the integrals. One is the far field approximation, in which the distance R from an origin in or near the obstacle to the far field observation point is much larger than any obstacle dimension. This allows the gradient of the Green's function to be well approximated by

$$\nabla \psi \simeq ik\hat{s}\psi \tag{5-4}$$

where $\hat{s}$ is a unit vector aligned along the scattering direction. Under far field conditions, the line integrals can be represented as surface integrals, and when combined with the other terms, another simplification results. It will be found that there can be no component of the surface field distribution along the scattering direction [12], whence the Stratton-Chu integrals can be written as

$$\bar{E}_s = ik\psi_0 \int_S \hat{s} \times [\hat{n} \times \bar{E} - Z_0 \hat{s} \times (\hat{n} \times \bar{H})] \, e^{ik\bar{r} \cdot (i - s)} \, dS \tag{5-5}$$

$$\bar{H}_s = ik\psi_0 \int_S \hat{s} \times [\hat{n} \times \bar{H} + Y_0 \hat{s} \times (\hat{n} \times \bar{E})] \, e^{ik\bar{r} \cdot (i - s)} \, dS \tag{5-6}$$

where $\bar{r}$ is now the position vector from the local origin to the surface patch dS, $Y_0 = 1/Z_0$ is the admittance of free space, and $\psi_0 = \exp(ikR)/4\pi R$ is the far field Green's function. Note that the scattered fields are represented in terms of the tangential components of the total fields on the surface, hence the desired scattered fields appear on both sides of the equation. Either of the two equations can be used to calculate the far scattered field because of the relationship:

$$\bar{H}_s = Y_0 \hat{s} \times \bar{E}_s \tag{5-7}$$

We will use Eq. (5-5) by way of illustration.

We can approximate the total fields within the integrals by making the tangent plane approximation. That is, we assign the surface fields the values that they would have had if the body had been perfectly smooth and flat at the surface patch of integration dS. This approximation can be made for any body material, but we shall assume the body to be perfectly conducting. In this case, the tangential components of the total fields are

$$\hat{n} \times \bar{E} = 0 \tag{5-8}$$

$$n \times \bar{H} = 2\hat{n} \times \bar{H}_i \tag{5-9}$$

where $\bar{H}_i$ is the incident magnetic field strength at the surface path. If the incident wave propagates in a direction given by the unit vector $\hat{i}$, with a magnetic intensity H_0 and a magnetic polarization along the unit vector $\hat{h}_i$, Eq. (5-5) becomes the physical optics integral:

$$\bar{E}_s = -i2kZ_0 H_0 \psi_0 \int_S \hat{s} \times [\hat{s} \times (\hat{n} \times \bar{H}_i)] \, e^{ik\bar{r} \cdot (i - s)} \, dS \tag{5-10}$$

where the surface S is now the illuminated portion of the body. In other words, the tangential fields on shaded portions of the body are assumed to be precisely zero.

In the computation of the scattering from complex objects in the high frequency region, the fields scattered by the many different components must be calculated and then added together before squaring to obtain the scattered power. This preserves the phase relationship between the various scatterers on the target so that interference effects may be faithfully represented, but it requires that we calculate quantities proportional to the square root of Eq. (3-1). Let us therefore extract the square root of Eq. (3-1). Furthermore, in light of the polarization dependence of the scattered fields, the signal sensed by a far field receiver will be proportional to the component of the scattered field aligned along the receiver polarization. Thus, it is convenient to redefine the RCS as

$$\sqrt{\sigma} = \lim_{R \to \infty} 2\sqrt{\pi} R \; \frac{\overline{E}_s \cdot \hat{e}_r}{E_0} \; e^{ikR} \tag{5-11}$$

where $\hat{e}_r$ is the electrical polarization of the receiver. Note that we have multiplied by the exponential $\exp(-ikR)$ to remove any dependence on the phase of the RCS on range. Note further that this expression may be a complex number, whereas the RCS as defined in Eq. (3-1) is a pure real number denoting amplitude only. When Eq. (5-10) is substituted into Eq. (5-11), we have the physical optics expression for the square root of the RCS,

$$\sqrt{\sigma} = -i \; \frac{k}{\sqrt{\pi}} \; \int_S \hat{n} \cdot \hat{e}_r \times \hat{h}_i \; e^{ikr \cdot (\hat{i} - \hat{s})} dS \tag{5-12}$$

The integral can be evaluated exactly only for a handful of cases that include flat plates, cylinders, and spherical caps viewed at axial incidence. In the case of flat plates, a coordinate system can be established in terms of two variables over the surface of the plate plus the position vector $\overline{r}_0$ of the origin of the coordinate system,

$$\overline{r} = \overline{r}_0 + x\hat{x} + y\hat{y} \tag{5-13}$$

Thus, the only term in Eq. (5-12) that varies over the surface is the phase term, all other terms being constant. The evaluation of the integral is particularly simple for a flat rectangular plate, the result being

$$\sqrt{\sigma} = -i \; \frac{kLW}{\sqrt{\pi}} \; \hat{n} \cdot \hat{e}_r \times \hat{h}_i \; e^{ik\overline{r}_o \cdot (\hat{i} - \hat{s})} \tag{5-14}$$

$$\times \frac{\sin\left[(1/2)\, k\overline{L} \cdot (\hat{i} - \hat{s})\right]}{(1/2)\, k\overline{L} \cdot (\hat{i} - \hat{s})} \cdot \frac{\sin\left[(1/2)\, k\overline{W} \cdot (\hat{i} - \hat{s})\right]}{(1/2)\, k\overline{W} \cdot (\hat{i} - \hat{s})}$$

where $\overline{W}$ and $\overline{L}$ are vectors aligned along the width and length of the plate, and have the attributes of length as well as direction. Note that this is a bistatic result because no restrictions have been placed on the directions of incidence and scattering.

By way of illustration, Fig. 5-2 shows the bistatic scattering pattern for a plate illuminated at an oblique angle. Note that there are two prominent lobes, one precisely in the forward direction and one in the specular direction. The lobes are equal in amplitude, which occurs only for flat structures like the plate. The strength of the forward scattered lobe for any obstacle is proportional to the projected area of the obstacle and, for the plate, the strength of the specular lobe is also proportional to the projected area. For scattering obstacles of other shapes, however, the specular lobe, if there is one, is less intense. As shown in Ch. 7, the strong forward scattering combines with the incident field to create a shadow behind the obstacle.

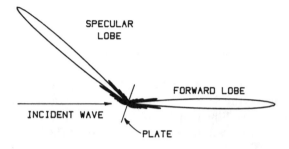

Figure 5-2 Bistatic scattering pattern of a flat plate

Figure 5-3 illustrates the bistatic scattering for a collection of incident angles in which the forward lobe has been omitted for clarity. Note that the main lobe is always centered on the specular direction. Although it may be difficult to detect differences in the amplitudes of the lobes from the small-scale patterns in this figure, the strength of the specular lobe decreases with the projection of the area of the plate in the specular direction, as suggested by the first two patterns at the upper left.

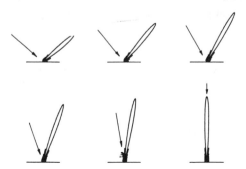

Figure 5-3 The main scattering lobe is always centered on the specular direction

The quadrilateral plate is an inconvenient surface element for modeling some surfaces. A more basic element is the triangular plate: any large, smooth surface can be approximated by a collection of small triangles placed edge-to-edge over the surface. Moreover, as shown by Gordon [13], the surface integral in Eq. (5-12) can be easily converted to a contour integral. When evaluated for polygonal plates, we obtain the expression:

$$\sqrt{\sigma} = -\frac{\hat{n} \cdot \hat{e}_r \times \hat{h}_i}{\sqrt{\pi} \, T} e^{ik\overline{r}_o \cdot \overline{w}} \sum_{m=1}^{M} (\hat{p} \cdot \overline{a}_m) \, e^{ik\overline{r}_m \cdot \overline{w}} \, \frac{\sin\left[\dfrac{1}{2} k\overline{a}_m \cdot \overline{w}\right]}{\dfrac{1}{2} k\overline{a}_m \cdot \overline{w}} \qquad (5\text{-}15)$$

where

σ = bistatic RCS of the plate

$\hat{n}$ = the unit normal of the illuminated plate surface

$\hat{e}_r$ = a unit vector along the electric polarization of a far field receiver

$\overline{r}_o$ = the position vector of an origin on or near the plate

$\overline{w}$ = $\hat{i} - \hat{s}$

$\overline{a}_m$ = a vector describing the length and orientation of the mth edge of the plate, arranged tip to tail around the perimeter

$\overline{r}_m$ = the position vector of the midpoint of the mth edge

T = the length of the projection of $\overline{w}$ onto the plane of the plate

$\hat{p}$ = $\hat{n} \times \overline{w} / |\hat{n} \times \overline{w}|$ = a unit vector in the plane of the plate perpendicular to $\overline{w}$

M = the number of plate edges

Note that the total return is comprised of a collection of $\sin(x)/x$ patterns, one due to each edge, and that the expression appears to become singular when the factor $T = 0$. This implies that there is no component of $(\hat{i} - \hat{s})$ in the plane of the plate and, consequently, the surface coincides with the surface of equal phase delay from the source to the far field point of observation. In other words, not even the phase varies over the surface and, hence, the integral Eq. (5-10) reduces simply to the area of the plate, A. Thus, in the event $T = 0$, Eq. (5-15) becomes

$$\sqrt{\sigma} = -\frac{ikA}{\sqrt{\pi}} \hat{n} \cdot \hat{e}_r \times \hat{h}_i \, e^{ik\overline{r}_o \cdot \overline{w}} \qquad (5\text{-}16)$$

For monostatic scattering, this becomes the familiar broadside return,

$$\sigma = 4\pi A^2 / \lambda^2 \qquad (5\text{-}17)$$

The physical optics integral is a bit more difficult to evaluate when applied to a right circular cylinder. The integration is best performed in cylindrical coordinates, for which the surface patch of integration is

$$dS = a \, d\phi \, dz \tag{5-18}$$

and in which the surface position vector $\overline{r}$ can be expressed in terms of axial and circumferential components,

$$\overline{r} = \overline{r}_0 + z\hat{z} + a\hat{n} \tag{5-19}$$

where a is the radius of the cylinder, $\hat{z}$ is a unit vector along the cylinder axis, $\hat{n}$ is an outward surface normal on the curved surface, and $\overline{r}_0$ is, as with the flat plate, the position vector to the origin of the coordinate system from some main origin associated with the mother target of which the cylinder is a component. It is assumed that the cylinder has end caps whose returns can be calculated from the flat plate expression and, hence, only the contribution from the curved surface is to be calculated here.

When Eqs. (5-18) and (5-19) are inserted in Eq. (5-12), the result is expressed in terms of two integrals, one with a variable of integration along the axial direction and the other with a variable in the circumferential direction. The expression is

$$\sqrt{\sigma} = -i \, \frac{ka}{\sqrt{\pi}} \, I_z \, I_\phi \, e^{ik\overline{r}_o \cdot (\hat{i} - \hat{s})} \tag{5-20}$$

where the axial and circumferential intergrals are

$$I_z = \int_{-l/2}^{l/2} e^{ikz\hat{z} \cdot (\hat{i} - \hat{s})} \, dz \tag{5-21}$$

$$I_\phi = \int_{-\pi/2}^{\pi/2} \hat{n} \cdot \hat{e}_r \times \hat{h}_i \, e^{ika\hat{n} \cdot (\hat{i} - \hat{s})} d\phi \tag{5-22}$$

where ϕ is the circumferential angle of the surface position vector from the plane containing the incident direction and the cylinder axis. The axial integral is easily evaluated, and the result for a cylinder of length l is

$$I_z = l \, \frac{\sin \left[(1/2) \, kl\hat{z} \cdot (\hat{i} - \hat{s}) \right]}{(1/2) \, kl\hat{z} \cdot (\hat{i} - \hat{s})} \tag{5-23}$$

Note that this result is the $\sin(x)/x$ function characteristic of uniformly illuminated rectangular apertures and we can perform the integration exactly, even for bistatic directions. However, the circumferential integral is not so easily evaluated.

The reason is, at least for bistatic scattering, that the limits of integration of the ϕ integral are not symmetricallly disposed with respect to the specular line along the side of the cylinder. This specular line is located in such a way that a surface normal erected anywhere along the line bisects the projections of the angles of incidence and scattering onto a plane perpendicular to the cylinder axis, and the limits of integration are the shadow boundaries due to the incident wave. An exact evaluation of the ϕ integral is available only when the limits of integration are from $-\pi/2$ to $\pi/2$, and this occurs only when the scattering direction lies in the plane of incidence (i.e., the plane containing the incident direction and the cylinder axis). Nevertheless, the integral can be evaluated approximately by means of the method of stationary phase and, for all except cylinders with very small diameters, it yields quite acceptable results.

It is common in physical optics problems to encounter integrals in which the phase of the integrand varies rapidly over the surface. The real and imaginary components vary more or less as sinusoids and, as such, the negative cycles of the variation tend to cancel the positive cycles. Over curved surfaces, however, there is usually a point at which the phase variation slows down and actually stops, and then begins varying in the opposite sense. This point of phase reversal is the dominant contribution to the integral, and it can be evaluated by the method of stationary phase.

We can expand the phase function in a Taylor series, and if all terms beyond the second derivative of that series are ignored, we obtain the approximation:

$$\int g(\phi)\, e^{if(\phi)}\, d\phi \simeq \left[\frac{2\pi}{-if''(\phi_0)}\right]^{1/2} g(\phi_0)\, e^{if(\phi_0)} \tag{5-24}$$

where the double prime indicates the second derivative. In this expression, ϕ_0 is the value of the coordinate that forces the first derivative $f'(\phi)$ to zero. In the case at hand, it is none other than the circumferential location of the specular line, at which point the phase is indeed stationary.

Therefore, let us express the outward surface normal $\hat{n}$ in terms of the angle ϕ,

$$\hat{n} = \hat{x}\cos\phi + \hat{y}\sin\phi \tag{5-25}$$

where ϕ is measured from the plane containing the cylinder axis and the incident direction. The angle $\phi = 0$ is the stationary phase point, and the ϕ-integral of Eq. (5-22) becomes

$$I_\phi = \hat{n}_0 \cdot \hat{e}_r \times \hat{h}_i\, e^{ika\hat{n}_0 \cdot (\hat{i} - \hat{s})}\, e^{-i\pi/4} \left[\frac{a}{\lambda}\, \hat{n}_0 \cdot (\hat{i} - \hat{s})\right]^{1/2} \tag{5-26}$$

where $\hat{n}_0$ is the outward surface normal erected anywhere along the specular line. When this result is substituted for the ϕ integral of Eq. (5-20) and Eq. (5-23) substituted for the z-integral, we obtain the bistatic scattering formula for the cylinder,

$$\sqrt{\sigma} = -il \left[\frac{2ka}{\hat{n}_0 \cdot (\hat{i} - \hat{s})} \right]^{1/2} \frac{\sin (1/2) \, kl\hat{z} \cdot (\hat{i} - \hat{s})}{(1/2) \, kl\hat{z} \cdot (\hat{i} - \hat{s})}$$
$$\times \ (\hat{n}_0 \cdot \hat{e}_r \times \hat{h}_i) \ e^{i k \overline{r}_0 \cdot (i - \hat{s})} \ e^{ika\hat{n}_0 \cdot (\hat{i} - \hat{s})} \ e^{-i\pi/4} \tag{5-27}$$

Note that the stationary phase integration yields a phase factor $\exp(-i\pi/4)$ because of the minus sign in the denominator of the radical of Eq. (5-24). If the quantity $\hat{n}_0 \cdot (\hat{i} - \hat{s})$ in Eq. (5-27) is negative, the extraction of the root will generate an additional phase factor $\exp(i\pi/2)$ which advances the phase to $\exp(i\pi/4)$. Thus, the sign of the exponent of the trailing phase term will always be opposite of that of the exponential $\exp[ika\hat{n}_0 \cdot (\hat{i} - \hat{s})]$.

For the case of backscattering, the scattering direction points back in the direction of incidence and Eq. (5-27) reduces to

$$\sqrt{\sigma} = -il \ \sqrt{ka\hat{n}_0 \cdot \hat{i}} \ \frac{\sin (kl\hat{z} \cdot \hat{i})}{kl\hat{z} \cdot \hat{i}} \ e^{i2k\overline{r}_0 \cdot i} \ e^{i2ka\hat{n}_0 \cdot i} \ e^{-i\pi/4} \tag{5-28}$$

If we take the amplitude of this expression and square it, we obtain the familiar equation for the monostatic radar cross section of the right-hand circular cylinder,

$$\rho = kal^2 \left| \cos \theta_i \ \frac{\sin (kl \sin \theta_i)}{kl \sin \theta_i} \right|^2 \tag{5-29}$$

where θ_i is the angle from broadside incidence (see Fig. 5-4). Note that the RCS rises linearly with the electrical circumference of the cylinder and with the square of its length.

When the cylinder axis and the directions of incidence and scattering all lie in the same plane as suggested in Fig. 5-4, the physical optics integral can be evaluated exactly, even if the integral itself is an approximation. In this event, the limits of integration (i.e., the shadow boundaries along the sides of the cylinder) in Eq. (5-22) are symmetrically placed on either side of the specular line. In this event, the ϕ integral can be evaluated exactly, and for the case of backscattering, Eq. (5-22) becomes

$$I_\phi = \{\pi \left[S_1 (2ka \cos \theta) + iJ_1 (2ka \cos \theta) \right] - 2\} \cos \theta \tag{5-30}$$

where S_1 and J_1 are the Struve function and the Bessel function of the first kind, respectively. It turns out, however, that this exact evaluation of the ϕ integral does not necessarily yield more accurate results than the stationary phase approximation.

A comparison of the results from the two ways for evaluating the integral in the case of backscattering is shown in Fig. 5-5. The two traces shown represent the broadside RCS of the cylinder normalized with respect to the square of the length, with the solid trace representing the results of the stationary phase evaluation. As suggested by Eq. (5-29), the RCS is a straight line as it varies linearly with the electrical circumference of the cylinder. The dashed line

᠕represents the solution in which the ϕ integral is given by Eq. (5-30), and it oscillates about the line representing the stationary phase approximation. The reason for the undulation is that the discontinuity in the assumed induced surface fields gives rise to a contribution from the shadow boundary that goes in and out of phase with the specular return from the near side of the cylinder as the cylinder grows electrically larger. Indeed, the periodicity of the undulation is appropriate to the returns from two scattering centers separated by the cylinder radius. It should be noted from Eq. (5-28) that the backscattering depends on the angle of incidence, but not on the polarization. Thus, physical optics gives the same results for any incident polarization, whether we use the exact evaluation of the physical optics integral or the stationary phase approximation.

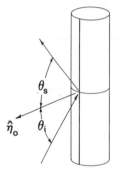

Figure 5-4 Cylinder bistatic scattering geometry for scattering in the plane of incidence. The specular scattering arises from a line down the side of the cylinder and the unit vector n_0 is a unit surface normal erected on that line

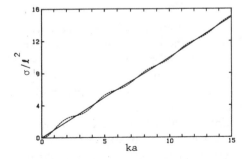

Figure 5-5 RCS of the cylinder as obtained from the physical optics approximation. The dashed trace is obtained from an exact evaluation of the *PO* integral and the solid trace is obtained from a stationary phase evaluation

By contrast, the exact solution of the wave equation for the circular cylinder depends on the incident polarization, as is shown in Fig. 5-6. The solid and dashed traces represent the solution when the incident electric and magnetic fields are parallel to the cylinder axis, called E- and H-polarization, respectively. The exact E-polarization solution exhibits undulations not unlike those of the dashed trace of Fig. 5-5, but the periodicity is quite different. The source of the undulations in the exact solution is a creeping wave which circles the rear of the cylinder boundary. However, because it actually traverses the rear of the cylinder, the phase of the creeping wave is delayed significantly more than that of the fictitious shadow boundary contribution and, hence, the period of the undulations is much shorter than those of the Struve function solution represented by the dashed trace of Fig. 5-5. The undulations in the dashed trace of Fig. 5-6 are due to precisely the same mechanism that creates the undulations for the sphere shown in Fig. 3-3.

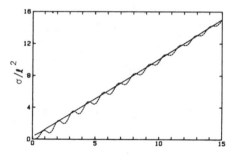

Figure 5-6 RCS of the cylinder as obtained from the exact (two-dimensional) solution. The solid trace is for E-polarization and the dashed for H-polarization

The errors in the physical optics approximation are worse for small cylinders than for large ones, as shown in Table 5-1. Neither form of the PO approximation works as well for H-polarization as it does for E-polarization, the error being on the order of 2.5 to 3.5 dB for $ka = 1.4$; (for E-polarization, it is on the order of 0.5 dB or less). The error in the PO approximation decreases to 0.15 dB or less for $ka = 9.7$, and diminishes even further with increasing size. Thus, the approximation is well within a decibel of the exact solution for cylinders as small as 1.5 wavelengths in diameter. Moreover, the stationary phase approximation yields slightly better accuracy than does the exact evaluation of the PO integral. Because the exact evaluation requires the generation of Struve functions, a significant undertaking, the stationary phase result is far more appealing.

Table 5-1
Errors of the Physical Optics Approximation

	Decibel Errors			
	stat. phase	stat. phase	Struve	Struve
ka	exact H	exact E	exact H	exact E
1.4	2.50	-0.61	3.35	0.25
4.9	0.44	-0.09	0.66	0.13
9.7	0.15	-0.03	0.08	-0.10

As with the right circular cylinder, the physical optics integral cannot be evaluated in a closed form solution for even a sphere for bistatic directions. Again, this is because the limits of integration are not symmetricallly disposed about the specular direction. For the same reason, we cannot even compute the RCS of a spherical cap in other than the backscattering and forward scattering directions. Nevertheless, it is of interest to examine the results for the special case of backscattering.

Upon evaluating the physical optics integral for the sphere, we obtain the result [14]:

$$\sqrt{\sigma} = \sqrt{\pi}\, a \left[\left(1 + \frac{1}{i2ka} \right) e^{-i2ka} - \frac{1}{i2ka} \right] \qquad (5\text{-}31)$$

where a is the radius of the sphere and the phase of the echo has been referenced to the center of the sphere. Note that this expression contains three terms, the first of which is identical to the geometric optics result obtained earlier. The second term is a correction to the first and agrees with the next term in the exact solution, also called the Mie series. The phase of these two terms is retarded by twice the radius of the sphere, and is therefore consistent with the location of the specular point at the front of the sphere (nearest the radar). The third term is due to the discontinuity of the assumed induced currents across the shadow boundary, and the phase of this term is consistent with the location of the shadow boundary. However, we must regard the term as fictitious because the induced currents do not drop suddenly to zero at the shadow boundary as assumed. Indeed, the result of Eq. (5-31) is more appropriate to the scattering from a hemispherical shell than to a complete sphere because, in that case, the induced fields change rapidly at the edge, which is also the shadow boundary.

For all the backscattering cases considered above, the polarization term $\hat{e}_r \times \hat{h}_i$ always yields a maximum when the receiver polarization is aligned along the incident polarization and yields precisely zero when aligned 90 degrees from that polarization. Hence, physical optics can yield no estimate of the cross-polarized return. Also, because the theory includes an erroneous contribution due to the shadow boundaries on otherwise smooth objects, we might be better off using the theory of geometric optics for smooth, doubly curved bodies. In addition to these failures, physical optics also gives increasingly erroneous results as the scattering direction moves further away from the specular direction, as will be shown in Ch. 6. The reason for this last failure is that the role of the edges bounding the surface has been ignored. These failures were recognized by Keller and Ufimtsev, and their pioneering efforts considerably advanced the ease with which RCS calculations can be performed.

5.4 GEOMETRICAL THEORY OF DIFFRACTION

Keller was well aware of the wide-angle failure of physical optics and introduced his classic geometrical theory of diffraction as a way of computing scattered fields well away from the specular directions. It was shown earlier that when a ray or electromagnetic wave impinges on a flat infinite surface, part of the wave or ray is reflected and part is transmitted through the surface. Whether the surface is dielectric or perfectly conducting, the reflected ray can propagate in only one direction in space; that direction is in the plane of incidence containing the incident ray and the surface normal, and it subtends the same angle with respect to the surface normal as does the incident ray. This unique specular direction of the reflected ray is a consequence of the doubly infinite size of the surface.

If the doubly infinite surface is now halved, thereby generating an edge, we may think of the reflected (or scattered) fields as arising from two sources, one being a surface contribution, as before, plus an edge contribution. Ignoring for the moment the surface contribution, let us consider the edge and the components of an incident ray parallel and perpendicular to it. We would expect reflected ray components along the edge to be constrained to a unique direction because that dimension of the edge is infinite. However, we would expect the transverse components of reflected rays to propagate in all directions perpendicular to the edge because the transverse dimension of the edge is zero. Thus, in contrast to the single, unique direction taken by the ray reflected by a surface, an edge-reflected ray can lie anywhere along a forward cone whose half-angle is precisely that subtended by the edge and the incident ray. This is the well known Keller cone sketched in Fig. 5-7. It represents the extension of Fermat's principle to diffraction by an edge, and the rays are understood to be diffracted rays.

If the edge singularity were somehow divorced from the surface it bounds, it would be a filament. If it were a conducting filament (i.e., a wire), we would expect the diffracted rays to have the same amplitude no matter where on the diffraction cone they might be. Indeed, this is the case. But how do we account for the presence of the surface that ends at the edge? Certainly, the diffracted rays on one side must be weaker than those on the other, especially for directions shaded from the incident ray by the surface itself. In other words, we expect the intensities of the diffracted rays to vary from one direction to another around the diffraction cone. What is that variation?

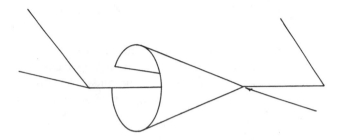

Figure 5-7 The Keller cone of diffracted rays

To get an answer, Keller turned to the exact solution of a two-dimensional canonical problem, the diffraction from an infinite wedge or half-plane, such as obtained by Sommerfeld [15,16]. In Sommerfeld's two-dimensional problem, the incident ray impinges on the edge at right angles, and Sommerfeld represented the diffracted rays as a spectrum of plane waves. His solution amounts to finding the coefficients of each of the elementary plane waves diffracted by the edge. Hence, for any scattering direction, the strength of the diffracted ray can be calculated. The Sommerfield two-dimensional solution for diffracted rays can be applied to Keller's three-dimensional problem by adjusting Sommerfeld's propagation constants to match the transverse propagation components of Keller's problem. The direction-dependent amplitudes of Sommerfeld's solution then make it possible to estimate the amplitude and phase of a diffracted ray anywhere on the Keller cone. This yielded the key ingredient of the theory: the diffraction coefficients, which depend on the polarization of the incident ray.

The other ingredients of Keller's GTD include the decay in field intensity away from the diffracting edge and the phase of the ray along its propagation path. The intensity of the diffracted ray thus has the form:

$$u = \frac{D e^{iks}}{[s(1+s/\rho_1)]^{1/2}} A e^{ik\psi} \tag{5-32}$$

where D is a diffraction coefficient depending on the polarization and angle of arrival of the incident ray and the direction of the scattered ray, s is the distance along the ray from the edge element to a far field observation point, A and ψ are the amplitude and phase of the incident ray, respectively, and ρ_i is the distance from the edge element to a caustic of the diffracted ray.

When Eq. (5-32) is applied to the components of the far diffracted electric field, the diffracted field can be expressed as

$$\overline{E}_d = - \frac{\Gamma\, e^{iks}}{\sin^3 \beta}\, [(\hat{\imath} \cdot \overline{E}_i)\,(X - Y)\hat{s} \times (\hat{s} \times \hat{\imath}) + Z_0\,(\hat{\imath} \cdot \overline{H}_i)\,(X + Y)\hat{s} + \hat{\imath}\,]$$

(5-33)

where Γ is a divergence factor accounting for the nature of the edge excitation (plane wave, spherical wave, *et cetera*) and the spreading of energy away from the edge and β is the angle subtended by the edge and the incident ray. In this expression, $\hat{\imath}$ is a unit vector aligned along the edge, and the diffraction coefficients are given by

$$X = \frac{(1/n)\,\sin\,(\pi/n)}{\cos\,(\pi/n) - \cos\,[(\psi_s - \psi_i)/n]}$$

(5-34)

$$Y = \frac{(1/n)\,\sin\,(\pi/n)}{\cos\,(\pi/n) - \cos\,[(\psi_s + \psi_i)/n]}$$

(5-35)

where n is the exterior wedge angle normalized with respect to π, and ψ_i and ψ_s are the angles of the transverse components of the incident and diffracted directions with respect to one of the surfaces meeting at the edge, as shown in Fig. 5-8. It should be noted that Eq. (5-34) and (5-35) are the form of the diffraction coefficients as suggested by Ufimtsev, although Ufimtsev did not specifically label them X and Y. The notation is not due to Keller either, but to Senior and Ushlenghi [17]. We use the difference in the coefficients for those components of the incident magnetic polarizations along the edge. To obtain the RCS due to an edge, we need only substitute the field of Eq. (5-33) into Eq. (3-1) or (5-11).

In applying Eq. (5-33) to a scattering obstacle with edges, we must first find those edge elements for which a generator on the local Keller cone pierces the far field observation point. We can imagine hundreds, and perhaps thousands, of little Keller cones erected on edges all over the target and dozens distributed along all curved edges. Only the contributions from those edges having Keller cone directions toward the far field point are included in the calculation, and all others are ignored. Thus, GTD is a kind of specular theory in precisely the way geometric optics is specular and, like geometric optics, GTD has certain failures.

For one, the diffraction coefficients are actually the result of a wide angle evaluation of an integral, and if the scattering direction approaches too close to a shadow boundary or reflection boundary, one or the other of the coefficients becomes singular. The results are obviously wrong because the far diffracted fields must remain finite in the real world. The difficulty can be traced to the wide angle approximation of the integral mentioned above, which is not valid in the transition regions of the reflection and shadow boundaries. The coefficients should be replaced by more accurate expressions than those of Eqs. (5-34) and (5-35), which shall be addressed below. In addition to the failure of the diffraction coefficients in the transition regions, there may be curved edges on the scattering obstacle for which the Keller cones erected at every point along the contour contain the scattering direction. In this case, an infinity of points contribute to the far diffracted field and we obtain an infinite result. It is serious enough to cause concern because it occurs, for example, whenever a ring, such as the open end of a tube, is presented normal to the incident wave. Such features are common on bodies of revolution of great practical and tactical importance.

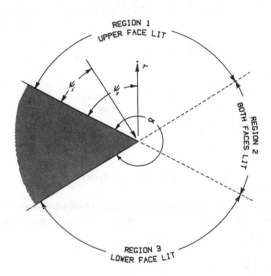

Figure 5-8 Geometry for wedge diffraction

Despite these drawbacks, GTD is a very powerful and convenient method for the computation of high frequency scattering, and Keller must be recognized for his significant contribution to electromagnetic theory. Benefiting from Keller's foundation of an extremely useful computational technique, others have taken steps to overcome what should be considered merely aberrations in the theory. Among the techniques that have improved GTD are the asymptotic theories.

5.5 A UNIFORM ASYMPTOTIC THEORY

Kouyoumjian and Pathak are generally credited with the development of the uniform theory of diffraction (UTD) [7], and Lee and Deschamps for developing the uniform asymptotic theory (UAT) [18,19]. Each strives to accurately represent the diffracted field, but using different approaches. Although Deschamps, Boersma, and Lee [19] contend that their UAT recovers the exact solution for the diffraction of a three-dimensional half-plane, here we shall summarize the procedure used in the UTD of Kouyoumjian and Pathak.

The singularities in the diffraction coefficents are overcome in the UTD essentially by the multiplication of the diffraction coefficients by a Fresnel integral. At the shadow boundary or the reflection boundary, the modifying Fresnel integral is zero while the diffraction coefficient is infinite, but the product of the two remains finite. The prescription given by Kouyoumjian and Pathak is

$$X = \frac{1}{2} \left[\cot \frac{\pi + (\psi_s - \psi_i)}{2n} \ F[kLa^+ (\psi_s - \psi_i)] \right.$$
$$\left. + \cot \frac{\pi - (\psi_s - \psi_i)}{2n} \ F[kLa^- (\psi_s - \psi_i)] \right] \tag{5-36}$$

$$Y = \frac{1}{2} \left[\cot \frac{\pi + (\psi_s + \psi_i)}{2n} \ F[kLa^+ (\psi_s + \psi_i)] \right.$$
$$\left. + \cot \frac{\pi - (\psi_s + \psi_i)}{2n} \ F[kLa^- (\psi_s + \psi_i)] \right] \tag{5-37}$$

where L is a function that depends on the nature of the source of the incident wave (i.e., plane wave, spherical wave, *et cetera*) and is given explicitly in [7]. The function F is the Fresnel integral:

$$F(Q) = -i \, 2 \sqrt{Q} \ e^{-iQ} \int_{\sqrt{Q}}^{\infty} e^{iz^2} \, dz \tag{5-38}$$

and it can be verified that when the argument Q is large, Eqs. (5-36) and (5-37) reduce exactly to the wide-angle coefficients given in Eqs. (5-34) and (5-35).

The function

$$a^{\pm} (\eta) = 2\cos^2 \left[\frac{1}{2} \ (2n \pi N^{\pm} - \eta) \right] \tag{5-39}$$

where the integers N are those that most nearly satisfy

$$2n \pi N^+ - \eta = \pi, \ 2n \pi N^- - \eta = -\pi$$

In the transition regions, the cotangent functions rise to infinity while the Fresnel integrals drop to zero in such a way that the product remains finite:

$$\cot \frac{\pi \pm \epsilon}{2n} \; F[kLa^{\pm}(\epsilon)] \simeq n \; \{\sqrt{2\pi kL} \; \mathrm{sgn}\epsilon - 2kL\epsilon \; e^{-i\,\pi/4}\} \; e^{-i\,\pi/4}$$

(5-40)

where ϵ is a small angle measured from the optical shadow or reflection boundary, as the case may be. The amplitude of this function is the same on either side of that boundary, but its sign changes, hence the diffraction coefficients are discontinuous across the boundary. The discontinuity is necessary to compensate for the abrupt change of the geometrical optics field from the illuminated region to the shadowed region, or from one side of the reflection boundary to the other side.

The Fresnel integral approaches unity for large arguments, hence its value can be taken as one for wide angles. Thus, for angles well away from either transition region, we may use the simpler expressions. Eqs. (5-34) and (5-35). When the scattering direction approaches one of the transition regions, only the affected diffraction coefficient needs to be calculated more carefully with the aid of either Eq. (5-36) or (5-37). This is because X becomes singular at the shadow boundary, while Y becomes singular at the reflection boundary, hence they never become singular simultaneously. Thus, it is necessary to use the more complicated expression involving the Fresnel integral only for the affected diffraction coefficient. Figure 5-9 compares the behavior of the diffraction coefficients of Eqs. (5-34) through (5-37) in the transition regions. It can be seen that the UTD indeed removes the singularities.

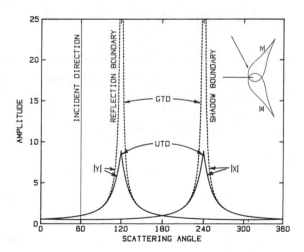

Figure 5-9 Comparison of the diffraction coefficients of *GTD* and *UTD* for an incident angle of 60° on the edge of a half-plane. Inset shows a polar diagram of the coefficients

However, the uniform theory of diffraction does not eliminate the difficulties of the caustics of GTD where an infinity of rays may converge. This is a particularly serious flaw, because many bodies of great practical interest are bodies of revolution. As such, many have ring discontinuities where different surfaces join each other, such as the junction between a cylinder and a cone, a very common feature of re-entrant bodies. The caustics due to such rings coincide with the body axis in the far field and, hence, we obtain the wrong answer for important aspect angles such as nose-on incidence. However, there is a way to fix the difficulty: the method of equivalent currents.

5.6 THE METHOD OF EQUIVALENT CURRENTS

The method of equivalent currents stems from the fact that any finite current distribution yields a finite result for the far diffracted field when that distribution is summed in a radiation integral. If the proper distribution can be found, then the axial caustics of the GTD can be avoided. Moreover, the method of equivalent currents extends the promise that the diffracted fields can be computed for scattering directions not on the Keller cone, a significant extension of Keller's theory. Thus, there are two important motivations for devising equivalent currents.

The notion of equivalent currents was employed by Millar in his studies of the diffraction by apertures [20], and Ryan and Peters used them to compute the fields along the axial caustics of bodies of revolution [3]. The basic approach is to postulate the existence of filamentary electric and magnetic currents I_e and I_m at each point around the singularity (the edge contour) and to sum them in the far field radiation integral,

$$\overline{E}_d = -ik\psi_0 \int [Z_0 I_e \, \hat{s} \times (\hat{s} \times \hat{t}) + I_m (\hat{s} \times \hat{t})] \, e^{ik\hat{r}\cdot\hat{s}} \, dt \qquad (5\text{-}41)$$

where ψ_0 is the far field Green's function defined earlier and $\hat{t}$ is a unit vector aligned along the contour. As we shall see, the equivalent currents are fictitious because they depend on the scattering direction as well as the direction of incidence, but they overcome the caustic difficulties. Perhaps, the strongest defense of them is that they yield what appears to be the correct result in directions where GTD does not.

The equivalent currents proposed by Ryan and Peters are

$$I_e \quad = i2 \, (X - Y) \, (\hat{t} \cdot \overline{E}_i)/kZ_0 \qquad (5\text{-}42)$$

$$I_m = i2 \, (X + Y) \, (\hat{t} \cdot \overline{H}_i)/kY_0 \qquad (5\text{-}43)$$

When these currents are used in Eq. (5-41), we can obtain the scattered fields anywhere, not just on the Keller cone. Later, Knott and Senior improved the equivalent current prescription by requiring that a stationary phase evaluation of the integral Eq. (5-41) yield the exact GTD result for directions on the

Keller cone [4]. This had the effect of dividing the currents in Eqs. (5-42) and (5-43) by $\sin^2\beta$, where β is the angle subtended by the incident ray and the local edge tangent where the currents are postulated. The two prescriptions become identical when $\beta = 90$ degrees, of course, such as for backscattering along the axis of a ring discontinuity.

In an attempt to extend the equivalent current notion to general bistatic cases, in which the scattering direction need not lie on the Keller cone, Knott and Senior interpreted the product:

$$\sin^2\beta = \sin\beta_i \sin\beta_s \tag{5-44}$$

where the subscripts i and s denote the angle subtended by the incident and scattering directions. For the cases they studied, however, β_i and β_s remained close enough to each other that the distinction between the two interpretations was small. When the equivalent currents are inserted in Eq. (5-41), the result is the contour integral:

$$\bar{E}_d = -2E_0\,\psi_0 \int_c \frac{e^{i\bar{k}r\cdot(\hat{i}-\hat{s})}}{\sin\beta_i \sin\beta_s} \times [(\hat{t}\cdot\hat{e}_i)\,(X-Y)\hat{s}\times(\hat{s}\times\hat{t}) + (\hat{t}\cdot\hat{h}_i)\,(X+Y)\hat{s}\times\hat{t}]\,dt \tag{5-45}$$

where $\hat{e}_i$ and $\hat{h}_i$ are unit vectors aligned along the incident electric and magnetic fields.

We have already seen that the diffraction coefficients X and Y depend on the projections of the directions of incidence and scattering onto the plane perpendicular to the edge. Thus, these coefficients vary around the edge singularity with the result that the integral cannot be evaluated in closed form for the general case. This need not deter us, of course, because there are ways to perform the integration, even if approximate. For one, the integral can always be evaluated numerically with a computer, although this is not a very effective method. Another option is to expand the diffraction coefficients in a Fourier series, which is usually effective for circular discontinuities. The method of stationary phase is particularly useful and, as expected, the result is that the scattering is due primarily to the "flash points" around the contour where incident and scattering directions subtend identically the same angle. In this case, the scattering direction lies on the Keller cones erected at the flash points. For special cases in which the phase is constant around the contour, the flash points spread out over the entire edge, and instead of a pair of bright flash points (one each at the near and far edges of a ring, for example), the entire ring lights up.

In examining these equivalent current postulates, Michaeli proposed a more rigorous development [5]. He considered the far field contribution due to two narrow surface strips (one on each face of a wedge) meeting at an edge element, and integrated the induced currents along the strips. In the direction along the surfaces perpendicular to the edge, only the asymptotic contribution of the end-point of integration (the edge) is retained, and upon comparing the result with the form of the surface integral Eq. (5-41), Michaeli deduced the form of the equivalent currents. He then related the surface integrals to the canonical solution for the wedge, and arrived at the equivalent current prescription:

$$I_e = \frac{i2\,(\hat{t}\cdot\bar{E}_i)D_e}{k\,Z_0\,\sin^2\beta_i} + \frac{i2\,(\hat{t}\cdot\bar{H}_i)\,D_{em}}{k\,\sin\beta_i} \tag{5-46}$$

$$I_m = -\frac{i2\,(\hat{t}\cdot\bar{H}_i)D_m}{k\,Y_0\,\sin\beta_i\,\sin\beta_s} \tag{5-47}$$

where the diffraction coefficients D_e, D_m, and D_{em} are

$$D_e = \frac{\dfrac{1}{n}\,\sin\dfrac{\phi_i}{n}}{\cos\dfrac{\pi-\alpha_1}{n} - \cos\dfrac{\phi_i}{n}}$$

$$+ \frac{\dfrac{1}{n}\,\sin\dfrac{\phi_i}{n}}{\cos\dfrac{\pi-\alpha_2}{n} + \cos\dfrac{\phi_i}{n}} \tag{5-48}$$

$$D_m = \frac{\sin\phi_s}{\sin\alpha_1}\cdot\frac{\dfrac{1}{n}\,\sin\dfrac{\pi-\alpha_1}{n}}{\cos\dfrac{\pi-\alpha_1}{n} - \cos\dfrac{\phi_i}{n}}$$

$$+ \frac{\sin(n\pi-\phi_s)}{\sin\alpha_2}\cdot\frac{\dfrac{1}{n}\,\sin\dfrac{\pi-\alpha_2}{n}}{\cos\dfrac{\pi-\alpha_2}{n} + \cos\dfrac{\phi_i}{n}} \tag{5-49}$$

$$D_{em} = \frac{Q}{\sin\beta_i}\left[\frac{\cos\phi_s}{\sin\alpha_1}\cdot\frac{\dfrac{1}{n}\,\sin\dfrac{\pi-\alpha_1}{n}}{\cos\dfrac{\pi-\alpha_1}{n} - \cos\dfrac{\phi_i}{n}}\right.$$

$$\left. - \frac{\cos(n\pi-\phi_s)}{\sin\alpha_2}\cdot\frac{\dfrac{1}{n}\,\sin\dfrac{\pi-\alpha_2}{n}}{\cos\dfrac{\pi-\alpha_2}{n} + \cos\dfrac{\phi_i}{n}}\right] \tag{5-50}$$

with

$$Q = 2 \, \frac{1 + \cos \beta_i \cos \beta_s}{\sin \beta_i \sin \beta_s} \, \sin \, \frac{1}{2} (\beta_s + \beta_i) \sin \, \frac{1}{2} (\beta_s - \beta_i) \qquad (5\text{-}51)$$

$$\sin \alpha_1 = [\sin^2 \beta_i - \sin^2 \beta_s \cos^2 \phi_s]^{1/2} / \sin \beta_i \qquad (5\text{-}52)$$

$$\sin \alpha_2 = [\sin^2 \beta_i - \sin^2 \beta_s \cos^2 (n\pi - \phi_s]^{1/2} / \sin \beta_i \qquad (5\text{-}53)$$

The diffraction coefficients in Eqs. (5-48) through (5-50) reduce exactly to the Keller coefficients X and Y when the scattering direction lies on the Keller cone, as indeed they must. For arbitrary directions, they are very complicated and there is no more hope of evaluating a contour integral containing them (via the equivalent currents) than there is with the simpler expressions devised by Ryan and Peters or Knott and Senior. Nevertheless, Michaeli's equivalent currents are more complete and were derived more rigorously. Note from Eq. (5-46), for example, that the equivalent electric current is due to the excitation of the edge element by the incident magnetic field as well as the incident electric field, a feature missing from the previous equivalent current theories. In addition, it would appear that Michaeli devised a new set of diffraction coefficients because those in Eqs. (5-48) through (5-50) are the generalized versions of the Keller coefficients X and Y. We will have more to say about this in a moment.

Whatever the case, the equivalent currents extend Keller's GTD in two important aspects. First, edge-diffracted fields remain finite in caustic directions. Although the caustics may be corrected in other ways, the equivalent currents represent a unified approach. Second, the scattering direction is no longer confined to a generator of the Keller cone, a significant extension. These two improvements are sufficient motivation to use the equivalent currents in any serious computation of the scattering by simple bodies, although the complexity of the diffraction coefficients may prohibit the implementation of the method for complex bodies comprised of hundreds or thousands of facets.

Despite Michaeli's extension of Keller's theory to arbitrary directions not on the Keller cone, the diffraction coefficients become singular in the transition regions of the shadow and reflection boundaries. That defect could be remedied by the development of a uniform theory like the UTD discussed above, but at the time of this writing, such a theory has not yet been put forth. An alternative is the incremental length diffraction coefficient (ILDC) developed by Mitzner [8], which remains finite in the transition regions (as discussed in sec. 5.8 below). However, because Mitzner's theory is based on Ufimtsev's PTD, we discuss the latter first.

5.7 THE PHYSICAL THEORY OF DIFFRACTION

Like Keller, Ufimtsev sought a more accurate representation of the scattered fields than yielded by physical optics and he relied on the canonical solution of the scattering by a wedge for his diffraction coefficients, but, unlike Keller, he retained in his solution the approximate physical optics result and sought instead a correction by which to improve the physical optics approximation. He therefore represented the scattered field as the sum of the physical optics contribution and an edge contribution, using the exact solution of the two-dimensional wedge problem to extract the latter. That is, if we have in hand the exact solution and subtract from it the physical optics (surface) contribution, what remains must be the contribution from the edge itself, there being no other scattering features present except the surface and the edge.

In developing his PTD, Ufimtsev claims to have considered the "non-uniform" induced edge currents in addition to the "uniform" induced surface currents of physical optics, but nowhere in his work will the reader find an explicit representation of the edge currents. He considered instead the scattered fields, not the surface currents, of the exact solution for a wedge. Nevertheless, we are seldom interested in the surface field, except as a means by which to compute the scattered field, hence his approach is defensible.

As in most two-dimensional problems, Ufimtsev recognized two distinct cases depending on whether the incident field is polarized parallel or perpendicular to the edge. Arbitrary polarizations, of course, can be handled as linear combinations of the two cases. He represented the total fields (incident plus scattered fields) as

$$E_z = E_{oz} \left[u \left(r, \psi_s - \psi_i \right) - u(r, \psi_s + \psi_i) \right] \tag{5-54}$$

$$H_z = H_{oz} \left[u \left(r, \psi_s - \psi_i \right) + u(r, \psi_s + \psi_i) \right] \tag{5-55}$$

where

$$u(r, \psi) = \frac{1}{2\alpha} \int_c \frac{e^{-ikr \cos \beta}}{1 - \exp \left[i \pi (\beta + \psi)/\alpha \right]} \, d\beta \tag{5-56}$$

where r is the distance from the edge to the point of observation, ψ_s is the angular coordinate of that point above one face of the wedge, ψ_i is the direction of arrival of the incident wave, similarly measured, α is the external wedge angle, and C is the Sommerfeld contour in the complex plane. Equations (5-54) through (5-56) represent Sommerfeld's classic solution of the wedge problem.

To obtain the edge-diffracted field, Ufimtsev subtracted the incident field and the physical optics approximation of the scattered field. We have seen that the physical optics field is a surface integral, and in the case of the wedge, the surface is infinite in extent. How, then, did Ufimtsev obtain a finite result? He did so by ignoring the contribution at infinity, but not without justification. We have seen that the result of evaluating the physical optics integral yields the contributions of the stationary phase points or the end-points of integration, or both. In the case of the wedge, one of the end-points is the edge itself, but there is no "other" end-point. Even if there were, we could force it to be so far away that the resultant scattered fields would be insignificant. This is exactly the procedure used by Michaeli in developing his equivalent currents. Thus, the physical optics contribution to be subtracted from the exact solution arises from the edge itself. As we shall see, its contribution will be in the form of a diffraction coefficient.

Depending on the direction of arrival of the incident wave, there are three possible physical optics contributions to subtract from the exact solution. One is if the upper face is illuminated, but not the lower face, another is if the lower face is illuminated, but not the upper, and the third is if both faces are illuminated. Thus, Ufimtsev's results have different forms in each of the three different regions shown in Fig. 5-8. It is a small annoyance to keep track of these three regions in implementing the PTD in practical computations, but it is a small price to pay for the other advantages of the theory.

Integral Eq. (5-56) in the exact solution must be evaluated and substituted in Eqs. (5-54) and (5-55), and this can be done by means of the familiar method of stationary phase. The result is none other than the Keller diffraction coefficients of Eqs. (5-34) and (5-35), at least for angles not too close to the shadow or reflection boundaries. A more exact evaluation of the integral in the transition regions is not necessary, however, because the physical optics contributions effectively eliminate the singularities. Ufimtsev's results can be summarized as follows:

$$E_z^s = E_{oz} f \; \frac{e^{i(kr + \pi/4)}}{\sqrt{2\pi kr}} \tag{5-57}$$

$$H_z^s = H_{oz} g \; \frac{e^{i(kr + \pi/4)}}{\sqrt{2\pi kr}} \tag{5-58}$$

where the diffraction coefficients are

$$f = \begin{cases} (X - Y) - (X_1 - Y_1) & 0 \le \psi_i \le \alpha - \pi, \\ (X - Y) - (X_1 - Y_1) - (X_2 - Y_2) & \alpha - \pi \le \psi_i \le \pi, \\ (X - Y) - (X_2 - Y_2) & \pi_i \le \psi_i \le \alpha. \end{cases} \tag{5-59}$$

$$g = \begin{cases} (X + Y) - (X_1 + Y_1) & 0 \le \psi_i \le \alpha - \pi, \\ (X + Y) - (X_1 + Y_1) - (X_2 + X_2) & \alpha - \pi \le \psi_i \le \pi, \\ (X + Y) - (X_2 + Y_2) & \pi \le \psi_i \le \alpha. \end{cases} \tag{5-60}$$

where the unsubscripted coefficients are the Keller coefficients of Eqs. (5-34) and (5-35). The subscripted coefficients are due to the uniform physical optics contributions,

$$X_1 = -\frac{1}{2} \tan\left[(\psi_s - \psi_i)/2\right] \tag{5-61}$$

$$Y_1 = -\frac{1}{2} \tan\left[(\psi_s - \psi_i)/2\right] \tag{5-62}$$

$$X_2 = \frac{1}{2} \tan\left[(\psi_s - \psi_i)/2\right] \tag{5-63}$$

$$Y_2 = -\frac{1}{2} \tan\left[\alpha - (\psi_s + \psi_i)/2\right] \tag{5-64}$$

Thus, as evidenced by Eqs. (5-59) and (5-60), Ufimtsev's diffraction coefficients are none other than Keller's, but modified by the presence of the physical optics "diffraction coefficients." These expressions are well worth considering in more detail.

First, the Keller coefficients should be expected to be singular along the reflection and shadow boundaries. Keller's prescription is obtained from the exact solution for an infinite wedge, and if we believe that solution, then the fields must be infinite along the reflection boundary. All the incident rays striking the surface must be reflected in the specular direction, and because there is an infinity of such rays, they will produce an infinite field. The same is true at the shadow boundary, marked by no incident field just inside the shadow zone to the full incident field just outside it. The diffraction coefficient must completely cancel the incident field over the entire half-space screened from the incident wave by the wedge surfaces. Hence, it is not unexpected that the diffraction coefficient rises to infinity at the shadow boundary as well as at the reflection boundary.

However, the physical optics contribution from an infinite half-space must also rise to infinity because of the infinite size of the surfaces. Indeed, Eqs. (5-61) and (5-63) become infinite along the shadow boundaries, while Eqs. (5-62) and (5-64) become infinite along the reflection boundaries. Thus, it would appear that the infinities of the exact solution could be handled adequately by the physical optics diffraction coefficients. This is, in fact, the case, and because the physical optics components have been subtracted by Ufimtsev from the exact solution, what remains is the diffraction from the edge alone. The distinction between the diffraction coefficients of Keller and Ufimtsev, therefore, is that Keller's coefficients inherently contain both surface and edge contributions, while Ufimtsev's coefficients contain only the edge contributions. Keller's become infinite in the transition regions while

Ufimtsev's remain finite. In fact, it can be shown that in the transition regions Ufimtsev's diffraction coefficients approach the value:

$$X - X_1 \quad \text{or} \quad X - X_2 \rightarrow - \frac{1}{2n} \cot \frac{\pi}{n} \tag{5-65}$$

$$Y - Y_1 \quad \text{or} \quad Y - Y_2 \rightarrow - \frac{1}{2n} \cot \frac{\pi}{n} \tag{5-66}$$

Figures 5-10 and 5-11 are graphical comparisons of the diffraction coefficients of Keller and Ufimtsev for the simple case of a half-plane illuminated at an angle of 60 degrees from the surface. In contrast to Fig. 5-8, in which the magnitude of the coefficients are displayed, we have retained the sign of the terms in these illustrations. Figure 5-10 compares Keller's X coefficient with Ufimtsev's $X - X'$; Keller's X becomes singular, as noted earlier, along the shadow boundary (240 degrees), and the coefficient changes sign from one side of the shadow boundary to the other. Ufimtsev's coefficient also changes sign at the shadow boundary, but is zero there. It is zero only because the internal wedge angle is zero, and takes on non-zero values at the shadow boundary in accordance with Eq. (5-65) for non-zero wedge angles. Keller's Y is compared with Ufimtsev's $Y - Y'$ in Fig. 5-11, and it can be seen that the behavior in the diffraction coefficients at the reflection boundary is similar to that at the shadow boundary.

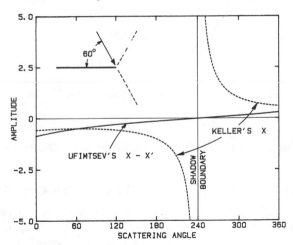

Figure 5-10 Comparison of Keller's X coefficient with Ufimtsev's X – X'

Either the PTD or the GTD can be used to estimate the fields scattered by finite bodies, of course, but the diffraction coefficients come from the solution for a semi-infinite structure. As such, difficulties will be experienced in problems solved by use of the GTD due to the singularities, while the estimates based on PTD will remain well behaved. However, because the PTD yields only the contributions due to edges, some other method, such as physical optics, must be used to obtain the surface contributions. It seems paradoxical to subtract the physical optics contribution to obtain well behaved diffraction coefficients, only to add them back in treating finite structures, yet the technique works out quite well. However, because both the GTD and the PTD rely on the exact solution of the two-dimensional wedge problem, they are applicable only to scattering directions on the Keller cone. We have seen how the equivalent currents of Michaeli extended the GTD to arbitrary directions. Let us now examine how Mitzner extended the PTD with his incremental diffraction coefficient.

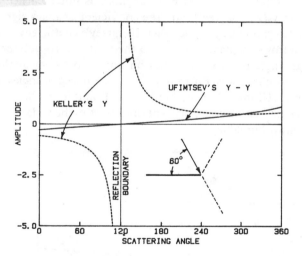

Figure 5-11 Comparison of Keller's Y coefficient with Ufimtsev's Y-Y'

5.8 THE INCREMENTAL LENGTH DIFFRACTION COEFFICIENT

Mitzner originally published his work in a government report of limited distribution [8], thereby denying access to the scattering community at large. However, distribution of the report is now unlimited, making it more available to researchers than previously.

Mitzner focuses immediately on an incremental form for his results with the understanding that the scattering from an edge of any contour can be obtained by integrating over the illuminated portions of the contour. He expresses the far diffracted field in terms of a dyadic diffraction coefficient $\bar{\bar{d}}$,

$$\bar{E}_d = E_i \frac{e^{i(kR-\pi/4)}}{\sqrt{2\pi} \; R} \; \bar{\bar{d}} \cdot \hat{p} \; dt \tag{5-67}$$

where $\hat{p}$ is a unit vector aligned along the incident electric polarization. In developing his ILDC, he establishes two pairs of unit vectors, one pair perpendicular and parallel to the plane of incidence, and one pair perpendicular and parallel to the plane of scattering. The plane of incidence is that plane containing the edge element dt and the incident direction, and the plane of scattering contains the edge element and the direction of scattering. These unit vectors are

$$\hat{e}_\perp^i = \hat{i} \times \hat{i} / |\hat{i} \times \hat{i}| \tag{5-68}$$

$$\hat{e}_\parallel^i = \hat{i} \times \hat{e}_\perp^i \tag{5-69}$$

$$\hat{e}_\perp^s = \hat{i} \times \hat{s} / |\hat{i} \times \hat{s}| \tag{5-70}$$

$$\hat{e}_\parallel^s = \hat{s} \times \hat{e}_\perp^s \tag{5-71}$$

where $\hat{i}$ is a unit vector along the edge. (Mitzner chooses his $\hat{i}$ in the opposite direction used by Michaeli [21]. He also uses the symbol α to denote the interior wedge half-angle, whereas we have used that symbol to denote the complete external wedge angle. This notational difference has been accounted for in the diffraction coefficients displayed below.)

Mitzner expresses the diffraction dyad in terms of components along these directions,

$$\bar{\bar{d}} = d_{\perp\perp} \; \hat{e}_\perp^s \; \hat{e}_\perp^i + d_{\perp\parallel} \; \hat{e}_\perp^s \; \hat{e}_\parallel^i + d_{\parallel\perp} \; \hat{e}_\parallel^s \; \hat{e}_\perp^i + d_{\parallel\parallel} \; \hat{e}_\parallel^s \; \hat{e}_\parallel^i \tag{5-72}$$

Similarly, we may express the incident field polarization as

$$\hat{p} = \hat{e}_\perp^i \cos\gamma + \hat{e}_\parallel^i \sin\gamma \tag{5-73}$$

where γ is the angle subtended by the incident polarization and the normal to the plane of incidence. We may therefore express the product $\bar{\bar{d}} \cdot \hat{p}$ as

$$\bar{\bar{d}} \cdot \hat{p} = d_{\perp\perp} \; \hat{e}_\perp^s \; \cos\gamma + d_{\perp\parallel} \; \hat{e}_\perp^s \; \sin\gamma + d_{\parallel\perp} \; \hat{e}_\parallel^s \; \cos\gamma + d_{\parallel\parallel} \; \hat{e}_\parallel^s \; \sin\gamma \tag{5-74}$$

To evaluate the components of Eq. (5-74), and thereby obtain expressions for the ILDC that can be compared to Michaeli's diffraction coefficients, we must apply Eqs. (3-46A) through (3-67) of Mitzner's report. There is an error in the last term of Mitzner's Eq. (3-46A), and the function $\sin[(\beta_s + \beta_i)/2]$ appearing therein must be replaced by $\sin[(\beta_s - \beta_i)/2]$. When this is done and the diffraction coefficients are expressed in Michaeli's form, we obtain

$$D_\perp = D_m - D_\perp' \tag{5-75}$$

$$D_\parallel = D_e - D_\parallel' \tag{5-76}$$

$$D_{\parallel\perp} = D_{em} \sin\beta_i - D_{\parallel\perp}' \tag{5-77}$$

where the primed diffraction coefficients are the physical optics terms,

$$D'_\perp = - U^+ \frac{\sin \phi_s}{\cos \alpha_1 + \cos \phi_i} - U^- \frac{\sin (n\pi - \phi_s)}{\cos \alpha_2 + \cos (n\pi - \phi_i)} \quad (5\text{-}78)$$

$$D'_\parallel = - U^+ \frac{\sin \phi_i}{\cos \alpha_1 + \cos \phi_i} - U^- \frac{\sin (n\pi - \phi_i)}{\cos \alpha_2 + \cos (n\pi - \phi_i)} \quad (5\text{-}79)$$

$$D'_{\parallel\perp} = - U^+ \left[\frac{Q \cos \phi_s}{\cos \alpha_1 + \cos \phi_i} - \cos \beta_i \right]$$

$$+ U^- \left[\frac{Q \cos (n\pi - \phi_s)}{\cos \alpha_2 + \cos (n\pi - \phi_i)} - \cos \beta_i \right] \quad (5\text{-}80)$$

In Eqs. (5-78) through (5-80), the quantities Q, α_1, and α_2 are defined in Eqs. (5-51) through (5-53), and the step functions are

$$U^+ = \begin{cases} 1, \text{ for "plus" face illumination} \\ \\ 0, \text{ otherwise} \end{cases} \quad (5\text{-}81)$$

$$U^- = \begin{cases} 1, \text{ for "minus" face illumination} \\ \\ 0, \text{ otherwise} \end{cases} \quad (5\text{-}82)$$

These step functions toggle on or off the appropriate physical optics coefficients, depending on whether a given wedge face is illuminated. Note that the components of Mitzner's ILDC are identical to those of Michaeli, modified by the removal of the physical optics coefficients. It can be shown that Mitzner's diffraction coefficients reduce exactly to those of Ufimtsev when the scattering direction lies on the Keller cone. Although the physical optics terms become singular along the reflection and shadow boundaries, Mitzner's ILDC remains finite in the transition regions because the singularities in the physical optics coefficients precisely cancel those in Michaeli's diffraction coefficients.

It should be clear that Mitzner's ILDC extends Ufimtsev's PTD to arbitrary directions in precisely the same way Michaeli's equivalent current approach extends Keller's theory. The singularities in Michaeli's coefficients can be attributed to a surface term which is cancelled in Mitzner's coefficients by the singularity in the physical optics term. The price we pay for this highly desirable result is that two separate computations are required for any finite edged body. One is the edge contribution as given in Eqs. (5-75) through (5-80), and the other is a physical optics integral over the illuminated portions of the body. Despite the very practical extension to arbitrary directions both these theories represent, the edge elements are assumed to scatter independently. Consequently, in and of themselves, they cannot handle a case of great practical importance: surface traveling waves.

5.9 THE SURFACE TRAVELING WAVE

Although the surface traveling wave is of primary importance for long, smooth structures illuminated at relatively low angles of incidence, and is therefore a high frequency phenomenon, the elements of the surface are strongly coupled by the fields propagating over the surface. Consequently, the surface elements do not scatter independently as assumed for the theories considered thus far in this chapter. The traveling wave is launched only if there is a component of the incident electric field tangential to the surface and in the plane of incidence. It travels close to the speed of light over the surface, much as a current wave is launched along thin wire antennas in the end-fire mode.

In fact, the analogy between the operation of end-fire antennas and the surface wave contribution to the scattered fields on long bodies is a convenient way to introduce the subject. Consider, for example, the antenna in the upper diagram of Fig. 5-12. In practice, the antenna is mounted relatively close to the ground and one end is fed by an RF source, as shown, while the far end is terminated to ground by an impedance chosen to minimize reflections. The wire supports a filamentary current directed along the wire, and although the current propagates at close to the speed of light, we may assume for generality a propagation velocity $v = pc$, where c is the speed of light. In most cases of practical interest, p is on the order of 99 to 100 percent the speed of light. When the assumed current is inserted in the radiation integral Eq. (5-41) (and with the magnetic current set to zero, of course), we obtain for the far radiated field:

$$E_r = -i\,kl\,\psi_0\,Z_0 I_0 \sin\theta \; \frac{\sin\left[\dfrac{1}{2p}\,kl\,(1-p\cos\theta)\right]}{\dfrac{1}{2p}\,kl\,(1-p\cos\theta)} \tag{5-83}$$

where I_0 is the amplitude of the current wave, assumed constant along the wire, and θ measures the angle of the point of observation away from the axis of the wire.

If the wire were somehow isolated in space (and ignoring how we would feed the antenna at one end and ground it at the other), the radiation pattern would be a figure of revolution. The polar pattern shown in the upper diagram of Fig. 5-12 is a half-slice through that spatial pattern, and note that the radiation intensity along the wire axis is zero because of the presence of the $\sin\theta$ factor in Eq. (5-83). The pattern is nothing more than a $\sin(x)/x$ function multiplied by a trigonometric function that forces the axial radiation to zero. The $\sin(x)/x$ function is characteristic of a uniformly illuminated aperture, and in the case of the long-wire antenna, the current is constant along the wire from one end to the other. When such a function is summed in a radiation integral, the result is essentially given by the end-points of integration. The presence of the electrical length kl in the argument of the $\sin(x)/x$ function shows that the lobe structure of resulting pattern becomes finer as the wire becomes electrically longer. The length of the wire, in fact, governs the spatial location of the main lobe: the longer the wire, the closer the lobe to the wire axis.

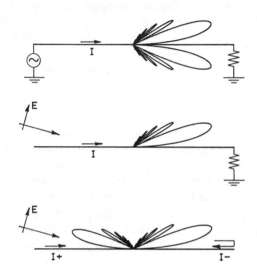

Figure 5-12 Polar patterns of surface traveling waves

We may remove the RF generator exciting the wire and replace it with an incident plane wave, as shown in the center diagram of Fig. 5-12, and obtain the same result. Although the intensity of the radiation will not be the same, the pattern will be unchanged, because the plane wave launches a current wave down the wire just as the signal generator did. Finally, we may remove the terminating impedance, as in the lower diagram, and more faithfully represent the behavior of an isolated wire. When this is done, however, the current wave will be reflected by the discontinuity at the end of the wire, and there will exist two current waves, one traveling in the forward direction and one in the backward direction. The backward traveling current wave will give rise to the same kind of pattern generated by the forward current wave, but its location in space will be in the opposite direction, as shown in the lower diagram. That diagram suggests somewhat less intense radiation in the backward direction than in the forward direction, accounting for less than perfect reflection of the forward current wave at the far end of the wire.

Thus, it is not the forward traveling current wave that gives rise to the surface wave contribution to the RCS of long, smooth metallic surfaces, but the backward current wave. The surface wave phenomenon is not constrained to slender objects like the wire, and is a common feature of the scattering of other structures as well, such as airfoils and missile bodies. In fact, even subsections of larger surfaces exhibit the effect due to surface discontinuities created by seams and gaps where skin sections of fuselages may be joined together. One of the great challenges of modern airframe construction is the elimination of gaps and seams to reduce or even eliminate the surface wave contribution. A particularly annoying problem is the discontinuity at the far end of the airframe (which could be the nose if the airframe is viewed from aspects in the rear quarter). There being no further structure beyond the end, the challenge is to somehow attenuate the forward traveling surface wave before it reaches the end of the surface. Typically, magnetic surface absorbers are used, as suggested in Ch. 9.

It is of significant diagnostic value to be able to estimate the angle at which the pattern of Eq. (5-83) attains its maximum value. One approach is to differentiate the expression with respect to the angle θ and to find those angles that force the first derivative to zero. This generates the transcendental equation:

$$\tan f - 2f(1 - f/kl) = 0 \tag{5-84}$$

where

$$f = \frac{1}{2} kl(1 - p\cos\theta) \tag{5-85}$$

There are a multitude of solutions of Eq. (5-84) and a value for the function f may be found for each sidelobe. Due to the presence of kl and p in the transcendental equation, each solution will depend on the electrical length kl and the relative phase velocity p. It is sufficient for diagnostic purposes to eliminate this dependence by ignoring the third term of Eq. (5-84) and setting $p = 1$. The solution may be effected by the Runge-Kutta method of successive approximations or by graphical means, and the first few solutions are

$$f = \pm\ 1.16556,\ \pm\ 4.60422,\ \pm\ 7.78988,\ \pm\ 10.94995,\ \ldots \tag{5-86}$$

Inserting the value of the first solution into Eq. (5-85), and converting from radians to degrees, we find that the first maximum in the pattern occurs at

$$\theta = 49.35 \sqrt{\lambda/l} \tag{5-87}$$

Despite the approximation used, this simple expression yields accurate estimates of the location of the first lobe for slender bodies as short as a wavelength: for a body one wavelength long, the location is given within 3 percent, and the accuracy improves as the body or surface becomes longer. For example, Eq. (5-87) gives the location with an error of only 0.2 percent for a body 15 wavelengths long.

The end-fire antenna analogy can even be used for curved bodies, such as an ogive. In this event, the axial velocity of the surface wave is somewhat slower than its velocity over the surface because of the greater distance traveled. In his classic analysis of the surface traveling wave [22], Peters suggested that an average phase velocity p could be assigned to the surface wave on the basis of the additional distance traveled, and the results agreed closely with measurement. What is not quite so easy to obtain is the amplitude of the reflection of the surface wave at the far end of the body, and therefore its contribution to the far scattered field, and subsequently the RCS.

For simple structures, it turns out that the repeated application of the GTD or PTD to account for higher order interactions between the edges of a plate yields very nearly the correct value. We might ask why this is so, since those theories treat only the localized diffraction by an edge, whereas the traveling wave is distributed over the surface. The answer has been suggested earlier: when a more or less uniform surface wave is integrated over the surface, the dominant contributions to the integral are the end-points of integration. Because these are none other than the edges of the plate to which the diffraction theories are applied, the apparent locations of the scattering centers are the same in both the diffraction theory and in the integration of the traveling current wave. It is important to note, however, that it is only by accounting for the higher order edge interactions that the diffraction theories emulate the surface traveling wave effect so well.

Even for a simple plate, the situation can become complicated very quickly if the angle of incidence on the plate edges is no longer normal. Waves are diffracted in the direction of the intersection of the local Keller cone and the surface of the plate, and those directions must be pursued until the surface ray is intercepted by another edge. The re-diffracted surface ray must be similarly traced, and so on, until all significant interactions have been accounted for. Although the procedure is probably prohibitively time consuming for arbitrary plates assembled together to represent a more complex target, it is, nevertheless, conceptually possible. The situation is not much better if one were to integrate the surface current wave for arbitrary plates, for the waves striking an oblique trailing edge would also be reflected in a direction closely related to the Keller cone, generating another surface wave. Thus, as important as the surface wave mechanism may be, there is no routine way to estimate its magnitude for arbitrary structures, as might be desired in the prediction of the RCS of complex bodies.

5.10 SUMMARY

In this chapter we have discussed the more important of the high frequency methods for predicting the RCS of simple structures. The great utility of these theories is that in the high frequency region, the target elements scatter the incident wave independently of one another. This makes it possible to assemble a collection of relatively simple shapes, such as flat plates, cylinders, and spheroids to model a complicated target. The analytical high-frequency formulas are relatively easy to derive for specific geometries, and we only need to sum the individual contributions coherently to obtain the total RCS of the target.

The simplest of those theories is geometric optics, which is best used for doubly curved surfaces. The formula is deceptively simple, and the same formula holds for bistatic directions, as well as in the backward direction, as long as the forward direction is avoided. The implementation of geometric optics requires only that there exist a specular point on the body and that the principal radii of curvature be specified there. We have shown that geometric optics fails when one or both radii of curvature is infinite, as is the case for flat and singly curved surfaces, and in this event the theory of physical optics proves most helpful.

The theory of physical optics uses the tangent plane approximation of the induced surface currents, and thus relies on estimates of the surface fields from geometric optics. The induced currents are integrated over the illuminated portions of the body to obtain the far scattered fields, and the integral, although itself an approximation of the actual scattered fields, can sometimes be evaluated exactly. For those cases when an exact closed-form evaluation cannot be obtained, the method of stationary phase often yields quite acceptable results. On the other hand, when applied to smooth bodies such as spheroids, physical optics yields an erroneous contribution due to the abrupt discontinuity in the assumed surface fields at shadow boundaries. Thus, for smooth objects of any appreciable size, we might be better off using the simpler theory of geometric optics.

However, the physical optics prediction fails by progressively wider margins as the scattering direction moves further from the specular direction. This is because the surface effects decay to the levels of edge effects, and the edge contributions are not well modeled in physical optics. The edge returns can be more accurately predicted in the non-specular regions with Keller's geometrical theory of diffraction, and the theory is attractively couched in terms of a pair of simple diffraction coefficients. Nonetheless, the GTD has its own set of difficulties, among them the singularities in the diffraction coefficients, the existence of caustics, and the restriction of the scattering direction to the Keller cone.

The first of these can be overcome with a uniform theory of diffraction and the UTD developed by Koumyoujian and Pathak was described. The singularities in the diffraction coefficients are multiplied by a Fresnel integral that drops to zero as the diffraction coefficients rise to infinity. The product remains finite and exhibits the proper change in sign from one side of the shadow or reflection boundary to the other. The caustic difficulties can be overcome by the method of equivalent currents, which postulates the existence of filamentary electric and magnetic currents along an edge singularity. When summed in the radiation integral, the equivalent currents guarantee finite fields at the observation point, provided we have well-behaved diffraction coefficients available. The equivalent currents also make it possible to compute the scattered field for arbitrary directions as well as those on the Keller cone.

We have shown that the diffraction coefficients of Ufimtsev's PTD are not singular, like those of Keller's GTD, and thus may be more attractive for RCS predictions. This is because Ufimtsev has, in essence, deleted the singularity in the Keller coefficients by subtracting a physical optics coefficient. As a result, the PTD contains only the edge contribution, while the GTD contains that contribution plus a surface contribution. Nevertheless, Ufimtsev's PTD is restricted, as is Keller's GTD, to directions lying on the Keller cone. Mitzner, with his incremental length diffraction coefficient, extended Ufimtsev's theory to arbitrary directions in much the same way Michaeli extended Keller's theory. Thus, the high frequency diffraction theories seem to have reached the stage where edge scattering can be accurately predicted.

Alas, this is not yet the case. The surface traveling wave, although a high frequency phenomenon, involves the entire surface as well as the edges bounding that surface, and the diffraction theories can properly account for only the localized edge diffraction. However, the repeated application of the diffraction theories to account for multiple interactions between the edges offers some promise of modeling the surface traveling wave effect. The incremental prescriptions of Mitzner and Michaeli hold well enough over edge sections whose contours change slowly, but those theories cannot be expected to work in the vicinity of corners where the edge takes an abrupt turn. Because the corner is a discontinuity where we would expect a strong reflection of a surface wave launched along the edge, a corner diffraction coefficient is badly needed. If one were available, then theories similar to those of Mitzner and Michaeli might be applied to account for multiple corner diffraction.

REFERENCES

1.   E.F. Knott, "A Tool for Predicting the Radar Cross Section of an Arbitrary Trihedral Corner," presented at the IEEE SOUTHEAST-CON '81 Conference, Huntsville, Alabama, 6-8 April 1981, IEEE Publication 81CH1650-1, pp. 17-20.

2. J.B. Keller, "Diffraction by an Aperture," *J. App. Phys.*, Vol. 28, No. 4, April 1957, pp. 426-444.

3. C.E. Ryan, Jr. and L. Peters, Jr., "Evaluation of Edge-Diffracted Fields Including Equivalent Currents for Caustic Regions," *IEEE Trans. Antennas Propag.*, Vol. AP-17, No. 3, May 1969, pp. 292-299. See also Correction in Vol. AP-18, March 1970, p. 275.

4. E.F. Knott and T.B.A. Senior, "Equivalent Currents for a Ring Discontinuity," *IEEE Trans. Antennas Propag.*, Vol. AP-23, No. 3, March 1984, pp. 252-258.

5. A. Michaeli, "Equivalent Edge Currents for Arbitrary Aspects of Observation," *IEEE Trans. Antennas Propag.*, Vol. AP-23, No. 3, March 1984, pp. 252-258. See also correction in Vol. AP-23, February 1985, p. 227.

6. P. Ia. Ufimtsev, "Approximate Computation of the Diffraction of Plane Electromagnetic Waves at Certain Metal Bodies: Pt. I. Diffraction Patterns at a Wedge and a Ribbon," *Zh. Tekhn. Fiz.* (USSR), Vol. 27, No. 8, 1957, pp. 1708-1718.

7. R.G. Kouyoumjian and P. H. Pathak, "A Uniform theory of Diffraction for an Edge in a Perfectly Conducting Surface," *Proc. IEEE*, Vol. 62, No. 11, November 1974, pp. 1448-1461.

8. K.M. Mitzner, "Incremental Length Diffraction Coefficients," Technical Report No. AFAL-TR-73-296, Northrop Corporation, Aircraft Division, April 1974.

9. R.A. Ross, "Radar Cross Section of Rectangular Plates as a Function of Aspect Angle," *IEEE Trans. Antennas Propag.*, Vol. AP-14, No. 3, May 1966, pp. 329-335.

10. D.K. Plummer, L.E. Hinton, E.F. Knott, and C.J. Ray, "Non-Metallic Chaff Study," Report NADC-79200-30, Prepared for the Department of the Navy, Naval Air Development Center by the Georgia Institute of Technology, Engineering Experiment Station, June 1980.

11. J.A. Stratton, *Electromagnetic Theory*, New York, McGraw-Hill, 1941, pp. 464-470.

12. G.T. Ruck, D.E. Barrick, W.D. Stuart, and C.K. Krichbaum, *Radar Cross Section Handbook*, Vol. 1, New York, Plenum Press, 1970, pp. 50-59.

13. W.B. Gordon, "Far Field Approximation of the Kirchhoff-Helmholtz Representation of Scattered Fields," *IEEE Trans. Antennas Propag.*, Vol. AP-23, No. 5, July 1975, pp. 864-876.

14. T.B.A. Senior, "A Survey of Analytical Techniques for Cross-Section Estimation," *Proc. IEEE*, Vol. 53, No. 8, August 1965, pp. 822-833.

15. A. Sommerfeld, "Mathematische Theorie der Diffracton," *Math. Ann.*, Vol. 47, 1896, pp. 317-374.

16.   A. Sommerfeld, "Lectures on Theorectical Physics," in *Optics*, Vol. 4, New York, Academic Press, 1964.

17.   T.B.A. Senior and P.L.E. Uslenghi, "High-Frequency Backscattering from a Finite Cone," *Radio Science*, Vol. 6, No. 3, March 1971, pp. 393-406.

18.   S.W. Lee and G.A. Deschamps, "A Uniform Asymptotic Theory of Electromagnetic Diffraction by a Curved Wedge," *IEEE Trans. Antennas Propag.*, Vol. AP-24, No. 1, January 1976, pp. 25-34.

19.   G.A. Deschamps, J. Boersma, and S.W. Lee, "Three-Dimensional Half-Plane Diffraction: Exact Solution and Testing of Uniform Theories," *IEEE Trans. Antennas Propag.*, Vol. AP-32, No. 3, March 1984, pp. 264-271.

20.   R.F. Millar, "An Approximate Theory of the Diffraction of an Electromagnetic Wave by an Aperture in a Plane Screen," *Proc. IEE*, Vol. 103, Part C, March 1956, pp. 177-185.

21.   E.F. Knott, "The Relationship between Mitzner's ILDC and Michaeli's Equivalent Currents," *IEEE Trans. Antennas Propag.*, Vol. AP-33, No. 1, January 1985, pp. 112-114.

22.   L. Peters, Jr., "End-Fire Echo Area of Long, Thin bodies," *IRE Trans. Antennas Propag.*, Vol. AP-6, No. 1, January 1958, pp. 133-139.

CHAPTER 6

CONCEPT OF RADAR CROSS SECTION

E. F. Knott

6.1 DEFINITIONS

When an obstacle is illuminated by an electromagnetic wave, energy is dispersed in all directions. The spatial distribution of energy depends on the size, shape, and composition of the obstacle, and on the frequency and nature of the incident wave. This distribution of energy is called *scattering*, and the obstacle itself is often called a *target* or a *scatterer*.

Bistatic scattering is the name given to the situation when the scattering direction is not back toward the source of the radiation; hence, *forward scattering* occurs when the bistatic angle is 180 degrees. It is called *monostatic scattering* when the source and receiver are located at the same point, as is the case with most radars. In many measurement situations, separate transmitting and receiving antennas are often used, which in a strict application of the definition represent a bistatic condition. However, the angle subtended at the target by the two directions to the antennas is usually small enough that the measurements are indistinguishable from the truly monostatic case.

Probably as an outgrowth of antenna research and design, this spatial distribution of scattered energy or scattered power is characterized by a *cross section*, a fictitious area property of the target. An antenna is often regarded as having "an aperture of effective area [which] extracts energy from a passing radio wave," and the power available at the terminals of a receiving antenna can be represented as the product of an incident power density and an effective area exposed to that power density [1]. In much the same way, the power reflected or scattered by a radar target can be expressed as the product of an effective area and an incident power density. In general, that area is called the *scattering cross section*. For directions other than back toward the radar, it can also be called the *bistatic cross section*, and, when the direction is back toward the radar, it is called the *backscattering cross section* or the *radar cross section*. In the early days of radar research, the term *echo area* was common and occasionally researchers defined "effective areas" that could be identified with the geometric area of an equivalent flat plate [2].

Whatever the case, symbol σ has been accepted almost universally to denote the radar cross section of a target and, occasionally, the bistatic cross section. Generally, the receiver of the scattered energy is assumed to be located far enough from the target so that the target is essentially a point scatterer. This implies that the distance to the target is much larger than any significant target dimension. If we assume that this "point scatterer" radiates energy isotropically, because scattered fields depend on the attitude at which the target is presented to the incident wave, the scattering cross section fluctuates, and the strength of this fictitious point scatterer must also fluctuate with target attitude. Thus, the scattering cross section is *not* a constant: it is a strongly angular-dependent property of the target.

The definition of the radar cross section is based upon this concept of isotropic scattering, assuming that the target is illuminated by a plane wave. For such a wave, the incident power density is [3]

$$W_i = \frac{1}{2} \ E_i H_i = \frac{1}{2} \ Y_0 |E_i|^2 \tag{6-1}$$

where E_i and H_i are the strengths of the incident electric and magnetic field intensities, respectively, and Y_0 is the admittance of free space (0.00165 mho). Thus, the net power intercepted by the target is

$$P = \sigma W_i = \frac{1}{2} \ \sigma Y_0 |E_i|^2 \tag{6-2}$$

where, as discussed above, σ is the scattering cross section of the target.

If that power is now radiated isotropically, the scattered power density at some large distance R from the target is

$$W_s = \frac{P}{4\pi R^2} = \frac{\sigma Y_0 |E_i|^2}{8\pi R^2} \tag{6-3}$$

However, the scattered power density can be written in terms of the scattered field strength E_s:

$$W_s = \frac{1}{2} \ Y_0 |E_s|^2 . \tag{6-4}$$

Equating (6-3) and (6-4), and solving for σ,

$$\sigma = 4\pi R^2 \frac{|E_s|^2}{|E_i|^2} \tag{6-5}$$

Because the incident wave is planar and since we have assumed the target to be effectively a point scatterer, the distance R should be allowed to become infinite. Thus, Eq. (6-5) should be written more precisely as

$$\sigma = \lim_{R \to \infty} 4\pi R^2 \frac{|E_s|^2}{|E_i|^2} \tag{6-6}$$

which is identical to Eq. (3-1). This is the basic definition of radar cross section [4], and an equally valid definition results when the incident and scattered electric fields are replaced by the incident and scattered magnetic fields.

Another definition can be established which retains the vector properties of electromagnetic fields such that the radar cross section can be expressed as

$$\sigma = (\lambda \hat{e}_r \cdot \overline{S})^2 / \pi \tag{6.7}$$

where λ is the radar wavelength, $\hat{e}_r$ is the polarization of the receiver and $\overline{S}$ is the vector scattering function:

$$\overline{S} = k\,Re^{-ikr} \frac{\overline{E}_s}{|E_i|} \tag{6.8}$$

where $k = 2\pi/\lambda$ is the free space wavenumber, $\overline{E}_s$ is the vector scattered field, and a harmonic time dependence $e^{-i\omega t}$ has been assumed. The exponential e^{-ikR} shifts the phase reference from the receiver to the target. Note that $\overline{S}$ is a dimensionless function and the presence of λ^2 in the numerator of Eq. (6-7) forces σ to be an area.

For three-dimensional geometries, the scattered field $\overline{E}_s$ decays as R^{-1} in the far field and, consequently the presence of R in the numerator of Eq. (6-8) removes the effect of range. For two-dimensional problems, which are not often encountered in nature but are extremely useful in the analysis of related three-dimensional problems, the fields decay as $R^{-1/2}$ away from the target. Consequently, the definitions in Eq. (6-7) and (6-8) are slightly different.

For the two-dimensional case, the body is an infinite cylinder, and the scattering cross section reduces to a scattering width, which can be thought of as a cross section per unit length of the cylinder. The scattering width is defined as [4]:

$$\sigma = \frac{4}{k} |P|^2 \tag{6-9}$$

where P is analogous to S in Eq. (6-8),

$$P = \frac{\sqrt{\pi k \pi}}{2} \frac{|E_s|}{|E_i|} e^{-i(k\rho - \pi/4)} \tag{6-10}$$

where ρ is the distance from the cylinder to a remote point where the scattering is observed. In two-dimensional problems, the scattering function P depends on whether the incident electric or magnetic field is aligned along the cylinder axis.

In general, the forward scattering from a target is stronger than the back-scattering, and at high frequencies — actually, when the body dimensions are significantly larger than the incident wavelength — the forward scattering cross section is proportional to the projected area of the target. The large forward scatter is often misunderstood. Aside from special cases, such as a refraction process that focuses energy in the forward direction, the forward scatter is subtracted from the incident field, thereby creating a shadow behind the target. This is demonstrated in Figs. 6-1 through 6-3.

Figure 6-1 is the far field bistatic pattern of an infinite circular metallic cylinder ten wavelengths in circumference. A circumference of ten wave-lengths is often considered at the edge of the "optics" region, where the term, as used here, does not refer to very small wavelengths, but to large body sizes as measured in wavelengths. As such, the scattering behavior is not very different from what might be observed at much shorter wavelengths. The cylinder whose pattern is shown in Fig. 6-1 is illuminated by a plane wave whose electrical polarization is aligned along the cylinder axis. In the backward direction, the scattering is approximately $|P| \simeq 3$, but in the forward direction it is more like $|P| \simeq 11$. Inserting these values in Eq. (6-10), we have

$$\sigma \simeq \pi a \quad \text{(backward direction)} \tag{6-11}$$

$$\sigma \simeq \frac{8\pi}{\lambda}\, a^2 \quad \text{(forward direction)} \tag{6-12}$$

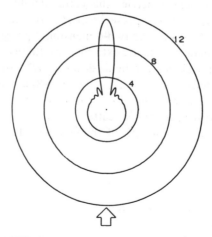

Figure 6-1 Bistatic scattering of an infinite cylinder 10λ in circumference. Quantity plotted is $|P|$

Note that the forward scattering width is proportional to the square of the cylinder diameter, while the backscattering varies linearly with the diameter.

For most of the angular coverage, the bistatic scattering varies little with the scattering direction, hence the bistatic scattering for all but a narrow forward sector is not very different from that in the backward direction, but the strong forward lobe creates a shadow behind the target, as shown in Figs. 6-2 and 6.3.

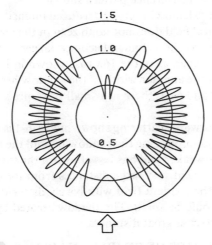

Figure 6-2 Total field in the vicinity of the infinite cylinder along a circular path 10λ from the cylinder axis

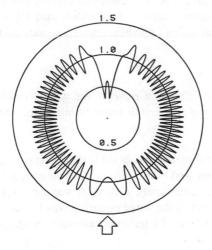

Figure 6-3 Total field in the vicinity of the infinite cylinder along a circular path 16λ from the cylinder axis

Figure 6-2 is a plot of the *total field* (incident field plus the scattered field) along the circular trajectory ten wavelengths from the cylinder axis. The variations in field strength are slowest in the backward and forward directions because, in those regions, the phase of the incident field is changing slowly along the circular trajectory. Near the sides, the phase of the incident field is changing much more rapidly, while the phase of the scattering is nearly constant. This creates the interference pattern shown.

Note that there is a "shadow" region of reduced intensity behind the cylinder. In general, the total field does not go to zero in the shadow region, but it does become smaller as the cylinder becomes larger. At a slightly greater distance from the cylinder, as in the 16λ case shown in Fig. 6-3, the shadow becomes deeper, but the angular width of the shadow becomes narrower. In both cases (Fig. 6-2 and 6-3), there is a small peak precisely in the forward direction.

Thus, the strong forward scattering shown in Fig. 6-1 is nearly out of phase with the incident field, producing the shadow behind the target. Conversely, the backscattering is weaker and has less effect on the total field. If the total fields were to be probed at very large distances from the cylinder, we would essentially measure the incident field with barely discernible ripples because the scattered field would be small. The shadow created by the bistatic fields would be hard to detect at great distances.

6.2 CHARACTERISTICS OF SIMPLE SHAPES

Because of the great variation in the radar cross section pattern from one aspect angle to another, it is convenient to display the data in logarithmic form. The unit commonly used is the decibel, and RCS values are usually expressed in dBsm, meaning "decibels above a square meter."

That is,

$$\text{dBsm} = 10 \log_{10} (\sigma) \tag{6-13}$$

where σ is expressed in square meters. We will have occasion to use this logarithmic scale of units as well as more familiar units of area.

6.2.1 The Sphere

The simplest three-dimensional body is the sphere. It is rotationally symmetric about the origin, and because its surface coincides with one coordinate of the spherical polar coordinate system, it was one of the first shapes for which an exact solution of the wave equation was found. The exact solution for the radar cross section of a perfectly conducting sphere is given in Ch. 4 and is repeated here. It is

$$\sqrt{\frac{\sigma}{\pi a^2}} = -\frac{i}{ka} \sum_{n=1}^{\infty} (-1)^n (2n + 1) (b_n - a_n) \tag{6-14}$$

$$a_n = \frac{j_n(ka)}{h_n^{(1)}(ka)} \tag{6-15}$$

$$b_n = \frac{ka\,j_{n-1}(ka) - nj_n(ka)}{ka\,h_{n-1}(ka) - nh'^{(1)}_n(ka)} \tag{6-16}$$

$$h'^{(1)}_n(ka) = j_n(ka) + iy_n(ka) \tag{6-17}$$

Here $j_n(ka)$, $y_n(ka)$, and $h'_n(ka)$ are the spherical Bessel functions of the first, second, and third kinds, respectively, of order n and argument ka. Note that the last is a complex number comprised of the first two. The argument $ka = 2\pi a/\lambda$ is the electrical circumference of the sphere whose radius is a. The square root of the radar cross section is given in Eq. (6-14) to display the relative phase of the return.

A plot of the radar cross section as a function of the electrical size of the sphere is shown in Fig. 6-4. The radar cross section is proportional to $(ka)^4$ for very small spheres, although this fact cannot be discerned from the figure. It attains a peak near $ka = 1$, then oscillates about the geometric optics value (πa^2) with damped excursions that become smaller as the sphere grows larger. In many cases, the geometric optics value is an acceptable approximation, even for electrical circumferences as small as 10 wavelengths.

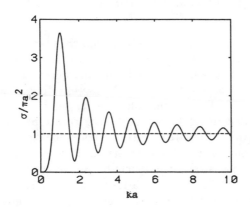

Figure 6-4 Radar cross section of a metallic sphere as a function of its electrical circumference

The oscillatory nature of return from the sphere and is due to a "creeping wave" that circles the rear of the sphere and is launched back in the direction of the radar, as shown in Fig. 6-5. The creeping wave travels an additional path length equal to the sphere diameter plus half the circumference, or a total additional distance of $(2+\pi)a$. The damped intereference pattern, therefore, should have peak-to-peak spacings in ka that occur when the path length difference is a wavelength, or for

$$\Delta ka = \frac{2\pi}{2+\pi} = 1.2220$$

Figure 6-5 A creeping wave spins around the back side of the sphere and comes back to the radar along with a direct reflection from the front

If the diagram is examined carefully the spacing between peaks is about $\delta ka \simeq 1.2101$, very close to that value. Because the creeping wave loses energy in proportion to the distance traveled behind the sphere, it becomes weaker as the sphere becomes larger. Consequently, the interference pattern becomes weaker as the electrical size of the sphere increases.

Like the circular cylinder, bistatic scattering from a sphere is nearly omnidirectional except in the forward direction. This is one disadvantage in using a sphere or a cylinder as a calibration standard, although their radar cross sections are known accurately enough to be used as calibration standards for RCS measurements. The bistatic scattering can be reflected from the target rotator back to the sphere (or cylinder), then back to the radar. The effect can be minimized by covering the target turntable with absorbing material, and by choosing a target mounting height as large as possible, consistent with the safety and integrity of the target.

6.2.2 The Prolate Spheroid

The radar return from a prolate spheroid oscillates much like that of a sphere with increasing frequency. Figure 6-6 is a comparison of the measured and predicted end-on returns from a 2:1 prolate spheroid as reported by

Moffatt and Kennaugh [5]; the theoretical curve is based upon an analysis of the impulse response of the spheroid. Figure 6-7 is a similar comparison for a 10:1 prolate spheroid as reported by Senior [6] using a less exotic approach. Note that Moffatt and Kennaugh normalize their results with respect to the square of the wavelength, while Senior normalizes his with respect to an undefined reference cross section σ_0. Nevertheless, there is an interference pattern in both cases, and the nulls in the patterns arise from creeping waves, and possibly traveling wave returns. As these go in and out of phase with the specular return from the near end of spheroid, they produce the alternating reinforcement or cancellation that generates the interference pattern. Note that because of the difference in the normalization factors, the returns in Fig. 6-6 rise with increasing spheroid size while those in Fig. 6-7 are steady.

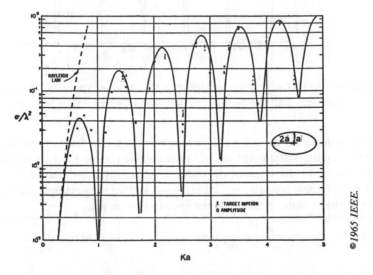

Figure 6-6 Calculated and measured axial echo area of a 2:1 prolate spheroid. Solid curve is predicted, and crosses and circles are measured data (from Moffatt and Kennaugh [5])

If the spheroid is large compared to the wavelength, the geometric optics approximation yields good results for estimating the radar cross section. The geometric optics approximation is

$$\sigma = \pi \rho_1 \rho_2 \qquad\qquad (6\text{-}18)$$

where ρ_1 and ρ_2 are the principal plane radii of curvature of the spheroid at the specular point. The radii of curvature are measured in mutually orthogonal

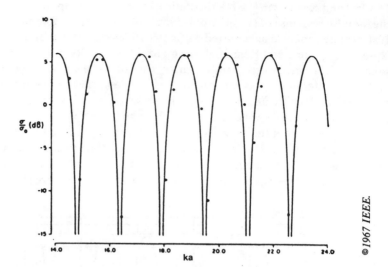

Figure 6-7 Axial radar cross section of a 10:1 prolate spheroid. The solid trace is theoretical and the points are measured values (from Senior [6])

planes containing the outward surface normal to the specular point, and the specular point is that point on the surface where the outward normal is aligned precisely in the direction of the radar illuminating the spheroid.

Using elementary calculus, it can be shown that the principal radii of curvature of the prolate spheroid at the specular point are

$$\rho_1 = -\frac{a^2 b^2}{(a^2 \cos^2 \alpha + b^2 \sin^2 \alpha)^{3/2}} \tag{6-19}$$

$$\rho_2 = -\frac{b^2}{(a^2 \cos^2 \alpha + b^2 \sin^2 \alpha)^{1/2}} \tag{6-20}$$

where $2a$ is the length of the spheroid, $2b$ is its diameter, and α is the aspect angle measured from end-on incidence. Inserting these values into Eq. (6-18). and normalizing the results with respect to πb^2 (the projected area when viewed end-on), we have

$$\frac{\alpha}{\pi b^2} = \left(\frac{ab}{a^2 \cos^2 \alpha + b^2 \sin^2 \alpha}\right)^2 \tag{6.21}$$

Figures 6-8 through 6-10 are prolate spheroid patterns as given by Eq. (6-21) for three different aspect ratios. Note that the patterns exhibit greater fluctuations for higher aspect ratios; in fact, the peak-to-peak excursion from end-on to broadside aspect varies with the fourth power of the aspect ratio, $(a/b)^4$. The simple prescription of Eq. (6-21) fails, of course, when the nose-on radius of curvature becomes less than about a wavelength.

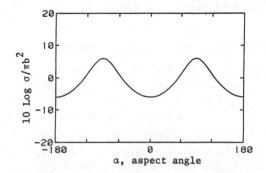

Figure 6-8 Geometric optics return from a 2:1 prolate spheroid

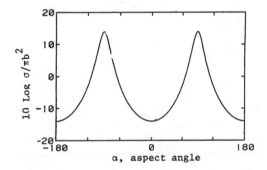

Figure 6-9 Geometric optics return from a 5:1 prolate spheroid

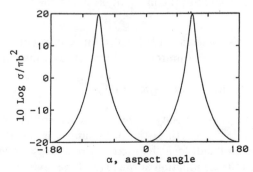

Figure 6-10 Geometric optics return from a 10:1 prolate spheroid

Even if the radius of curvature is large enough, Eq. (6-21) does not account for the contributions due to an induced surface traveling wave. Crispin and Maffett [7] show that the induced surface wave contribution can be as much as three orders of magnitude larger than the end-on echo predicted by geometric optics. Consequently, the surface wave mechanism is an extremely important contributor.

6.2.3 The Surface Wave (Traveling Wave) Mechanism

When a long body is illuminated near end-on incidence, the incident field induces surface currents that build up to relatively large levels at the far end of the body. In order for this build-up to occur, there must be a component of the incident electric field along the surface in the direction of propagation, as shown in Fig. 6-11. The electric field component induces surface currents that flow along the body in the longitudinal direction. The magnitude of the current increases toward the rear of the body and unless they are absorbed there, or flow around some smooth termination, the currents are reflected back toward the front of the body.

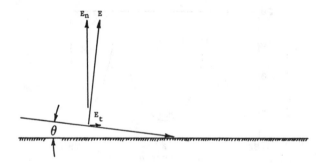

Figure 6-11 The surface wave mechanism is due to surface currents induced by a longitudinal component of the incident electric field along the surface

Peters analyzed this mechanism [8] using an equivalent traveling wave antenna concept. For a long, thin rod, the form of his approximate result is

$$\sigma = A \left[\frac{\sin \theta}{a - p \cos \theta} \sin \{1/2 \, kl \, (1 - p\cos\theta)\} \right]^2 \tag{6-22}$$

where p is the relative phase velocity of the induced currents along the body, $kl = 2\pi l/\lambda$ is the electrical length of the body, and θ is the aspect angle as measured from end-on incidence. The factor A includes a reflection coefficient associated with the termination at the rear of the body, and the effect of the surface conductivity or surface impedance of the body. Peters reported that the strength of the first maximum in the pattern given by Eq. (6-22) was nine square wavelengths for a steel rod 39λ long and $\lambda/4$ in diameter, but 28.2 square wavelengths for a silver rod of identical size. Thus, the surface conductivity plays a much more important role for this scattering mechanism than it does for the optical type of reflection that occurs for fatter bodies.

Peters applied his traveling wave theory to an ogive (a body formed by the rotation of an arc of a circle around its chord) by estimating the phase velocity to be the average axial velocity of currents induced on the ogive surface at a relative phase velocity of unity. A comparison of the measured and predicted RCS pattern is shown in Fig. 6-12 (taken from Peters' paper) for an ogive 39λ long and an apex half-angle of 15 degrees. The surface wave theory predicts higher returns than were measured, and a pattern with deeper nulls, but the locations of the traveling wave lobes are predicted quite well. As the aspect angle swings out toward the broadside aspect angle, the geometric optics prediction (of the specular returns) is seen to be quite accurate.

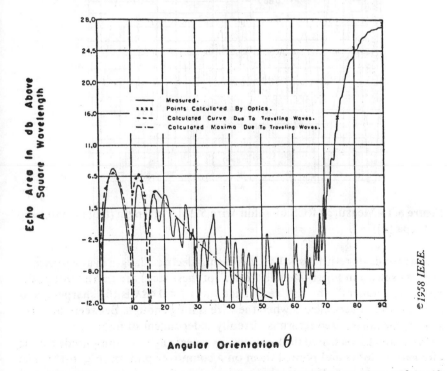

Figure 6-12 Comparison of measured and predicted RCS of a 39λ, $15°$ (half-angle) metallic ogive (from Peters [8])

6.2.4 Thin Wires

The infinitely thin wire has been analyzed in detail by several authors, and the experimental work of Chang and Liepa [9] is a good example. Chang and Liepa measured the backscattering of thin wires ranging in length from 0.301 to 5.422 wavelengths. The wires were actually steel rods, 0.0625 inch in diameter, measured at a frequency of 2.37 GHz. The broadside lobe of the wire is the strongest anywhere in the pattern, but the traveling wave lobe can also be strong (see Fig. 6-13 as an example).

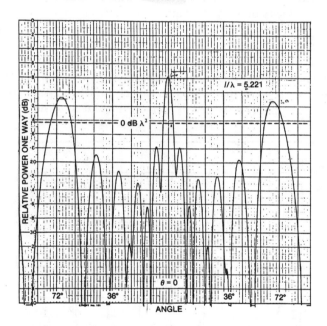

Figure 6-13 Measured RCS of a thin-wire 5.221λ in length (from Chang and Liepa [9])

The broadside return rises with increasing electrical length in a sequence of steps, as shown in Fig. 6-14. The "risers" of these steps are sharp local peaks called "resonances"; the first resonance (near $l = 0.48\lambda$ is the sharpest, and eventually the risers die out when the wire is long enough. Between the resonances, the normalized return is virtually independent of frequency.

Chang and Liepa traced the locations (in aspect angle) and amplitudes of the wire pattern lobes and plotted them on a composite pattern (Fig. 6-15). The upper curve of Fig. 6-15 traces the location of the first traveling wave lobe, and the nearby dashed line is the location predicted by Eq. (5-87). Even for wires as short as a wavelength or two, the prediction is quite accurate.

Because the wire essentially has only one dimension, length, only that component of the incident electric field along the wire can induce currents along the wire. Thus, the wire is invisible if the incident electric field is perpendicular to the wire, and the radar cross section of the wire is maximum when the field is in the plane of incidence containing the line of sight and the wire axis.

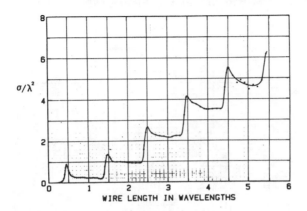

Figure 6-14 Measured broadside RCS of a thin-wire as a function of the wire length in wavelengths (from Chang and Liepa [9])

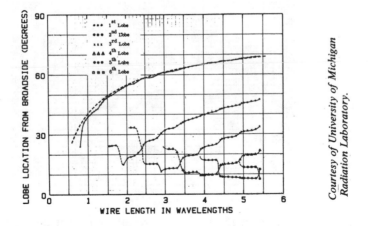

Figure 6-15 Measured lobe locations for a thin wire (from Chang and Liepa [9])

6.2.5 The Circular Cone

The radar return from a metallic cone when observed from within the backward half-cone is primarily due to diffraction from the base of the cone. The scattering from the tip is small, and as the viewing angle increases, there is a broadside flash from the slanted side of the cone. Radar waves are also diffracted across the base of the cone. Figures 6-16 and 6-17 are examples of

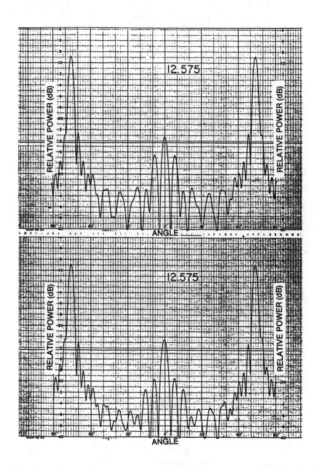

Courtesy of University of Michigan Radiation Laboratory.

Figure 6-16 RCS of a 15° half-angle metallic cone for horizontal incident polarization. The cone was bare for the upper pattern and had an absorber pad cemented to its base for the lower pattern (from Knott and Senior [10])

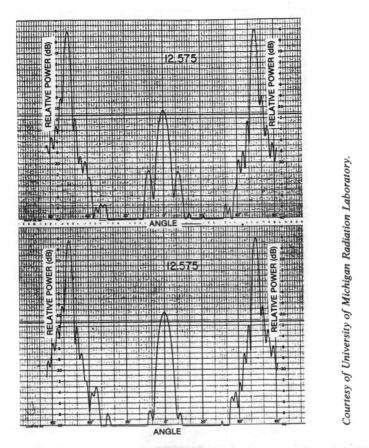

Courtesy of University of Michigan Radiation Laboratory.

Figure 6-17 RCS of the 15° cone for vertical incident polarization. The base was 12.575λ in circumference (from Knott and Senior [10])

the radar cross section patterns of a metallic cone having a 15 degree half-angle, and Fig. 6-18 and 6-19 are for a 40 degree half-angle cone. These patterns were recorded at the University of Michigan [10] for horizontal and vertical polarization, meaning that the incident electric vector was in and perpendicular to the plane of incidence containing the cone axis and the line of sight, respectively. These eight patterns represent only a small fraction of a much larger set of 328 patterns in which the frequency was varied.

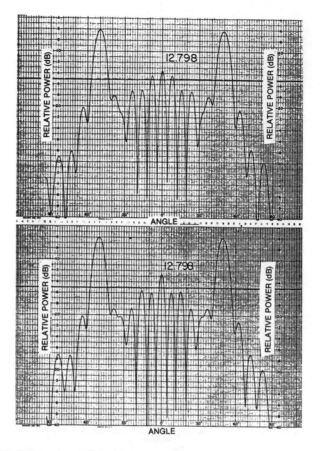

Courtesy of University of Michigan Radiation Laboratory.

Figure 6-18 RCS of a 40° half-angle metallic cone for horizontal incident polarization. The cone was bare for the upper pattern and had an absorber pad cemented to its base for the lower pattern (from Knott and Senior [10])

The upper patterns in these four figures are of the bare metallic cone, and the lower patterns are of the cone with an absorber pad cemented to its base. The purpose of the pad is to suppress the diffraction of waves across the base of the cone. There are significant differences in the upper and lower patterns in the nose-on regions, demonstrating that multiple diffraction across the shadowed base does in fact occur.

The base of the 15 degree cone was 12.575 wavelengths in circumference, while that of the 40 degree cone was 12.798 wavelengths. The patterns were calibrated by a sphere of known size, and the heavy line drawn across the

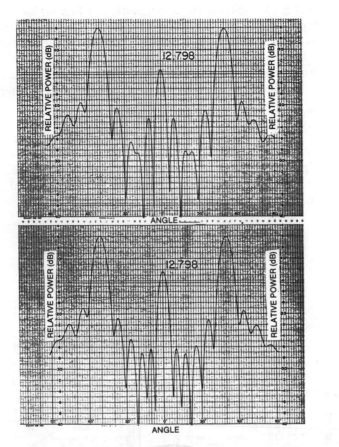

Courtesy of University of Michigan Radiation Laboratory.

Figure 6-19 RCS of the 40° cone for vertical incident polarization. The base of the cone was 12.798λ in circumference (from Knott and Senior [10])

patterns represents a return of ten square wavelengths. Note that the broadside flashes of the 15 degree cone lie at ±75 degrees, and that those of the 40 degree cone lie at ±50 degrees, as expected. Because of the difference in the flare angle, the 40 degree cone was much shorter than the 15 degree cone, hence its broadside lobe is much wider and return is lower.

Figures 6-16 and 6-18 are for horizontal polarization, in which the incident electric field lies in the plane of incidence defined by the cone axis and the line of sight. Figures 6-17 and 6-19 are for vertical polarization, in which the incident electric vector is perpendicular to the plane of incidence. In the nose-on region, the returns are usually higher for horizontal polarization than for vertical.

The lobes at ±13 degrees in the upper pattern of Fig. 6-17 are apparently due to multiple diffraction across the base of the cone, because they virtually disappear when the absorber pad is applied to the base (lower pattern). Cross-base diffraction tends to be less significant for horizontal polarization than for vertical. Thus, at least for these simple shapes, the scattering signatures can be modified by treatments that are not directly exposed to the incident wave.

6.2.6 Flat Plates

Multiple diffraction also occurs for flat plates. This is due to the edges of the plate and an example is given in Fig. 6-20 for a square plate five wavelengths along an edge. The figure is a comparison between measured and predicted plate returns, in which the geometric theory of diffraction was used to estimate the multiple diffraction [11].

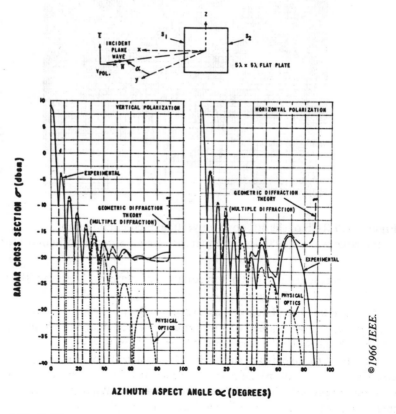

Figure 6-20 RCS patterns of a 5λ by 5λ square plate (from Ross [11]

In the region at within 20 or 30 degrees to either side of normal incidence, the simple theory of physical optics performs quite well in predicting the returns. Beyond this region, the single diffraction predicted by GTD (not shown) improves the prediction, but still falls short. The inclusion of multiple diffracted waves across the front and rear faces of the plate further improves the estimate, but the theory still fails near edge-on incidence, where the predicted returns tend to rise to infinity. Aside from these regions, however, the theory does quite well.

The surface traveling wave pehomenom is not restricted to ogives and rods, as evidenced by the lobe at 69 degrees on the right-hand side of Fig. 6-20. This pattern is for horizontal polarization, which is the appropriate polarization to excite the traveling wave. Equation (5-87) predicts that the lobe should occur 22 degrees from edge-on incidence (i.e., at 68 degrees from normal in incidence), hence Peters' analysis can be extended to cases other than thin rods.

6.2.7 Corner Reflectors

Corner reflectors are re-entrant structures formed by the intersection of two or three mutually perpendicular flat faces. The trihedral corner has three faces, and has a broad pattern due to a triple internal bounce mechanism. The dihedral corner has two faces, and its pattern in the plane perpendicular to the dihedral axis is also broad due to a double-bounce mechanism. Off that axis, however, the return is more like the flat plate patterns of Fig. 6-20.

The RCS pattern of a 90 degree dihedral corner is shown in Fig. 6-21. It was measured at a frequency of 9.4 GHz and had square faces 17.9 cm along a side. Thus, the faces were 5.6 wavelengths along a side, not much larger than the square plate whose patterns are shown in Fig. 6-20. The broad central part of the pattern is due to the interaction between the two faces, with the incident wave being reflected twice, once from each face. The peaks at either side of the pattern are the direct returns from the individual faces, and the ripples in the central region are due to the sidelobes of the individual face patterns.

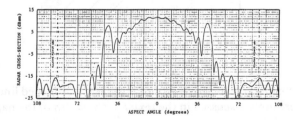

Figure 6-21 RCS pattern of a 90° dihedral corner with square faces 17.9 cm along a side measured at 9.4 GHz

   The broad double-bounce contribution can be reduced by angling the faces
at some angle other than 90 degrees. Opening up the dihedral angle to 100
degrees produces the pattern shown in Fig. 6-22. The double-bounce return

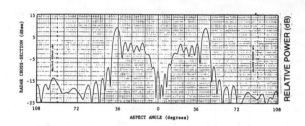

Figure 6-22 RCS pattern of a 100° dihedral corner with square faces 17.9 cm
   along a side measured at 9.4 GHz

has been greatly reduced, and most of the pattern is due to the superposition of
a pair of patterns like those of Fig. 6-20, but displaced from one another.
Based on an analysis of the double-bounce mechanism [12], the dihedral angle
required for a given reduction can be estimated as in Fig. 6-23.

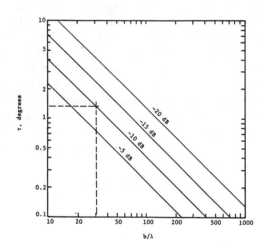

Figure 6-23 Dihedral corner RCS reduction chart. The dashed lines represent
   a specific example discussed in the text; *b* is the width of the narrower face
   (from Knott [12])

The reduction in the double-bounce radar cross section is given approximately by

$$R \simeq (kb \sin \tau \cos \gamma)^2 \tag{6-23}$$

where the two dihedral faces have been assumed to have a common length along the dihedral axis, b is the width of the smaller face, and

$$\gamma = \arctan \frac{b}{a} \tag{6-24}$$

where a is the width of the larger face. Here, τ is the angle from which the faces must vary from perpendicularity. The presence of k in Eq. (6-23) implies that the reduction depends on frquency, hence the desired reduction at the lowest anticipated frequency must be selected, thereby ensuring that it will be at least that amount at all higher frequencies.

Similar reductions can be obtained for trihedral corner reflectors, but because there are three internal bounces instead of two, the faces need not be angled away from each other by as large an angle as indicated in Fig. 6-23. The earliest known exploration of this effect was reported by Sloan Robertson [13].

6.3 THE HIERARCHY OF SCATTERING SHAPES

The previous discussion and illustrations show that the radar cross section of a body depends on the operation of several kinds of mechanisms, some of them simple, but many of them complex. These mechanisms can be categorized according to their strength, which in turn is closely related to their frequency dependence, as shown in Table 6-1.

The "largest" scatterers in this hierarachy are the corner reflectors listed in the top of the table. The large echo is due to the mutual perpendicularity of the two or three flat faces comprising the corners. Rays that impinge on these re-entrant structures are reflected back in the direction from which they came for a wide range of viewing angles. Corner reflectors have been installed on drone aircraft to make them easier to track with radar systems, and they are sold commercially for installation on small boats to enhance the radar detectability. Extended to the optical region, arrays of tiny corner reflectors are built into bicycle reflectors, automobile tail lights, roadside signs, and highway markers. They are also inadvertently created on ships and ground vehicles simply because the designers of such structures, unaware of their radar properties, often force two or three surfaces to meet at right angles in an apex. Note that the echo strength of a corner reflector rises with the square of the frequency of the incident wave.

The next three simple shapes (flat plate, cylinder, sphere) listed in Table 6-1 continue downward through the list hierarchy of echo characteristics, and the effect of surface curvature can be clearly seen. In each of the three cases, the body is oriented for a specular return, another way of saying that there is a point on the surface where the outward surface normal points directly back to the radar. In the case of the flat plate, the entire surface is specular, while for the cylinder, it is a "bright" line running from one end of the cylinder to the other. The specular point is, in fact, a point for the doubly curved surface of the sphere.

Table 6-1
Hierarchy of Scattering Shapes

Geometry	Type	Freq. Dep.	Size Dep.	Formula	Remarks
	Square trihedral corner retro-reflector	F^2	L^4	Maximum $$\sigma = \frac{12\pi\,a^4}{\lambda^2}$$	Strongest return; high RCS due to triple reflection
	Right dihedral corner reflector	F^2	L^4	Maximum $$\sigma = \frac{8\pi a^2 b^2}{\lambda^2}$$	Second strongest; high RCS due to double reflection, tapers off gradually from the maximum with changing θ and sharply with changing ϕ.
	Flat plate	F^2	L^4	Maximum $$\sigma = \frac{4\pi a^2 b^2}{\lambda^2}$$ Normal Incidence	Third strongest; High RCS due to direct reflection, drops off sharply as incidence changes from normal.
	Cylinder	F^1	L^3	Maximum $$\sigma = \frac{2\pi a b^2}{\lambda}$$ Normal Incidence	Prevalent cause of strong, broad RCS over varying aspect (θ), drops off sharply as azimuth (ϕ) changes from normal. Can combine with flat plate to form dihedral corner reflector.
	Sphere	F^0	L^2	Maximum $$\sigma = \pi a^2$$ Normal Incidence	Prevalent cause of strong, broad RCS peaks other than those due to large openings in target body. Energy defocused in two directions.

Table 6-1
Hierarchy of Scattering Shapes (continued)

Geometry	Type	Freq. Dep.	Size Dep.	Formula	Remarks
	Straight edge normal incidence	F^0	L^2	$f(\theta,\phi)L^2$ θ - aspect θ_{int} - interior dihedral angle between faces meeting at edge	Limiting case of 2-dimensional curved plate mechanism as radius shrinks to 0. Prevalent cause of strong, narrow RCS peaks from supersonic aircraft.
	Curved edge normal incidence	F^{-1}	L^1	$f(\theta,\theta_{int})\ a\lambda/2$ $a \geq \lambda$	Limiting case of 3-dimensional curved plate mechanism as principal radius shrinks to 0. The function f is the same as in mechanism 3.
	Apex	F^{-2}	L^0	$\lambda^2 g(a,\beta,\theta,\phi)$ a,β - interior angles of tip θ,ϕ - aspect angles	Limiting case of previous mechanism as a shrinks to 0. For a=ß, the tip is that of a cone. For a=0, the tip is the corner of a thin sheet, or fin.
	Discontinuity of curvature along a straight line, normal incidence	F^{-2}	L^0	$\dfrac{\lambda^2}{64\pi^3} \left(\dfrac{1}{a}\right)^2 \left\{1+\left(\dfrac{dy}{dx}\right)^2\right\}^{-3/2}$ $a \geq \lambda$ (1/a) - jump in reciprocal of dy/dx - slope of surface w.r.t. incident ray	Strongest of an infinite sequence of discontinuities. Very weak mechanism which together with 6 shares dominance of nose-on RCS of cone sphere.
	Discontinuity of curvature of a curved edge	F^{-3}	L^{-1}	$f(\theta,\phi)\dfrac{\lambda^3 b}{a^2}\left\{1+\left(\dfrac{dy}{dx}\right)^2\right\}^{-3/2}$ $f(\theta,\phi)$ - function of aspect b - radius of edge $>\lambda$	Important mechanism for traveling wave backscatter where RCS of discontinuity is augmented by gain of traveling wave structure. Dependences are based on dimensional considerations.
	Discontinuity of curvature along an edge	F^{-4}	L^{-2}	$g(\theta,\phi)\lambda\left(\dfrac{1}{a}\right)^2\{1+\left(\dfrac{dy}{dx}\right)^2\}^{-3/2}$ $g(\theta,\phi)$ - function of aspect	Important mechanism for traveling wave backscatter where RCS of discontinuity is augmented by gain of traveling wave structure. Dependences are based on dimensional considerations.

Note that the plate, the cylinder, and the sphere are flat, singly curved, and doubly curved bodies, respectively. The dependence of their radar echo amplitudes on the frequency of the incident wave is F^2, F^1, F^0, respectively. Consequently, the dependence on the body size L ranges from L^4 for the flat plate to L^2 for the sphere. We emphasize that these dependencies are only for specular orientations of the shapes.

When one of the radii of surface curvature goes to zero, an edge is created and another triad (straight edge, curved edge, tip) of cases can be listed in the hierarchy of radar echoes. These are shown in Table 6-1 for the cases where the remaining radius of curvature is infinite, finite, and zero, corresponding to a straight edge, a curved edge, and a vertex or apex. Much as in the previous cases for surfaces, there is a "specular" condition that the edge (just as previously for a surface) must be oriented such that there is a point on the edge where the outward normal points directly toward the radar. For the straight edge, the entire edge "lights up," while there is only a single "bright point" for the curved edge. The point respresented by the vertex represents the case of the sphere as the radius vanishes.

Finally, there is a collection of echo sources whose returns are very small, as shown near the bottom of Table 6-1. These all involve discontinuities in the surface curvature, and it is understood that these shapes are oriented such that *no* portion of the surface is where the outward surface normal points toward the direction of the radar. It is this nonspecular condition which leads to the very small values of the radar echo.

Obviously, if we tried to minimize the return of a complex body, we should build the body with only edges and vertexes, and no surfaces. Because such a body does not exist, the best practical option is to orient the body such that no surface or edge has a surface normal pointing toward the radar. Unfortunately, we cannot always predict where the radar will be, hence favorable surface orientations cannot be routinely specified. Even if they could be, it is not likely that the required orientations would always be achieved.

6.4 COMPLEX SHAPES

The RCS of a complex body is obviously more complicated. A complex target, such as an aircraft, contains several dozen significant "scattering centers" and myriad less significant scatterers, such as seams, ports, and rivet heads. Because of this multiplicity of scatterers, the net RCS pattern typically exhibits a rapid scintillation with aspect angle due to the mutual interference as the various contributions go in and out of phase with each other. Figure 6-24, a measured pattern of a scale model of a commercial jetliner, illustrates these characteristics [14]. The larger the target in terms of wavelengths, the more rapid these scintillations become.

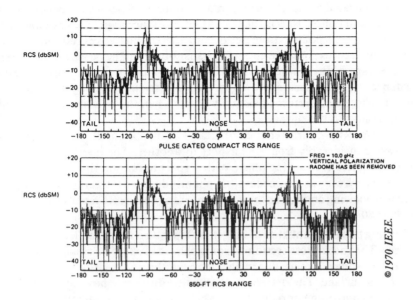

Figure 6-24 Measured RCS of a 1/15 scale model of a Boeing 737 commercial jetliner at 10 GHz and vertical polarization (from Howell [14])

In the nose-on region, the pattern is dominated by reflections from jet intake ducts. Ducts, like corner reflectors, are re-entrant structures with large radar cross sections. If there is a radar antenna in the nose of the craft, this, too, will contribute to the nose-on return because antennas are efficient reflectors along their boresight directions. A few degrees away from nose-on incidence, the leading edges of the wings become major scatterers. If the radius of curvature of the leading edge is large compared to the wavelength, the return is large for any polarization of the incident wave. If not, the leading edge return is stronger when the incident electrical polarization is parallel to the edge than it is when perpendicular to the edge. If the wing is seen from slightly below or slightly above, the trailing edge return can be stronger when the electrical polarization is *perpendicular* to the trailing edge than it is when parallel to the edge.

As the aspect angle swings near the broadside view, the sides of the fuselage and the engine pods become dominant sources. At broadside incidence, the vertical fin stands out. If viewed from considerably above the plane of the wings, the scattering from the dihedral corner formed by the upper wing surface and the fuselage is strong. Similarly, a dihedral corner may be formed by the vertical fin and the horizontal stabilizer. Somewhere between the broadside and tail-on aspects the trailing edges of the wings light up, especially for the electrical polarization parallel to the trailing edges. At the tail-on aspect, the returns from the engine exhaust ducts will be large. At any intermediate aspect angle where a relatively long surface is seen near grazing incidence, large returns may be observed due to the traveling wave phenomenon. These large returns primarily occur when the polarization is nearly perpendicular to the surface.

Thus, the RCS characteristics of a large complex object, like an aircraft, will vary considerably from one aspect to another. The interference due to several reflections changing phase with respect to one another can swing the net return by one or two orders of magnitude over a fraction of a degree change in viewing angle. The total dynamic range of the RCS pattern can be as much as 80 dB.

Turning to a different kind of target, a ship is also a large collection of echo sources. The radar cross section of some ships can exceed a square mile (64.1 dB above a square meter), due in part to the multipath environment provided by the sea surface. The designers of ships traditionally have had very little interest in controlling or reducing ship echoes, and this is obvious because of their fondness for bringing decks and bulkheads together at right angles. In addition, many topside surfaces are vertical, thereby forming efficient dihedral corner reflectors with the mean sea surface.

Figure 6-25, taken from Skolnik [15], shows the RCS patterns of a large ship at two different frequencies. These patterns were measured as the ship steamed in a circle while in the beam of a multifrequency shore-based radar. The three traces in each diagram represent the 20, 50, and 80 percentile levels, and plotted statistics were derived from the returns over a 2 degree aspect angle "window." Note that the patterns are not very frequency dependent and the mean return attains local maxima at the broadsides, bow-on, and stern-on aspect angles.

The significant scatterers on a ship depend on the range between the radar and the ship. For all except broadside incidence, the hull is not the dominant scatterer, even when the ship's hull is not occluded by the horizon. The superstructure and masts are the dominant scatterers, and every ship has a large assortment of fixtures and equipment located topside. Examples are vents, cleats, hoists, railings, capstans, lockers, pipes and conduits, stiffening plates, ladders, catwalks, and hatches. A ship is a very complicated target, and even as its masts disappear over the horizon, its radar antennas — which may have large radar cross sections — are the last major contributors to disappear.

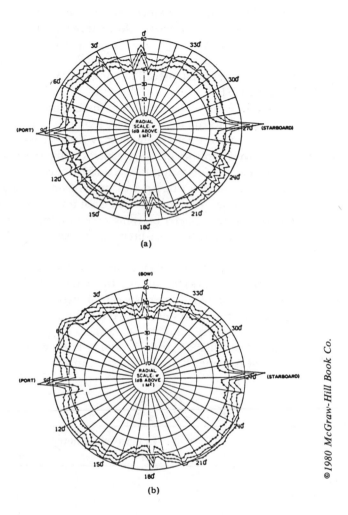

(a)

(b)

Figure 6-25 Upper diagram is the RCS pattern of a large naval auxiliary ship measured at 2.8 GHz, and lower diagram is for 9.225 GHz, both for horizontal incidence polarization (from Skolnik [15])

6.5 NATURAL TARGETS

Thus far we have discussed simple and complex man-made targets, primarily metallic ones. Occasionally, we desire information on natural targets, such as animals and vegetation. Although the echo properties of vegetation and terrain are of vital concern for many radar applications, we cannot discuss them here. The reader is referred instead to books like those of M. W. Long [16] and N. C. Currie [17].

A particular concern is the radar echo from insects, and we often wonder how they affect radar performance. An insect flying through a radar beam may constitute a false alarm, even though it may have a small echo. Insects may also interfere with laboratory measurements. During the course of some indoor RCS measurements,for example, the author once noted fluctuating receiver signals even though no target was present on the target support pedestal. After spending several hours trouble-shooting the instrumentation system, he discovered a small spider creeping about on the support column. The animal was removed and the measurements proceeded without further disruption.

A systematic series of measurements was once conducted at the University of Texas by Hajovsky, *et al.* [18] to establish the magnitude of insect echoes. the experimenters collected ten different species of insect and measured them with a simple CW measurement system at 9.4 GHz, similar to that discussed in Ch. 13. A variety of sizes were represented, ranging from 5 mm in length to about 20 mm and, not surprisingly, the echo magnitude increases with the size of the insect. The measurements are summarized in Table 6-2.

Table 6-2
Insect RCS at 9.4 GHz*

	Length mm	Diameter mm	σ_L dBsm	σ_T dBsm
Blue-winged locust	20	4	−30	−40
Army worm moth	14	4	−39	−49
Alfalfa caterpillar butterfly	14	1.5	−42	−57
Honey bee worker	13	6	−40	−45
California harvester ant	13	6	−54	−57
Range crane fly	13	1	−45	−57
Green bottle fly	9	3	−46	−50
Twelve-spotted cucumber beetle	8	4	−49	−53
Convergent lady beetle	5	3	−57	−60
Spider (unidentified)	5	3.5	−50	−52

*From Reference [18]

Not unexpectedly, the RCS depends on the aspect angle, and the experimenters measured their natural targets for head-on and broadside incidence. The longitudinal (broadside) RCS exceeds the transverse (head-on) RCS by as much as 15 dB for long, thin insects, while the difference between the longitudinal and head-on values is 2 dB or less for insects whose widths are not much different from their lengths. Although the original results are presented in square cm, we have converted the values to dBsm for comparison with other RCS values presented in this chapter.

The University of Texas experimenters positioned their targets very close to the radiating antenna in order to measure such small echo values, but they did not report the actual distances used. Moreover, the insects were drugged to prevent them from moving during the measurements. Hajovsky, *et al.*, noted that the insects could not be killed to keep them from moving because the carcasses would dry out and the dielectric constant would change during the test. Subject motion was also a problem for Blacksmith and Mack, who attempted to measure ducks and chickens at 400 MHz [19].

Blacksmith and Mack had intended to place their subjects on a rotatable foamed plastic support column so that the RCS could be measured as a function of aspect angle, in the tradition of conventional RCS measurements. However, the birds repeatedly flew off the column and the conventional approach was abandoned. Blacksmith and Mack settled for measurements at a single aspect angle, and the birds were eventually trained to sit still long enough for a measurement. The measured echoes changed when the animals stretched their necks, and it also changed when the birds shifted from a standing position to a squatting position. The experimenters' results, which are reproduced in statistical form in Fig. 6-26, suggest that the RCS of an adult duck is about –12 dBsm.

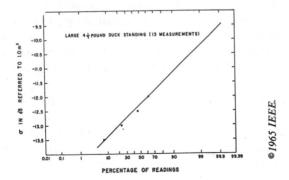

Figure 6-26 RCS of adult ducks measured by Blacksmith and Mack [19]

Being pragmatic experimenters, Blacksmith and Mack observed that a carefully planned experimental program could take advantage of the animals' natural growth. By starting with very young birds, the RCS could be measured periodically as they grew, and the echo strength could then be related to the size of the bird. This would also reduce the cost of acquiring subjects of different size because the same birds could serve as test objects over the duration of the program. The authors further suggest that at the end of the test program the advice of Rombauer and Becker [20] is "especially interesting and useful."

6.6 SUMMARY

In this chapter we have attempted to acquaint the reader with the practical concept of radar cross section. The radar cross section applies to the "scattering" of energy by an obstacle back in the direction of the source of radiation, but, in general, energy is scattered in all directions. For arbitrary directions, the common term used in the "scattering cross section," and the bistatic scattering by a cylinder was used as an illustration. It is the forward scattering from an obstacle that combines with the incident wave to produce a shadow behind the target and, hence, the forward scattering is usually stronger than the backscattering.

A creeping wave can be launched at the shadow boundary of smooth objects, and it skirts the rear of the body only to be launched back toward the radar at the opposite shadow boundary. The creeping wave goes in and out of phase with the specular return from the front of the object as the object becomes larger (or the wavelength smaller) to produce undulations in the net return as a function of frequency. For all practical purposes, the creeping wave becomes negligible for obstacles more than ten or fifteen wavelengths in size. A more important scattering mechanism is the surface traveling wave. It is much like a creeping wave, but is launched along illuminated parts of the body instead of in the shadow zone. As such, it may build up from the near end of the body to the far end, and is reflected by the abrupt termination at the far nd of edged bodies. We showed that the surface traveling wave can exist even on wires as short as a wavelength.

In fact, the discontinuities due to the edges of obstacles are prime reasons for the scattering of incident energy. The dominant sources of return for the cone, for example, are the flash points at the base; the tip is an insignificant scatterer until the aspect angle nears the broadside region. The return from a flat plate is dominated by the edge contributions for aspects beyond that bounded by the main lobe.

Scattering features can be arranged according to the strength and frequency dependence of the echo, and it was shown that corner reflectors and flat plates at normal incidence are at the top of the list. Surface scatterers may be ordered according to the two radii of curvature of the surface, and the edge scatterers may be ordered according to the radius of the edge. The specular echo from doubly curved surfaces and from straight edges at normal incidence are both independent of frequency. Non-specular echoes decay rapidly with increasing frequency, and examples are the returns from curved edges at glancing incidence and apexes.

Most targets are more complex than this, and the resulting RCS pattern has many scintillations due to the rapid phase changes between the contributions of the individual scatterers as they go in and out of phase with each other. It does not take many scattering centers to generate a complicated target, and two examples were shown. The actual scintillations for a ship cannot be measured, and the patterns presented were two-degree averages. Ships are very large, and their echo areas are perhaps better presented in units of square miles instead of square meters.

In addition to the characteristics of man-made targets, two examples of natural targets were given. The echoes from insects are best measured in square cm, and 1 cm^2 is not atypical. The RCS of chickens and ducks is on the order of 0.1m^2. Thus, the RCS values of scattering obstacles cited in this chapter covered 12 orders of magnitude, from a millionth of a square meter for small insects to a million square meters for ships.

REFERENCES

1. John D. Kraus, *Antennas*, McGraw-Hill, 1950, p. 41.
2. Sloan D. Robertson, "Targets for Microwave Radar Navigation," *Bell Sys. Tech. J.*, Vol. 26, 1947, pp. 852-869.
3. Julius A. Stratton, *Electromagnetic Theory*, New York, McGraw-Hill, 1941, p. 281.
4. J.J. Bowman, T.B.A. Senior, and P.L.E. Uslenghi, editors, *Electromagnetic and Acoustic Scattering by Simple Shapes*, Amsterdam, North-Holland, 1969, p. 7.
5. D.L. Moffatt and E.M. Kennaugh, "The Axial Echo of a Perfectly Conducting Prolate Spheroid," *IEEE Trans. Antennas Propag.*, Vol. AP-13, May 1965, pp. 401-409.
6. T.B.A. Senior, "Axial Backscattering by a Prolate Spheroid," *IEEE Trans. Antennas Propag.*, Vol. AP-15, July 1967, pp. 587-588.
7   J.W. Crispin, Jr., and A.L. Maffett, "Radar Cross Section Estimation for Simple Shapes," *Proc. IEEE*, Vol. 53, No.8, August 1965, pp. 833-848.
8. Leon Peters, Jr., "End-Fire Echo Area of Long, Thin Bodies," *IRE Trans. Antennas Propag.*, Vol. AP-6, January 1958, pp. 133-139.

9.   S. Chang and V.V. Liepa, "Measured Backscattering Cross Section of Thin Wires," University of Michigan Radiation Laboratory Report No. 8077-4-T, May 1967.

10.   E.F. Knott and T.B.A. Senior, "CW Measurements of Right Circular Cones," University of Michigan Radiation Laboratory Report No. 011758-1-T, April 1973.

11.   R.A. Ross, "Radar Cross Section of Rectangular Flat Plates as a Function of Aspect Angle," *IEEE Trans. Antennas Propag.*, Vol. AP-14, No. 3, May 1966, pp. 329-335.

12.   E.F. Knott, "RCSR Guidelines Handbook," Georgia Institute of Technology, Engineering Experiment Station, Report on Contract No. 0039-73-C-0676, April 1976, AD A099566.

13.   Sloan D. Robertson, "Targets for Microwave Radar Navigation," *Bell Sys. Tech. J.*, Vol. 26, 1947, pp. 852-869.

14.   N.A. Howell, "Design of Pulse Gated Compact Radar Cross Section Range," 1970 G-AP International Program and Digest, IEEE Publication 70c 36-AP, September 1970, pp. 187-195.

15.   Merrill I. Skolnik, *Introduction to Radar Systems*, New York, McGraw-Hill, 1980, p. 45.

16.   M.W. Long, *Radar Reflectivity of Land and Sea*, Dedham, MA, Artech House, 1975, 1983.

17.   N.C. Currie, editor, *Techniques of Radar Reflectivity Measurement*, Dedham, MA, Artech House, 1984.

18.   R.G. Hajovsky, A.P. Dean and A.H. LaGrone, "Radar Reflections from Insects in the Lower Atmosphere," *IEEE Trans. Antennas Propag.*, Vol. AP-14, No. 2, March 1966, pp. 224-227.

19.   P. Blacksmith, Jr., and R.B. Mack, "On Measuring the Radar Cross Sections of Ducks and Chickens," *Proc. IEEE*, Vol. 53, No. 8, August 1965, p.1125.

20.   I.S. Rombauer and M.R. Becker, *The Joy of Cooking*, New York, Bobbs-Merrill, 1953, pp. 389-424.

CHAPTER 7

RADAR CROSS SECTION REDUCTION

E. F. Knott

7.1 BACKGROUND

It has become apparent in the last few years that the development of increasingly effective hostile detection systems threatens to reduce the missions with which many types of weapons platforms carry out their effectiveness, and attention is now being given to methods of increasing survivability by reducing detectability. Because the specific configuration of any platform is determined by many factors involved in its mission, the final design represents a compromise between conflicting requirements. The purpose of this chapter is to survey some of the radar cross section reduction (RCSR) options available.

This chapter is intentionally tutorial, with emphasis on concepts rather than application to specific systems. There are several reasons for this. For one, access to many research and development efforts, both past and future, is limited in the interest of national security — the information is too dificult to obtain, and even if access were to be granted, we would not be able to disseminate the information. Besides, there are other media for the exchange of technical information on specific systems. Examples are the Radar Camouflage Symposia held in Dayton, Ohio in 1975 and in Orlando, Florida in 1980 [1,2].

Moreover, complex targets like ships and aircraft can be represented as collections of basic target elements, such as flat plates, cone frusta, spheroids, edges, vertexes, *et cetera*. As such, we typically isolate the dominant sources of target echo and fix our attention on a limited number of individual elements instead of the composite target.

We emphasize that radar cross section reduction is a study of compromises in which advantages are balanced against disadvantages, and this fact should become apparent. A reduction in RCS at one viewing angle is usually accompanied by an enhancement at another when target surfaces are reshaped or reoriented to achieve the reduction. However, if radar absorbing materials (RAM) are used, the reduction is obtained by the dissipation of energy within the material, thus leaving the RCS levels relatively unchanged in other directions. On the other hand, the use of RAM is a compromise paid for with added weight, volume, and surface maintenance problems. Thus, each approach involves its own form of trade-off. This is another reason why we cannot deal with specific systems: the trade-offs cannot be made without information that is not currently available.

No matter which technique is employed, each decrement in RCS is obtained at successively higher cost. The first 50 percent reduction is usually quite inexpensive, while the next 10 percent is a little more expensive, the next more costly still, until a 95 percent total reduction may be prohibitively costly, in terms of dollars as well as trade-offs in weight, size, and configuration. The cut-off point for deciding how much RCS reduction to incorporate depends on a host of variables, such as relative effectiveness, incremental cost, and the platform's mission. Because each particular target poses its own particular problems, we cannot establish the optimum RCS design approach for the general case. However, the concepts and principles illustrated here should make it easier to arrive at such a decision when specific systems are considered.

7.2 THE FOUR BASIC METHODS OF RCSR

There are only four basic techniques for reducing radar cross section. They are
1. Shaping;
2. Radar absorbing materials;
3. Passive cancellation; and
4. Active cancellation.
Each of these methods has its advantages and disadvantages.

7.2.1 Shaping

The objective of shaping is to orient the target surfaces and edges so as to deflect the reflected energy in directions away from the radar. This cannot be done for all viewing angles within the entire sphere of solid angles because there will always be viewing angles at which surfaces are seen at normal incidence and there the echoes will be high. The success of shaping depends on the existence of angular sectors over which low radar cross section is less important than over others.

Typically, a forward cone of angles is of interest and, hence, large cross sections can be "shifted" out of the forward sector toward the broadside sectors. This can be accomplished by sweeping airfoils back at sharper angles, for example. The forward sector includes the elevation plane as well as the azimuth plane, and if a target is hardly ever seen from above, echo sources, such as engine intakes, can be placed on the top side of the target where they may be hidden by the forward portion of the body when viewed from below.

For more "boxy" structures, such as ships and ground vehicles, internal dihedral and trihedral corners can be avoided by bringing intersecting surfaces together at acute or obtuse angles. Because of the presence of the sea surface, vertical bulkheads on ships, in particular, form efficient corners, and the effect can be reduced by tilting the bulkheads away from the vertical. However, this is virtually impossible to do with existing vessels and the cost of implementing the concept in future ships must be balanced against increased mission effectiveness.

7.2.2 Radar Absorbing Materials

As the name implies, radar absorbing materials reduce the energy reflected back to the radar by means of absorption. Radar energy is absorbed through a kind of ohmic loss, not unlike the way a resistor dissipates heat when an electrical current passes through it. The loss is actually the conversion of microwave energy into heat, and although most absorbers do not absorb enough energy to become hot, or even detectably warm, when illuminated by a radar, this is nevertheless the mechanism by which they operate. If the transmitter were brought close enough to the radar absorbing materials, they would indeed become warm.

The basic feature of RAM lies in the fact that substances either exist or can be fabricated whose indices of refraction are complex numbers. In the index of refraction, which includes magnetic as well as electrical effects, it is the imaginary part that accounts for the loss. At microwave frequencies, the loss is due to the finite conductivity of the material, as well as a kind of molecular friction experienced by molecules in attempting to follow the alternating fields of an impressed wave. It is customary to lump the effects of all loss mechanisms into the permittivity and permeability of the material because the engineer is usually only interested in the cumulative effect.

Carbon was the basic material used in the fabrication of early absorbers because of its imperfect conductivity. In fact, there are many commercial materials now being marketed whose designs have not changed substantially for more than 20 years. Most are intended for experimental and diagnostic work, including the construction of indoor microwave anechoic chambers, but these materials are not easily applied to operational weapons platforms. They are usually too bulky and fragile in operational environments.

Instead, magnetic absorbers are the more widely used for operational systems. The loss mechanism is primarily due to a magnetic dipole moment, and compounds of iron are the basic ingredients. Carbonyl iron has been used extensively, as have oxides of iron (ferrites). Magnetic materials offer the advantage of compactness because they are typically a fraction of the thickness of dielectric absorbers. However, magnetic absorbers are heavy because of their iron content. The basic lossy material is usually embedded in a matrix or binder such that the composite structure has the electromagnetic characteristics appropriate to a given range of frequencies.

7.2.3 Passive Cancellation

Passive cancellation, also known as impedance loading, received a great deal of attention in the 1960s, but the method is severely limited. The basic concept is to introduce an echo source whose amplitude and phase can be adjusted so as to cancel another echo source. This can be accomplished for relatively simple objects, provided that a loading point can be identified on the body. A port can be machined in the body, and the size and shape of the interior cavity can be designed to present an optimum impedance at the aperture. Unfortunately, even for simple bodies, it is extremely difficult to generate the required frequency dependence for this built-in impedance, and the reduction obtained for one frequency in the spectrum rapidly disappears as the frequency is changed.

Furthermore, typical weapons platforms are hundreds of wavelengths in size and have dozens, if not hundreds, of echo sources. Clearly, it is not practical to devise a passive cancellation treatment for each of these sources. Moreover, the cancellation can revert to a reinforcement with a small change in frequency or viewing angle. Consequently, passive cancellation has been discarded as a useful RCS reduction technique.

7.2.4 Active Cancellation

Also known as active loading, active cancellation is even more ambitious than passive loading. In essence, the target must emit radiation whose amplitude and phase cancels the reflected energy. This implies that the target must be "smart" enough to sense the angle of arrival, intensity, frequency, and waveform of the incident wave. It must also be smart enough to know its own echo characteristics for that particular wave and angle of arrival, fast enough to generate the proper waveform and frequency, and versatile enough to adjust and radiate a pulse of the proper amplitude and phase. It is not known whether this technique is being considered for new weapons systems.

7.2.5 The Penalties of RCSR

Most of the time, the requirement for reduced radar echo conflicts with conventional or traditional requirements for structures. As a result, the final system design is a compromise that inevitably increases the cost of the overall system, from initial engineering through production. Cost is only one penalty of RCSR; others are

1. Reduced payload;
2. Reduced range;
3. Added weight; and
4. Increased maintenance.

The relative importance of each factor depends on the mission of the particular platform involved, of course, and these factors change from one system to another. Not surprisingly, there are often cases when radar cross section reduction cannot be justified. In one study, for example, Georgia Tech calculated the detection range for a hypothetical sea target. The detection range was decreased less than 10 percent despite drastic changes in the target to reduce its radar echo. One reason for this was that the assumed threats were very sensitive and the target was detected as it came over the horizon, treated or not. Thus, there are cases when radar cross section reduction is not warranted.

7.3 THE RCSR NUMBERS GAME

Before discussing some practical ways to reduce the RCS of a given target, it is instructive to consider a few illustrative numerical exercises. If a target can be resolved into a collection of N discrete scatterers or scattering centers, then the net radar return at a given frequency is

$$\sigma = \left| \sum_{n=1}^{N} \sqrt{\sigma_n}\, e^{i\phi_n} \right|^2 \tag{7.1}$$

where σ_n is the RCS of the nth scatterer and ϕ_n is the relative phase of that particular contribution due to its physical location in space. Equation (7-1) will be recognized as a coherent sum because the phase of the contribution of each scatterer is included in the phasor summation.

This coherent sum is for a particular aspect angle, because not all of the scattering elements on a target are visible at the same time. Some are shielded from the radar, for example, by other parts of the target itself. Consequently, the number of contributing scatterers N will change from one aspect angle to another as certain features come into, or disappear, from view.

Moreover, the amplitudes of the individual returns (σ_n) are aspect angle sensitive, changing with the viewing angle. Therefore, the radar cross sections are functions, not merely fixed numbers. Some of the individual contributions are quite broad in their aspect angle dependence, while others are quite narrow.

Similarly, the individual phase angle ϕ_n depends on the relative distance between the scatterer and the radar. The phase angle depends on the radar frequency as well as on distance, and consequently the coherent sum changes from one radar frequency to another. Also, because the distance of each scatterer depends on the target aspect angle, the relative phase of each contributor changes with aspect.

For large targets, such as those that are at least a few dozen wavelengths in size, the phase angle variations produce rapid flunctuations in the total return. However, because the returns from individual contributors vary relatively slowly with aspect angle, the mean return can be surprisingly steady over aspect intervals of a few degrees. These mean values are more useful in characterizing the gross target returns than the detailed structure of a given target measured at a certain frequency.

This gives rise to the notion of forming the noncoherent sum. In many cases it can be safely argued that all phase angles are equally likely. Those cases are when the target is large, the wavelength is short, or both. If all phase angles are equally likely, then a good statistical characterization of the target is the noncoherent sum,

$$\sigma = \sum_{n=1}^{N} \sigma_n \tag{7.2}$$

The RCS pattern predicted by Eq. (7-2) does not have the rapid fluctuations characteristic of the more accurate Eq. (7-1), thus, scintillation in the pattern is largely lost.

Table 7-1
Effect of a Dominant Scatterer

	Untreated	Reduce σ_1 by 10 dB	Reduce σ_1, σ_2 by 10 dB	Reduce σ_1, σ_2, σ_3 by 10 dB
σ_1	200	20	20	20
σ_2	20	20	2	2
σ_3	20	20	20	2
TOTAL	240	60	42	24
dB reduction	0.0	6.0	7.6	10.0

The noncoherent sum is particularly useful in illustrating the need to first reduce the returns of the dominant scatterers before considering the less dominant ones. Assume, for example, that a very simple target has but three contributors, and that the return from one of them is ten times stronger than the returns from the other two. For the sake of numerical comparison, assume its return is 200 square meters, while the returns from the remaining two are only 20 square meters. Using the noncoherent sum to evaluate the effect of reducing the returns of the three, we can construct Table 7-1, showing the effect of reducing the returns of one or all of the contributors.

Note that a 10 dB reduction in the return of the dominant scatterer amounts to only a 6 dB reduction in the total return. Thus, although a dominant scatterer is present, the reduction in the total return of the target will always be less than the amount by which that individual return is reduced. We see from the table, for example, that a net reduction of 10 dB was accomplished only by reducing the returns of all three scatterers by 10 dB.

Can we do better by working harder on the dominant scatterer? The answer is yes. Suppose we can reduce the dominant scatterer by 15 dB, but the others by only 10 dB (due to the nature of the particular echo sources, for example), then our efforts have the results shown in Table 7-2. This table shows that even though we considerably improved the treatment of the dominant scatterer (from 10 dB reduction to 15 dB), the reduction in the total return improved by only 1.1 dB (from a 6 dB reduction to a 7.1 dB reduction). Clearly, the task is more difficult than might have been assumed at the outset. If one of the less important scatterers is given a 10 dB reduction in addition to a 15 dB reduction of the dominant scatterer, the net reduction is 9.3 dB. Treatment of the third scatterer brings about a 13.7 dB total reduction.

Table 7-2
Effect of Working Harder on the Dominant Scatterer

	Untreated	Reduce σ_1 by 15 dB	Reduce σ_1, σ_2 by 15 dB, σ_2 by 10 dB	Reduce σ_1 by 15 dB, σ_2 by 10 dB, σ_3 by 10 dB
σ_1	200	6.3	6.3	6.3
σ_2	20	20.0	2.0	2.0
σ_3	20	20.0	20.0	2.0
TOTAL	240	46.3	28.3	10.3
dB reduction	0	7.1	9.3	13.7

Going a step further, we perform some total eliminations to examine the very best that could be done if only either the dominant or lesser contributors were absent. The results are shown in Table 7-3. Clearly, the total elimination of one or the other cannot produce a 10 dB reduction. Eliminating σ_1 results in a 7.8 dB reduction, while the elimination of σ_2 and σ_3 produces barely 1 dB of reduction.

Table 7-3
Effects of Selective Elimination of Scatterers

	Untreated	Eliminate σ_1	Eliminate σ_2 and σ_3
σ_1	200	0	200
σ_2	20	20	0
σ_3	20	20	0
TOTAL	240	40	200
dB reduction	0.0	7.8	0.8

What if all three scatterers had been of the same amplitude? Table 7-4 like Table 7-1, shows that *all* the scatterers must be reduced by 10 dB in order to effect a 10 dB reduction in the total. Consequently, as we might have guessed, it is important to concentrate first on the dominant scatterers on a target because a reduction there has the greatest payoff. However, we must always be aware that the job is much more difficult when there are many echo sources, all of about the same magnitude. In that event, treatments must be devised for many scatterers instead of just a few.

Table 7-4
Effect of Reduction When All Have the Same Amplitude

	Untreated	Reduce σ_1 by 10 dB	Reduce σ_1, σ_2 by 10 dB	Reduce σ_1, σ_2, σ_3 by 10 dB
σ_1	20	2	2	2
σ_2	20	20	2	2
σ_3	20	20	20	2
TOTAL	60	42	24	6
dB reduction	0.0	1.6	4.0	10.0

7.4 FLARE SPOT IDENTIFICATION

The very first task that needs to be accomplished in an RCSR effort is the identification of "flare spots." This is a term used in the industry for the locations of dominant echo sources on the target. Obviously, we must know where these strong echo sources, or flare spots, are before addressing the problem of how to remove or suppress them.

Flare spots can be identified or located in several ways: examples are measurements or RCS prediction programs, and simply good engineering guesses. The experienced RCS specialist can usually pick out, without the benefit of measurements or RCS prediction codes, the features of a weapons system that will be large sources of echo. For an aircraft, dominant sources are the jet engine intake and exhaust ducts, the leading and trailing edges of airfoils, the radar antenna, external stores, the cockpit canopy, and assorted protuberances in the aircraft, such as airspeed indicators and communications antennas.

Depending on the viewing angle, other sources may or may not be important. Near the broadside aspects, the fuselage, engine pods, external fuel tanks, and the vertical stabilizer are large echo sources. When seen from above or below, the broad surfaces of airfoils are large scatterers. Interactions can occur between certain surfaces that meet at right angles, such as the dihedral corner formed by the horizontal and vertical stabilizers, but this echo source is strongest for aspect angles to the side and above the aircraft.

However, no matter how experienced the RCS specialist may be, his intuition and experience in the location of flare spots need to be verified. In addition, the strengths of the echo sources need to be established so that an approach can be developed for reducing the echo strength. If a model or target exists, it can be measured on an RCS range using special instrumentation. If there is no model or target available, any one of several RCS prediction codes can be used. The latter is often the only way to conduct preliminary assessments of flare spots for new weapons system development, because the system configuration is often no more than a few engineering models for which no physical model has been fabricated.

If a target model exists, short pulse (high resolution) measurements can be made to aid in the identification of scattering centers. The generation and reception of short pulses complicates the design of radar instrumentation, but helps in diagnostic work. The design complications arise because a short pulse has large bandwidth, and the system components must also have the bandwidth required to pass the entire spectrum of the pulse. Furthermore, the wideband signals increase system noise levels and reduce overall system sensitivity.

Examples of high resolution measurements are shown in Figs. 7-1 through 7-3 for very simple targets[3]. The pulse width used in these measurements was typically 0.5 ns, and a receiver bandwidth of 4 GHz accommodated the wide spectrum. Such a pulse width implies a range resolution of a few inches or better.

Figure 7-1 shows the return from a pair of 2in diameter spheres and a jack of about the same size. The three targets were arranged in a row, 12in apart, and were rotated in aspect angle. At each aspect angle the return is plotted as a function of time, which is essentially the same as distance because of the linear relation between time and distance. At each aspect angle (measured between the radar line of sight and the axis along which the three targets were mounted), three distinct echoes can be seen, one from each target. (The large spike along the left side of the diagram represents a reference return in both time and amplitude.)

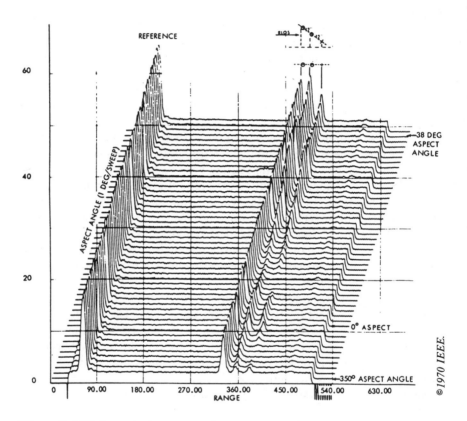

Figure 7-1 High resolution echoes from two spheres and a jack as a function of aspect angle (from Cisco, *et al.* [4])

Note that the echo from the leading sphere is essentially constant for all of the aspect angles displayed from about 350 degrees (−10 degrees) to 40 degrees. The second peak, which is the return from the second sphere, becomes weak as the second sphere passes into the shadow of the first near an aspect angle of zero degrees. An interaction between the two spheres leads to a slight enhancement of the return from the second sphere at aspects between 30 and

40 degrees. The return from the jack, the most distant target, disappears completely as it moves into the shadow of the two spheres in front of it, but can be seen in the deep shadow at zero degrees. This is due to diffraction of the incident wave around the sides of the two leading spheres.

If the display had been presented for extended aspect angles, the peaks would have converged toward a central location at aspects of ±90 degrees because all three targets would have been aligned broadside to the incident wave. However, Cisco, *et al*. chose not to report the data.

Figure 7-2 contains the same kind of displays for a hollow cylinder 35in long and 11in diameter, capped at one end. The display on the left side of the figure shows three distinct returns, the dominant being due to internal reflections off the distant capped end. The small echo preceding this large return is due to a ring represented by the open end of the cylinder nearest the radar, and a small echo is also seen to follow (in time) the large return from the rear face. This small echo is probably due to multiple internal reflections involving diffraction from the open end back toward the capped end of the cylinder.

The display on the right of Fig. 7-2 shows what happens when a pad of absorber is placed inside the cylinder covering the capped end. The large internal reflection is reduced by as much as 20 dB, and the multiple-diffracted internal rays are eliminated, but the return from the open end is unchanged.

These patterns show what might be expected for the intake of a jet engine. Internally, an engine is much like an end-capped cylinder, because the first rotor stage of the engine (the compressor stage) effectively closes off the interior of the intake to the incident radar wave. If the compressor blades were covered with absorbing material, the return from the compressor stage could be reduced. However, due to the typical speeds of jet turbines, this is impractical.

The high resolution traces need not be displayed in the format shown in Figs. 7-1 and 7-2. Using a suitable recording medium, such as film, range-time-intensity (RTI) plots can be made. With this kind of display, the amplitudes are not plotted as line drawings but in the form of intensity (on film) as functions of range and aspect angle. Such a display is shown in Fig. 7-3 for a metallic cone.

The cone was 28in long and had a flare angle of 15 degrees. The aspect angle variation in this particular figure runs from zero degrees (nose-on) at the bottom of the diagram to 180 degrees (base-on) at the top. For purposes of illustration, the three primary scattering centers on the cone — the tip and the two flash points at the base — are identified as S_1, S_2, and S_3. As the aspect angle varies, these scattering centers move in or out in range, tracing out sinusoidal trajectories.

The tip return is too weak at nose-on incidence to register on the diagram. In fact, the base is by far the dominant echo source. As the aspect angle swings

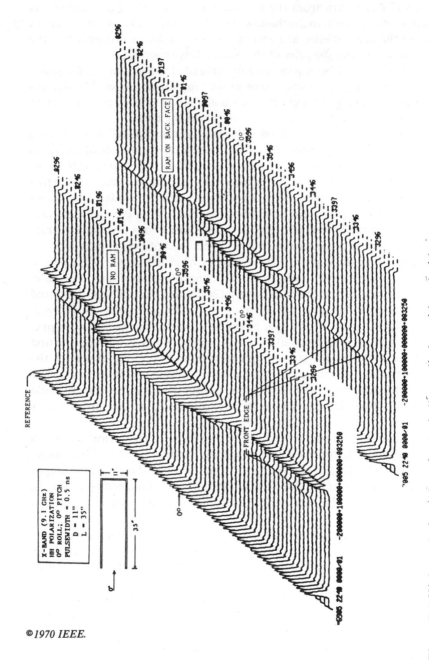

Figure 7-2 High resolution echoes from a hollow cylinder, with and without absorber on the rear face (from Cisco, *et al.* [4])

past the cone-half angle, the far flash point S_3 is shadowed from the incident wave and does not reappear until the aspect has swung past the specular flash at 75 degrees.

By contrast, the near flash point S_2 remains strong and visible throughout the entire 180 degree rotation. Near base-on incidence, the two flash points merge because they move to a position of equal distance from the radar at the base of the cone itself. Near the specular flash and base-on incidence, other returns can be seen in addition to those from the base and tip. For the base-on condition, these are due to multiple waves that flash across the base and excite the opposite edge. At the specular aspect, they are due to creeping waves that spin around the shadowed side of the cone.

If all targets were as simple as those represented in Fig. 7-1 through 7-3, high resolution range plots would be sufficient for diagnostic purposes. For large complex targets at high frequencies, however, additional information is helpful. There may be several scatterers located at the same range and we need to know how they can be separated.

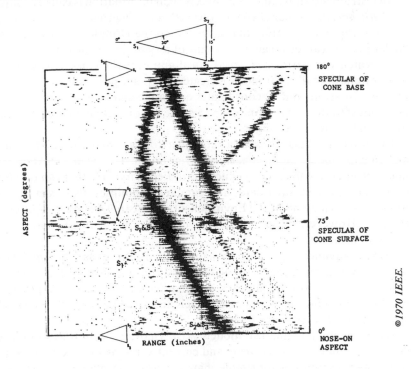

Figure 7-3 RTI plot of the echo from a cone (from Cisco, *et al.* [4])

One method is to use synthetic aperture techniques for data processing and display. The *synthetic aperture radar* (SAR) was initially used for radar terrain mapping, and it utilizes the motion of the radar platform (the aircraft) to create an effective antenna aperture that is very large along the direction of motion. This large effective aperture allows high resolution in the direction parallel to the aperture, and the resolution depends on the Doppler shift of the received signal due to the relative motion between the radar and the echo source.

The same principle can be applied if the radar is held stationary and the target is moved instead. Specifically, the return from a scatterer or a rotating target shifts the frequency of the signal received from the scatterer. The amount of the Doppler shift is proportional to the relative radial velocity (toward or away from the radar) and the radar frequency. Because the relative velocity depends on the scatterer location from the center of rotation and on the angular position in azimuth, we have a way of extracting cross-range location.

An instrumentation radar that collects such information is called an *inverse synthetic aperture radar* (ISAR), "inverse" implying that it is the target motion, not the radar motion, that gives rise to the Doppler shifted signals. A "radar image" can be created by means of a two-dimensional Fourier transform, which was performed with optical apparatus in earlier days, but can now be done digitally. The image is created in a plane perpendicular to the axis of target rotation, and its coordinates are range and cross range. The processing must be done over a finite aspect angle sector, but images can be created for virtually any aspect angle.

The instrumentation for such a data collection system is discussed in Ch. 13, and an example of radar imagery is given in Ch. 11. Scientific-Atlanta announced a new commercial product line in February 1983 that can be used to collect and process such data, and similar capabilities also exist at Point Mugu, California and at the Environmental Research Institute of Michigan (ERIM). However, those facilities are of little value if there is no physical model to measure.

In many advanced system designs, this is precisely the case. The model itself is often nothing more than a set of drawings, thus physical measurements are impossible. Therefore, we must resort to mathematical models that predict the radar returns. These mathematical models are based largely on the analytical formulations discussed in Ch. 3-5.

Several such mathematical models exist. Georgia Tech, for example, has developed models over the years and current versions include basic bistatic scattering prescriptions for simple shapes such as flat plates, cylinders, cone frusta, spheroids, and edges. The Northrop Corporation has also developed RCS prediction codes that have been widely used in industry. Versions of these codes are called GENSCAT, MISCAT, and HELISCAT, and newer versions are being developed. The Naval Research Laboratory (NRL) is also developing RCS prediction codes.

Some of these programs include multipath effects known to influence the returns from marine and land targets. Georgia Tech's program includes the multipath effect, NRL's program will eventually include it, and the Northrop programs do not, since those codes were developed for isolated targets in free space.

It should be noted that these codes are based on high frequency approximations of the scattering by simple objects, and that not all the natural scattering phenomena can be accounted for. The codes are extremely useful for development purposes, but should not be relied upon as the final word; at some point in the development process, the results should be verified by tests of real, physical models. Computer programs are useful tools; nothing more, nothing less.

Nevertheless, part of their utility is that the returns from the various echo sources on the body can be "remembered," catalogued, ordered, and listed. This allows the RCSR specialist to examine the list, concentrate on the most important scatterers on his target, and gauge their relative strength. Without the short-pulse high-resolution radar imaging systems described above, this can be a difficult task at times, becoming impossible, of course, if there is no target to measure.

In many development programs where computer predictions and theoretical analyses are inadequate, it is possible to devise experimental programs for isolated portions of the target. An experimental program requires the fabrication of a test model, but it is not necessary, nor desirable, to build a model of the entire target. First, not many RCS ranges have the high resolution capability mentioned earlier, making diagnostic measurements more difficult to perform and interpret. Second, models are very expensive to build; thus the smaller and less complicated the model, the better.

Therefore, for development and optimization of RCS reduction treatments for specific target features, only those features should be incorporated in a physical model. However, this poses a dilemma: separating one portion of a target from the main body usually creates additional echo sources on both the target and the isolated component. A solution is to build a "test body" that emulates the local geometry of the scattering structure, but from which all other scattering features are deleted. An example is shown in Fig. 7-4.

The upper diagram illustrates what might be constructed for developing treatments for a blade antenna. Because the antenna depends upon a local metallic ground plane for operation, part of the target in the vicinity of the antenna needs to be included in the test model. However, because only treatments of the antenna and local structure are to be evaluated, all other sources of echo must be suppressed. To eliminate reflections from the front of the test body, the forward portion of the test body is drawn to a sharp tip because the return from a sharp tip is small in a forward cone of aspect angles. The conical portion of the test body is smoothly faired into the main central portion to eliminate returns from the junction between the two. The rear portion of the body is terminated by a smooth rounded section to eliminate surface waves that might otherwise be reflected by sharp corners or edges at the rear of the body.

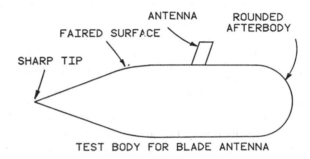

TEST BODY FOR BLADE ANTENNA

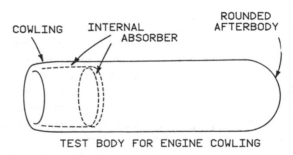

TEST BODY FOR ENGINE COWLING

Figure 7-4 Examples of test bodies

The lower diagram shows what an engine cowling test body might look like. Again, a rounded afterbody minimizes reflections of the surface wave. Because only the lip is desired in the model, the remainder of the internal duct is coated with radar absorbing material. Depending on the test frequency and the relative strength of the lip return, the internal absorber ought to be very high quality, because the return from a large flat panel, even of absorber material, may be high enough to mask the lip return.

The test bodies should be fabricated so that various RCSR test treatments can be applied or fitted easily to the test body. For a matrix of tests to be performed on a radar cross-section measurement range in relatively close succession, this means that several interchangeable test fixtures ought to be fabricated for fast mounting and dismounting on the body. Some experimenters design calibration features directly into the test body to aid in establishing reflection levels without a separate calibration measurement. Such features need to be carefully implemented, however, so that they do not interfere with the low RCS characteristics of the test body. Typically, the calibration feature is a small corner reflector installed somewhere on the rear.

It is common practice to build scale models, and even full-scale versions, of test bodies and test targets made of fiberglass. One fabrication technique is to build plaster molds on which the fiberglass and resin can be overlayed in sections. After final sanding and machining, the sections are assembled on a lightweight framework. Sometimes, a foamed plastic inner core can be used. Fiberglass models can be much lighter than metal ones, making the model handling and measurement much easier and less hazardous.

The outer surface of fiberglass models must be metallized. This can be done with several coats of silver paint, or by electroplating, depending on the physical size of the model. For silver paint alone, enough layers must be sprayed on the model that the surface conductivity is acceptably high. Unfortunately, there are no standards for an "acceptable" surface conductivity; some shops strive for a resistance of less than 0.1 ohm measured by an ordinary ohm-meter between two probes 12in apart on the surface, while others are content with as much as 0.5 ohm.

Such high resistance can probably be tolerated for most measurements not involving the surface wave mechanism. However, the surface resistance may be high enough to prevent the build up of surface traveling waves launched toward the rear of the body at and near grazing angles. It has been noted, for example, that Peters [4] measured traveling wave returns for silver and steel rods 39λ long at 23.85 GHz. The amplitude of the surface wave return for the silver rod was 28.2 square wavelengths, but for the steel rod it was only nine square wavelengths.

This ratio in the amplitudes of the returns is slightly more than 3:1, while the ratio of the conductivities ranges from about 7:1 to 50:1, depending on the particular steel alloy of the rod that Peters used. Peters does not mention what kind of steel he used, but it is clear that the traveling wave return decreases with conductivity. We do not have enough information to know the rate of decrease (for example, with the square root or the cube root of the conductivity ratio), but it definitely increases.

We conclude that the surface conductivity is very important in any physical modeling and testing that involves the surface wave mechanism. If the surface conductivity is built up using several layers of silver paint, it is important that the layers not be permitted to oxidize before subsequent layers are added.

7.5 SHAPING AS A MEANS OF RCSR

Of the four basic methods for reducing the return from a body, the uses of shaping and absorbers have been the most effective by far. The retrofitting of RCSR treatments largely involves only absorbers because the basic shape of a target can rarely be changed to take advantage of shaping. In the development of new weapons systems, on the other hand, both techniques can be implemented to advantage. As mentioned earlier, RCSR treatments represent trade-offs between structural and electromagnetic requirements, and we rarely have the latitude to perform a "full out" treatment without compromising other important performance characteristics of the vehicle.

Whatever the compromises, the most logical approach to RCS requirements in new system development is to first take advantage of shaping without resorting to absorbing materials. The shape must be chosen to optimize the mission of the vehicle, and the mission itself determines the blend of RCSR *versus* structural characteristics. This usually requires, for example, iterations of RCS and aerodymanic performance predictions for a selection of shapes, later verified by experimental testing. If the final shape configuration does not meet RCS requirements, we can always seek to refine the RCS performance by selective appliation of absorber treatments.

Shaping can best be exploited if threat sectors are established. This is because shaping usually does nothing more than shift the regions of high echoes from one aspect angle sector to another. The RCS reduction achieved over one sector is accompanied by an RCS enhancement over another. What rationale can be taken if all threat directions are equally likely? It is a question that has not been satisfactorily answered.

We might think that the surface of the vehicle could be reshaped or enclosed by a surface having a lower RCS, an example being a square flat plate enclosed by a cylinder or sphere just large enough to accommodate the plate, as depicted in Fig. 7-5. The echoes from these three targets can be estimated and compared using the analytical prescriptions discussed in Ch. 5. These prescriptions are

$$\sigma = \begin{cases} \dfrac{l^2}{\pi} \left[kw \cos\theta \; \dfrac{\sin(kw\sin\theta)}{kw\sin\theta} \right]^2 & \text{(flat plate)} \quad (7.3) \\[2ex] kal^2 & \text{(cylinder)} \quad (7.4) \\[1ex] \pi a^2 & \text{(sphere)} \quad (7.5) \end{cases}$$

where l is the vertical dimension of the plate or cylinder, $w\,(=l)$ is the width of the plate, a is the radius of the cylinder or sphere, and θ is the aspect angle as shown in Fig. 7-5. In order to enclose a flat plate whose width is equal to its length, the radius of the cylinder must be $a = l/2$ and the radius of the sphere must be $a = l/\sqrt{2}$. Note that the RCS of the cylinder and sphere are independent of θ because these two objects are rotationally symmetric about the axis normal to the plane containing the direction of incidence.

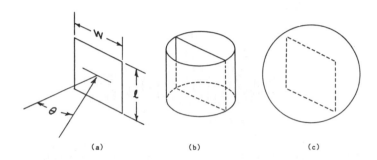

(a) (b) (c)

Figure 7-5 Radar cross section of (a) flat plate can be reduced by replacing it or enclosing it with (b) cylinder or (c) sphere

The radar cross section patterns are plotted for an arbitrary plate length of 25λ in Fig. 7-6. Note that the RCS has been normalized to the square of the plate length. The pattern of the flat plate shows the characteristic $\sin x/x$ behavior of uniformly illuminated apertures and has a large specular value (at $\theta = 0$) for broadside incidence. The amplitude falls off by more than 30 dB as the aspect angle swings toward ± 15 degrees, making it clear that a substantial reduction in RCS is available if the plate can be oriented so that it is never seen broadside. In sharp contrast, the cylinder return is constant and lies some 20 dB below the specular plate return. Similarly, the return from the sphere is constant and lies nearly 30 dB below the specular plate echo. Thus, dramatic reductions in RCS are potentially available for certain target orientations.

However, the display of Fig. 7-6 also illustrates another principle: that a reduction of RCS at one angle is usually accompanied by an enhancement at another. For example, the normalized flat plate pattern at aspects of ± 10 degrees is 10 dB, while that of the cylinder is 19 dB; consequently, the 20 dB reduction obtained at 0 degrees by enclosing the plate by a cylinder is offset by a 9 dB increase at ± 10 degrees. Moreover, the plate return is reduced only over an angle (in this case) of about 4 degrees, while it is enhanced over an angle of nearly 30 degrees. Thus, while the enhancement is not as great as the reduction, it is more persistent. Whether this is acceptable or not depends in large measure on the specific details of the particular RCS reduction task at hand and upon mission requirements; thus, in some cases, it may not be acceptable at all.

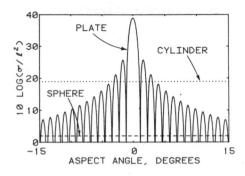

Figure 7-6 RCS patterns of flat plate, circular cylinder, and sphere

Thus, there arises a dichotomy: is it better to have a lower, but broader RCS pattern, or a higher RCS concentrated over a narrow sector of viewing angles? The answer depends on the vulnerability of the vehicle as a function of aspect angle. If the threat sector is confined to a range of angles, then a few large spikes in the RCS pattern are acceptable, provided that those spikes lie outside the threat sector. However, if the threat sector covers all directions, then the answer depends on the detection process.

The probability of detection increases as the RCS increases, but it decreases as the width of the specular flash narrows. Whether either effect dominates depends on the details of a particular mission scenario. A homely example is the rock strewn and brush studded scene in a western film where a US marshal is pursuing an outlaw. Somewhere in those hills crouches the desperado; the hero slowly sweeps his eyes over the landscape, looking for a sign, trying to see through the brush and over the rocks.

By happenstance, the outlaw swings his shiny Winchester through a specular orientation as he adjusts his position behind a rock. The flash of reflected sunlight is brief but bright, and the outlaw has revealed his general position. Although the outlaw remains hidden, the marshal now has only a few dozen square yards to concentrate on, instead of several dozen acres. The search is now narrowed to this relatively small area, even though the rifle flash may not be seen again

In this case, the very narrow angle but intense flash of sunlight from the rifle altered the detection strategy from a wide angle sweep to a narrow angle scan.

Although the probability of detection may have been small at first, the shift in detection strategy changed abruptly. Even if the rifle flash is never seen again, the new strategy will likely ensure that the marshall flushes the outlaw out from behind his rock.

The story may well have taken a different direction had the lawman not seen the flash. He could have developed any of a dozen different strategies, depending on the writer's whim, ranging from a waiting game (the lawman has a full canteen of water, but the outlaw's canteen is empty) to setting the brush afire. Thus, the dichotomy of shaping depends on the mission scenario given that all threat directions are equally likely.

This is the case for marine targets. The threat radar of a ship would be from another ship or for an airborne platform, both of which would be near the horizon. The most likely viewing angles are restricted to a region from the horizon upward to a small angle (a degree or two at most) above the horizon. Thus, there is a favored range of angles in the elevation (vertical) plane. In the azimuth plane, however, all angles are equally likely, so that the objective of shaping in the horizontal plane will be to reduce the large broadside, bow, and stern echoes. With the probable viewing angles confined within a few degrees of the horizon, the optimum shape of a ship would be a right circular cone with its tip pointed skyward. This is absurd from a practical viewpoint, of course, but it represents a shaping objective from an electromagnetic standpoint. Note that the cone half-angle should be chosen at a value of half the most likely elevation angle of the threat [5].

Most assuredly, the Navy will never commission a ship that looks like an upside-down ice cream cone, but it might consider the one advantageous feature of the cone that is attractive in the marine environment: its slanted surface. Unlike airborne targets, ships are doomed to spend their lives on the ground plane formed by the sea surface. Although that surface at times may be so rough that it can scarcely be considered a ground plane, it enchances the signal reflection from metal surfaces near or on it, particularly if those surfaces are vertical.

When a vertical surface intersects the ground plane formed by the mean sea surface, it forms an efficient corner reflector. As we saw in Ch. 6, corner reflectors have high RCS levels because they reflect a radar beam back in the direction from which it is coming. Any vertical surface forms a dihedral corner reflector with the mean sea surface, hence, at broadside aspect angles, the radar echo is large. This is also true of bulkheads and uprights on the ship that meet the decks at right angles.

The large echo area of a corner reflector is due to this perpendicularity, which sends the incident wave back in the direction it came from after two internal bounces, one off the sea surface and the other off the bulkhead. (For a trihedral corner, this happens after three internal bounces.) One way to defeat this process is to tilt the vertical member away from the vertical, thereby sending the reflected wave in some direction other than back to the radar.

An analysis of the dihedral corner shows that the double-bounce mechanism dominates the radar echo [6] and this contribution can be reduced by a specified amount if the electrical sizes of the dihedral corner sides are known. Specifically, we assign the symbols a and b to the wider and narrower faces of the dihedral corner, as shown in Fig. 7-7. The angle γ_0 in Fig. 7-7 is that subtended by the narrower face when seen from the outer edge of the wider face. The double-bounce return attains its maximum value when both faces are fully illuminated by the reflection of the incident wave off the opposite face, and this occurs when the incident wave arrives at an angle γ_0, measured upward from the plane of the wider face.

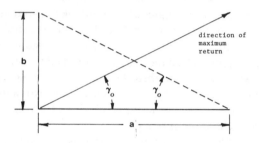

Figure 7-7 Dihedral corner geometry

We now tilt the smaller face by some angle τ away from perpendicular to the larger face. When this is done, the maximum of the double-bounce return is reduced by the factor given in Eq. (6-23). A plot of this reduction is given in Fig. 7-8 for three different dihedral face width ratios.

A more useful form of this result is given in Fig. 6-23, which may be used as follows. First, determine the lowest frequency for which the given RCS reduction must hold, thereby fixing the wavelength λ. Then find the width of the smaller face in wavelengths. Enter the chart with this value of $(b/\lambda$ and read the required angle for the given cross section reduction value. By way of example, assume the return from a dihedral, whose smaller face is 10ft wide, which must be reduced 10 dB for all frequencies above 3 GHz. At 3 GHz the wavelength is 3.94in, hence $b/\lambda = 30.5$. Entering the chart at $b/\lambda = 30.5$ and $R = -10$ dB, we read $\tau = 1.35$ degrees, whence the dihedral angle must be either 91.35 or 88.65 degrees. On some structures, the construction tolerances may be such that they automatically provide a cross section reduction of this magnitude.

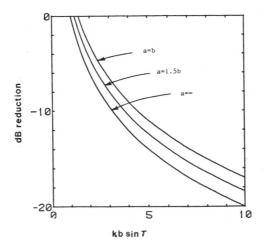

Figure 7-8 Optimum dihedral angle may be selected knowing *kb* and the reduction required

The Navy has recently been persuaded that this technique is practical for surface ships, and studies are being conducted to evaluate its cost effectiveness. Examples of possible implementations are shown in Fig. 7-9. The original structure or design is the hypothetical profile depicted in Fig. 7-9a and might represent a transverse section of the ship taken somewhere amidship.

To take advantage of the benefits of slanted surfaces, the profile could have been designed and built as suggested in Fig. 7-9b, where only the vertical surfaces have been sloped. Note in this case that the hull of the vessel has been canted outward at the top while topside vertical surfaces have been canted inward. The use of screening to provide essentially the same effect is shown in Fig. 7-9c, but observe that some of the deck surfaces have been narrowed. An alternative shielding option is shown in Fig. 7-9d, where the point of attachment of the uppermost screen has been moved to provide a kind of tunnel, presumably recovering the use of the deck space that had been narrowed in Fig. 7-9c. Thus, screening could be a convenient form of retrofit procedure for reducing RCS.

Airborne targets are quite different from ships, of course. First, they are often seen against the background of free space, where there are no clutter returns to compete with or to mask the target return. Second, they are much smaller and, consequently, have much lower radar echoes. Third, their very mission depends on their ability to fly, hence a favorable shape from the electromagnetic viewpoint may not be compatible with a favorable shape from the aerodynamic viewpoint. Fourth, and finally, shaping is out of the question for production vehicles.

This being the case, we examine some shaping concepts for hypothetical airborne platforms that are still in the engineering and design stage. Let us begin with the planform of the wings.

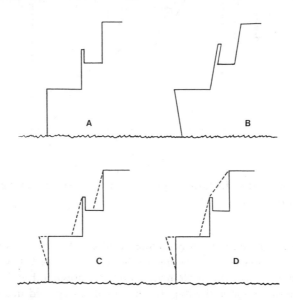

Figure 7-9 RCS reduction options: (a) original profile; (b) upright surfaces tilted away from the vertical; (c) deployment of screens; (d) improved deployment of uppermost screen

Figure 7-10 shows what might be a conventional planform for an airborne subsonic craft. The trailing edges in this particular case are perpendicular to the longitudinal vehicle axis, while the leading edges have a slight taper aft. Thus, the leading edges will be large contributors to the radar return in the forward aspect angle region, especially at an angle to either side of nose-on where the leading edge is perpendicular to the radar line of sight.

When viewed from small angles below the yaw plane, the trailing edges are significant scatterer for vertical polarization. The maximum return occurs when the trailing edges are normal to the line of sight and, consequently, those returns are strongest at nose-on incidence.

The leading edge returns can be shifted further away from the nose-on aspect by giving the wing planform a greater sweep angle, as in Fig. 7-11. Here, the trailing edges are also swept back and, hence, their contributions are also moved away from the nose-on aspect. As with the non-swept trailing edges in

Fig. 7-10, the maximum trailing edge returns occur when the aircraft is seen from below at a slight depression angle.

The leading edge returns can be shifted even further in aspect angle by the planform in Fig. 7-12. This increases the length of the edge, thereby increasing the amplitude of its contribution. One way to reduce the amplitude to its previous value is to curve the edge, as shown in the figure. However, the curvature increases the aspect angle coverage of the return attributable to the leading edge.

A long straight edge, such as the delta wing shown in Fig. 7-13, will have a large but narrow lobe, well removed from the nose-on region. However, the sweep of this planform allows the build-up of a surface traveling wave contribution with returns that will be seen in the nose-on region. The traveling wave builds up from front to back along the leading edge, and the effect occurs primarily for horizontal polarization.

The magnitude of traveling wave contribution is due, in part, to the sharp discontinuity at the wing tips, and its effect can be reduced by gently rounding the wing tips, as shown in Fig. 7-14. In this planform, the trailing edges have been curved and inclined forward to reduce their contributions, and to shift them out of the nose-on region.

Implicit in the design evolution in Figs. 7-10 to 7-14 was the strategy to reduce the RCS in a forward cone of angles. Consequently, the returns from the leading and trailing edges of the wing were swept back sharply to shift their contributions toward the broadside sector. However, it is also important in some cases to reduce the broadside return as well.

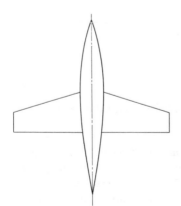
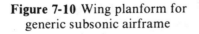
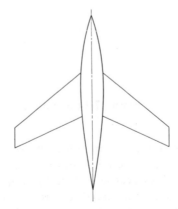

Figure 7-10 Wing planform for generic subsonic airframe

Figure 7-11 Planform with a greater sweep angle

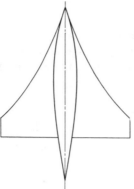

Figure 7-12 Curved leading edge at
a greater sweep angle

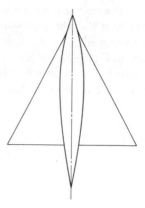

Figure 7-13 Delta wing

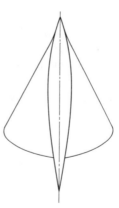

Figure 7-14 Modified delta wing with rounded corners

The dominant broadside returns are from the fuselage, engine pods, and fuel pods. These echoes are large because the scatterers are large, doubly curved surfaces seen from specular directions. It was shown in Ch. 6 that the return from a curved edge is proportional to the product of the wavelength and the curvature of the edge, while the return from a doubly curved surface is proportional to the product of the two radii of curvature of the surface at the specular point. This suggests the fuselage shaping treatments shown in Fig. 7-15.

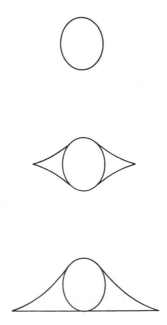

Figure 7-15 Fuselage shape treatments

The upper diagram shows a transverse section of a generic fuselage, while the center diagram shows a fairing drawn to a chine running along the sides of the fuselage. The return from the edge presented by the chine is considerably lower than that of the generic fuselage and can even be faired into the wing planform of Fig. 7-14.

This treatment is essentially intended for viewing angles in the yaw plane. When the fuselage is seen at large depression angles in the broadside sector, however, the lower fairing will have a large return for specular incidence angles. The depression angle coverage can be extended nearly to the nadir by making the bottom of the fuselage flat, as in the lowermost diagram of Fig. 7-15. The advantage of the chine is preserved and the flat underside may even contribute some additional lift to the aerodynamic performance of the vehicle.

A collection of shaping concepts are shown for a generic vehicle in Fig. 7-16. Here, we see the advantages of a sharp, straight leading edge with a long taper aft, rounded corners, and a flat underbody. In addition, the vertical fins are canted inward to reduce their broadside contribution and the nose is brought to a sharp point. For the particular vehicle shown, the engine intakes have been placed on the top side of the vehicle to reduce their contributions when seen from depression angles at the forward aspects.

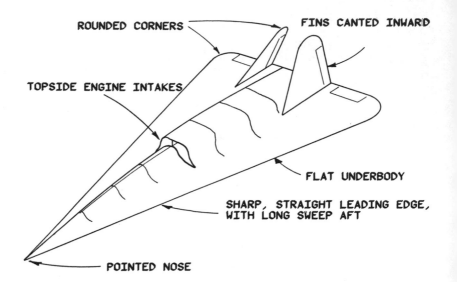

ROUNDED CORNERS

FINS CANTED INWARD

TOPSIDE ENGINE INTAKES

FLAT UNDERBODY

SHARP, STRAIGHT LEADING EDGE,
WITH LONG SWEEP AFT

POINTED NOSE

Figure 7-16 A collection of shaping options for an air-breathing vehicle

7.6 SOME CIVILIAN APPLICATIONS

President Jimmy Carter announced in August 1980 that the US was actively developing *stealth* technology. It was a fact well known within a few government agencies, but the announcement fired the curiosity of the civilian population. It also induced a few entrepreneurs to capitalize on the public awareness, and at least one journalist published a topical account of civilian interest.

Bedard [7] reported some informal studies sponsored by *Car and Driver* magazine about the possibilities of eluding police radars. Citing earlier experiments (also published by *Car and Driver*) using resonant absorbing materials tuned to the police radar frequencies, the radar detection range of a Porsche 9115C was reduced by only about 20 percent. In a subsequent experiment involving 14 different cars and trucks, Bedard reported that a radar able to spot "an Allied moving van 1 ½ miles away was unable to see a Corvette 600 feet away." He concluded that a Corvette is a "natural-born stealth car."

Car and Driver's reports on microwave absorbers inspired a few capitalists to attempt to market kits. The commercial products and their producers were identified as

"The Barrier"
Free Space
1656 Mission Drive
Solvang, CA 93463

"The Shadow"
C & J Manufacturing Company
29 Rear West Third Avenue
Columbus, OH 43201

"The Radar Eater"
(company name and address not given)

The three products were of similar construction, consisting of a conventional plastic stone shield lined with absorber. Unfortunately, the radar shields were stiff and did not conform to the contours of the vehicles that they were installed on. They tended to wrinkle up, giving the cars a "prune" look.

Moreover, Bedard went on, the kits were not very effective in reducing the detection range of a Mazda RX-7. He concluded that "unless you really like the prune look for the front of your RX-7, $300 for less than 100 feet of reduction [in detection range] is not a very good deal."

In search of more information, Bedard located an engineer employed by an auto company in Detroit who was willing to discuss stealth cars, but whom Bedard declined to identify. The engineer was given the pseudonym, "Deep Thwart," and he found that the detection range could be reduced by tilting the radiator and air conditioning condenser of the vehicle 5 degrees from the vertical. This is because significant amounts of radar energy propagate through the grillwork of the car, and are reflected by the radiator and condenser.

Deep Thwart is alledged to have installed a movable metal shield in the front of one car to deflect the incident radar beam away from the radar. The shield was tilted inward (backward at the top) and was reportedly effective. As with all RCSR treatments, there was a price to be paid: Deep Thwart's evaluation of the metal shield was, "The car pretty well disappeared, but it was hell to see where I was going."

Educated by his interviews with Deep Thwart, Bedard re-examined his absorber kits. He found that they covered only the sloping surfaces of the RX-7, the very portions of the car that did not require the application of absorbing materials. The vertical face of the radiator was exposed to the airstream below the bumper, an area not covered by the absorber. Even if an absorber shield were placed there, the cooling system would have to be redesigned for proper exposure to the airstream while being hidden from the radar.

Thus, the civilian section has encountered the same problems experienced by the military services in RCS reduction treatments. Solutions to those problems first require a thorough understanding of the scattering mechanisms and then a redesign of the system allowing the vehicle to accomplish its mission. That design, like military designs, involves trade-offs, and the cost of the treated vehicle will be higher than that of an untreated vehicle. Finally, and probably initially as well, an assessment should be made of the cost benefits. Once again, Bedard discovered there are times when "$300 for less than 100 feet of reduction in detection range is not a very good deal."

7.7 SUMMARY

In this chapter we presented the four basic techniques available for RCS reduction and, of the four, the use of shaping and radar absorbers are by far the most effective. Shaping is typically available only for systems still in the design stages because it can seldom be exploited for vehicles already in production.

In addressing the RCSR problem, we must have tools available for identifying flare spots. Experimental methods can be used when there are models available, otherwise computer codes based on well known analytical procedures can be used. However, these codes are rarely all-encompassing, and measurements must still be performed at some stage of design development. Even if the dominant contributors to the radar return can be treated, the existence of several other less dominant echo sources can make the overall reduction goal difficult to achieve.

Finally, in our survey of Bedard's article, we saw that the civilian sector has encountered the same kinds of problems experienced by the military. The resolution of those problems requires background knowledge and an understanding of the scattering mechanisms. Even so, we must attempt to assess the cost benefits of RCSR because the expense may not at all be worth the benefits.

REFERENCES

1.   "1980 Radar Camouflage Symposium," Air Force Avionics Laboratory, Technical Report AFWAL-TR-81-1015, March 1981.
2.   "Proceedings of the 1975 Radar Camouflage Symposium," Air Force Avionics Laboratory Report AFAL-TR-75-100, December 1975.
3.   D.O. Cisco, R. W. Clay, C. S. Liang, G. W. Grover, and W. A. Pierson, "Wide Bandwidth (Sub-Nanosecond) Scattering Measurements," presented at the 1970 G-AP International Symposium, Columbus OH, IEEE Publication 70c 36-AP, pp. 196-204.
4.   Leon Peters, Jr., "End-Fire Echo Area of Long, Thin Bodies," *IRE Trans. Antennas Propag.*, Vol. AP-6, January 1958, pp. 133-139.
5.   E.F. Knott, "RCSR Guidelines Handbook," Final Technical Report on Phase Two, Contract N00039-73-C-0676, Georgia Institue of Technology, Engineering Experiment Station, April 1976.

6.   E.F. Knott, "RCS Reduction of Dihedral Corners," *IEEE Trans. Antennas Propag.*, Vol. AP-25, May 1977, pp. 406-407.
7.   Patrick Bedard, "Stealth Cars," *Car and Driver*, Vol. 26, No. 6, December 1980, pp. 140 *et seq*.

CHAPTER 8

ANALYSIS OF REFLECTION FROM MULTILAYER DIELECTRICS

M. T. Tuley

8.1 INTRODUCTION

While shaping can often provide dramatic RCSR results over limited aspect angles, many situations require absorption of the incident electromagnetic energy if design goals are to be achieved. Thus, a knowledge of the design and application of radar absorbing materials (RAM) is vital to the engineer whose task it is to minimize the radar signature of a vehicle.

The design of RAM, particularly specular RAM, is principally the design of single or multilayer lossy dielectrics with specified scattering properties. To separate the basic theory underlying RAM design from the applications aspects, the study of RAM in this book has been broken into two sections.

This chapter provides the analytical background required for evaluation of the scattering from dielectric surfaces. Chapter 9 details many of the RAM types which have been formulated, comments on typical applications for each type, and surveys some of the commercial materials which are available.

The study of RAM should begin with the microscopic theory of radar absorption, but we will instead approach this topic with a macroscopic view of electromagnetics. While the loss mechanisms through which RAM operates are microscopic in nature (i.e., on the atomic and crystal lattice levels), the analysis is most easily handled by taking a classical transmission line approach for calculating the reflection and transmission properties of RAM. In effect, then, the design of RAM is simply the design of a lossy distributed network which matches the impedance of free space to that of a conducting body to be shielded.

The remainder of this introduction discusses the material properties useful in calculating RAM performance and introduces the concept of reflection coefficient in terms of normalized and intrinsic impedances. Section 8.2 discusses reflection from flat dielectric multilayers, taking the simpler case of normal incidence, and then sec. 8.3 generalizes the approach to include oblique incidence scatter. Section 8.4 presents the wave matrix formulation for reflection calculations through the analysis of a two-port circuit. In sec. 8.5, normal incidence scattering from cylindrical dielectric multilayers is introduced and the computational difficulties of the associated calculations are discussed. Finally, an approximate reflection analysis useful in initial design calculations is presented in sec. 8.6.

Radar absorbing materials are based on the fact that there are substances which absorb energy from electromagnetic fields passing through them. Such materials have indices of refraction which are complex numbers. In the index of refraction, which includes magnetic as well as electric effects, it is the imaginary component that accounts for the loss in a material. The term "loss" refers to the dissipation of power or energy, quite analogous to the way energy is consumed by a resistor when electrical current passes through it. The loss is actually the conversion of electrical energy into heat, and although most absorbers do not dissipate enough energy to get detectably warm when illuminated by a radar, this is nevertheless the mechanism by which they operate. At microwave frequencies, the loss is due to a number of effects on the atomic and molecular level. However, for most practical electric absorbers a majority of the loss is due to the finite conductivity of the material, while for most magnetic absorbers at microwave frequencies, magnetization rotation within the domains is the principal loss mechanism. In any event, it is customary to group the effects of all loss mechanisms into the permittivity (ϵ) and permeability (μ) of the material because the engineer is usually interested only in the cumulative effect.

Several common usages exist for expressing the complex permittivity and permeability. Generally, we shall deal with the relative permittivity, ϵ_r, and relative permeability, μ_r, which are normalized by the free space values, ϵ_0 and μ_0. The complex notation for ϵ_r and μ_r is normally given as

$$\epsilon_r = \epsilon_r' + i\,\epsilon_r''$$
$$\mu_r = \mu_r' + i\,\mu_r'' \tag{8-1}$$

where the real (energy storage) part of each parameter is denoted by a prime and the imaginary (loss) part is denoted by a double prime. Because the

$D = \epsilon E$
$E = E_0 e^{j\omega t}$
$\therefore D = j\omega \epsilon E$

Curl $H = \dot{D} + \dot{i}$

conductivity σ of electric absorbers is often the major loss mechanism, it is convenient to express the effect of the conductivity in terms of ϵ''_r. For that case, ϵ''_r and σ are related by

Curl $H = \sigma E + j\omega \epsilon E = E(\sigma + j\omega \epsilon)$

$\dfrac{\sigma}{\omega \epsilon} = \dfrac{\text{Conduction current density}}{\text{displacement current density}}$

$$\epsilon_r'' = \sigma/\omega\epsilon_0 \tag{8-2}$$

where ω is the radian frequency. Equivalently, in polar notation

$$\epsilon_r = |\epsilon_r| e^{i\delta_m}$$
$$\mu_r = |\mu_r| e^{i\delta_m} \tag{8-3}$$

where δ and δ_m are the electric and magnetic loss tangents given by

$$\tan\delta = \epsilon_r''/\epsilon_r'$$
$$\tan\delta_m = \mu_r''/\mu_r' \tag{8-4}$$

The index of refraction n is the ratio of the wavenumber describing wave propagation within the material to the free space wavenumber, and is to the geometric mean of the relative permittivity and permeability:

$$n = k/k_0 = \sqrt{\mu_r\,\epsilon_r} = \dfrac{V_0}{V} \geq 1 \tag{8-5}$$

where k is the wavenumber in the material, and $k_0 = \omega\sqrt{\mu_0\epsilon_0}$ is the free space wavenumber. Similarly, μ_r and ϵ_r also define the intrinsic impedance, Z, of the material,

$$Z = Z_0 \sqrt{\mu_r/\epsilon_r} \tag{8-6}$$

where Z_0 is the impedance of free space, 120π, which is approximately 377 ohms.

The intrinsic impedance is the impedance value seen by a normally incident wave on a semi-infinite slab of a material. In practical applications, a layer of dielectric will often be backed by a conducting surface. For that case a transmission line analysis can be performed to find the effective input impedance at the front face of the layer.

For a flat metallic surface coated with a layer of dielectric material, the normalized input impedance η is given by

$$\eta = \sqrt{\mu_r/\epsilon_r}\,\tanh(-ik_0\,d\,\sqrt{\mu_r\,\epsilon_r}) = \dfrac{Z}{Z_0}\tanh(-i\beta d) \tag{8-7}$$

where $\beta = k = k_0\sqrt{\mu_r\epsilon_r}$ and $\sqrt{\dfrac{\mu_r}{\epsilon_r}} = \dfrac{Z}{Z_0}$

where d is the thickness of the dielectric layer. This formula applies to a wave striking the surface at normal incidence, and it becomes more complicated when the wave arrives at oblique angles. The normalized impedance can be used to calculate the reflection coefficient R

$$R = \dfrac{\eta - 1}{\eta + 1} \tag{8-8}$$

R, like η, is a complex number, but has a magnitude between zero and one.
In discussing reflection coefficients, it is customary to ignore the phase

angle and to refer only to the "voltage" amplitude $|R|$, so that the power reflection in decibels is

$$|R| \text{ (dB)} = 20 \log_{10} |R| \tag{8-9}$$

In the discussion which follows, the terms "reflection coefficent" and "reflectivity" will be used essentially interchangeably.

The objective of RAM design is to produce a material for which $|R|$ remains as small as possible over as wide a frequency range as possible. It should be noted that unless the material has some loss, the amplitude of the reflection coefficient will be entirely controlled by the phase and amplitude relationship between the portion of the incident wave reflected at the front surface and the portion returning via reflections at the backing surface. In some cases, we may take advantage of the phase shift upon reflection to provide resonant energy cancellation. This is inherently a narrowband RCSR technique. However, it is also often taken advantage of with lossy materials to provide improved performance at certain frquencies. The Dallenbach layer, which is discussed in detail in Ch. 9, is an example of an absorber which makes use of a combination of loss and resonant cancellation.

8.2 NORMAL INCIDENCE SCATTERING FROM FLAT DIELECTRIC MULTILAYERS

Calculation of the reflection of a normally incident plane wave from an infinite flat multilayer structure is a straightforward problem involving application of boundary conditions derived from Maxwell's equations to the general solution for the electric and magnetic fields in each layer. The functional form of the fields, complex exponentials, and the stepping procedure required for multilayers make implementation of the equations on a computer (or programmable calculator) desirable. Note that existing transmission line design computer programs can often be used for absorber design, either as they are or with slight modifications.

The basic geometry to be considered is that of a finite number of dielectric layers stacked against a metallic backing plate, as shown in Fig. 8-1. The layer properties may differ from one layer to the next, or they may be the same. It is assumed that impedance sheets of zero thickness may be sandwiched between layers, as suggested in Fig. 8-2. The sheets can be characterized by a resistance value R in ohms per square, or by a conductance, G mhos per square, where $G = R^{-1}$. For cases such as circuit analog absorbers, where the sheets can provide a complex impedance, the resistance R may be directly replaced by the impedance Z, or the conductance G by the admittance Y. To minimize confusion in notation in the following analysis, G will be used for the sheet admittance to differentiate from the dielectric layer intrinsic admittance, which is denoted with a Y.

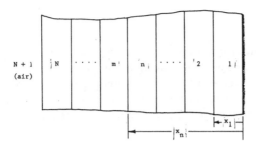

Figure 8-1 A sequence of N layers

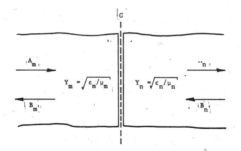

Figure 8-2 Resistive sheet sandwiched between two dielectric layers

The approach used to analyze the scattering is to postulate the form of the electric and magnetic fields in the dielectric layers on either side of the resistive sheet and to specify the boundary conditions which these fields must satisfy. This allows the coefficients of the field representation on one side of the sheet to be related to those on the other side. Because the layers in Fig. 8-1 are numbered outward from the backing plane, it is convenient to have x increasing to the left. Thus, the positive traveling wave will be associated with the B coefficients in Fig. 8-2.

The electric and magnetic field structure in a given layer is taken to be

$$E = A e^{-ikx} + B e^{ikx} \tag{8-10}$$

$$H = Y (A e^{-ikx} - B e^{ikx}) \tag{8-11}$$

where A and B represent the amplitudes of forward and backward propagating waves, and Y is the layer intrinsic admittance. The boundary conditions to be satisfied at the interface are

$$GE^+ = GE^- = J$$
$$H^+ - H^- = J \tag{8-12}$$

where the plus and minus superscripts denote the fields on opposite sides of the sheet, and J is the current flowing in the sheet.

If the resistive sheet location or the boundary between two layers is represented by x_n, and subscripts per Fig. 8-1 are appended to the quantities in Eq. (8-10) and (8-11) to identify the two media, then the application of Eq. (8-12) yields two equations,

$$A_m e^{-ik_m x_n} + B_m e^{ik_m x_n} = A_n e^{-ik_n x_n} + B_n e^{ik_n x_n} \tag{8-13}$$

$$Y_m (A_m e^{-ik_m x_n} - B_m e^{ik_m x_n}) = (G + Y_n) A_n e^{-ik_n x_n}$$
$$+ (G - Y_n) B_n e^{ik_n x_n} \tag{8-14}$$

These equations may be used to find A_m and B_m in terms of A_n and B_n,

$$A_m = \frac{e^{ik_m x_n}}{2Y_m} [A_n (Y_m + Y_n + G) e^{-ik_n x_n} + B_n (Y_m - Y_n + G) e^{ik_n x_n}] \tag{8-15}$$

$$B_m = \frac{e^{-ik_m x_n}}{2Y_m} [A_n (Y_m - Y_n - G) e^{-ik_n x_n} + B_n (Y_m + Y_n - G) e^{ik_n x_n}] \tag{8-16}$$

The stepping procedure begins with the assignment of arbitrary values to A_1 and B_1, the coefficients of the fields in the first layer, which is adjacent to the metal sheet. For a metallic backing, the total electrical field must vanish on the sheet, thus by Eq. (8-10), at $x = 0$, $B_1 = -A_1$. The arbitrary assignment $A_1 = 1$, $B_1 = -1$ satisfies this particular condition. If there is no metallic backing (i.e., if the backing is free space), there will be no wave traveling to the left, hence $B_1 = 0$ at $x = 0$. Next, the transformer relations (8-15) and (8-16) are used at the first interface located at $x = x_1$, and A_2 and B_2 are calculated. (In a computer code, a pair of variables may be replaced by updated values representing the change as a boundary is traversed.) The sequence is iterated until the $N+1$ layer is reached, which is free space outside the structure.

Because the stepping is initiated using arbitrary values for A_1 and B_1, the final results of A_{N+1} and B_{N+1} are in error by precisely the same amount as A_1 and B_1, because the transformations across the boundaries are linear operations. It can be assumed without loss of generality that outside the structure the incident wave has unit amplitude and the reflected wave has an amplitude R (for the reflection coefficient). Therefore, all the coefficients could have been corrected by normalizing with respect to A_{N+1}, had it been known at the outset. Because the normalization constant is now known, the reflection coefficient R, associated with the structure is simply

$$R = \frac{B_{N+1}}{A_{N+1}} \tag{8-17}$$

8.3 OBLIQUE INCIDENCE SCATTERING FROM FLAT DIELECTRIC MULTILAYERS

The preceding discussion details the simplest case of scattering from planar multilayer structures, that of normal incidence. This section generalizes to the more complex case of oblique incidence. The geometry is similar to that defined in Figs. 8-1 and 8-2, except that the directions of the propagating waves are not necessarily normal to the layer boundaries, an example of which is illustrated in Fig. 8-3. For this case, the form of the wave will be

$$E = A \ e^{ik(-x\cos\theta \ + \ z\sin\theta)} + B \ e^{ik \ (x\cos\theta \ + \ z\sin\theta)} \tag{8-18}$$

where x is normal to the layer boundary and positive upward, and z is to the right in Fig. 8-3. Note that Eq. (8-18) reduces exactly to Eq. (8-10) for $\theta = 0$.

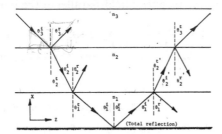

Figure 8-3 Wave propagation and reflection in a dielectric multilayer for oblique incidence

Along with the previous boundary conditions on the z component of the fields, there is an additional requirement, given by Snell's law, that

$$k_m \ \sin\theta_m = k_n \ \sin\theta_n \tag{8-19}$$

Obviously, if k_m or k_n is complex, implying lossy media, the sine of the angles in general must also be complex for the equality to hold. The complex angle is a result of the fact that the planes of constant phase and planes of constant amplitude no longer coincide, and so a plane wave no longer exists [1]. While the concept of a "complex angle" may be difficult to grasp, we may allow θ to be complex, so that

$$\theta = \theta' + i \ \theta'' \tag{8-20}$$

Then

$$\sin \theta = \cosh \theta'' \sin \theta' + i \sinh \theta'' \cos \theta' \tag{8-21}$$

The net result of having a complex angle is that the imaginary component introduces an attenuation factor in the propagation of the transmitted wave, in addition to the usual attenuation associated with the imaginary component of the complex wavenumber.

For the case of oblique incidence two cases must be considered. The first, for the electric field parallel to the interface, provides a formula for the coefficients of

$$A_m = \frac{e^{ik_m x_n \cos\theta_m}}{2 Y_m \cos\theta_m} [A_n (Y_m \cos\theta_m + Y_n \cos\theta_n + G) e^{-ik_n x_n \cos\theta_n}$$
$$+ B_n (Y_m \cos\theta_m - Y_n \cos\theta_n + G) e^{ik_n x_n \cos\theta_n}]$$

$$(8\text{-}22)$$

$$B_m = \frac{e^{ik_m x_n \cos\theta_m}}{2 Y_m \cos\theta_m} [A_n (Y_m \cos\theta_m - Y_n \cos\theta_n - G) e^{-ik_n x_n \cos\theta_n}$$
$$+ B_n (Y_m \cos\theta_m + Y_n \cos\theta_n - G) e^{ik_n x_n \cos\theta_n}]$$

The second case is for the magnetic field parallel to the interface, for which

$$A_m = \frac{e^{ik_m x_n \cos\theta_m}}{2 Y_m \cos\theta_m} [A_n (Y_m \cos\theta_n + Y_n \cos\theta_m + G \cos\theta_n \cos\theta_m) e^{-ik_n x_n \cos\theta_n}$$
$$+ B_n (Y_m \cos\theta_n - Y_n \cos\theta_m + G \cos\theta_n \cos\theta_m) e^{ik_n x_n \cos\theta_n}]$$

$$(8\text{-}23)$$

$$B_m = \frac{e^{ik_m x_n \cos\theta_m}}{2 Y_m \cos\theta_m} [A_n (Y_m \cos\theta_n - Y_n \cos\theta_m - G \cos\theta_n \cos\theta_m) e^{-ik_n x_n \cos\theta_n}$$
$$+ B_n (Y_m \cos\theta_n + Y_n \cos\theta_m - G \cos\theta_n \cos\theta_m) e^{ik_n x_n \cos\theta_n}]$$

As in the case of normal incidence, a stepping procedure is used, where the boundary conditions at the innermost boundary are used to determine the relationship between A_1 and B_1. Each successive layer is then stepped through until free space is reached. However, for the case of oblique incidence, it is generally the incidence angle on the outermost layer that is given, while the first angle required in the computation is that in the innermost layer. Thus, a double stepping procedure is required, where we must begin at layer $N + 1$ (free space) and step inward, calculating the value of each θ using Snell's law. These values can be stored and then recalled, as needed, when the program steps out from the inner layer outward to calculate the A's and B's. As before, the reflection coefficient is given by Eq. (8-17). Note from Fig. 8-3 that this procedure calculates the specular reflection coefficient, not the backscatter reflection coefficient. Only in the case of normal incidence will the two coincide.

8.4 THE WAVE MATRIX APPROACH TO SCATTERING FROM FLAT DIELECTRIC MULTILAYERS

An equivalent approach to calculation of scattering from flat multilayer dielectrics is the wave matrix approach. The basic analysis can be conducted in terms of either cascade matrices, relating the output side of a two-port to its input side, or in terms of scattering matrices, which relate the incident and reflected scattering coefficients Collin [2] gives an excelletn treatment of the cascade matrix approach in terms of reflection and transmission coefficients. The description given here utilizes the scattering matrix parameters because those are generally the most easily measured characteristics of a two-port network. This treatment follows Kerns [3], and it should be noted that Collin's matrix notation and Kerns' differ.

The shunt element circuit of Fig. 8-4 might represent a circuit analog sheet or a resistive sheet in an absorber layup. The reflected waves (b_1, b_2) at each side of its two ports are related to the incident values (a_1, a_2) by the scattering matrix $[S]$, where

$$\begin{bmatrix} b_1 \\ b_2 \end{bmatrix} = [S] \begin{bmatrix} a_1 \\ a_2 \end{bmatrix} = \begin{bmatrix} S_{11} S_{12} \\ S_{21} S_{22} \end{bmatrix} \begin{bmatrix} a_1 \\ a_2 \end{bmatrix} \tag{8-24}$$

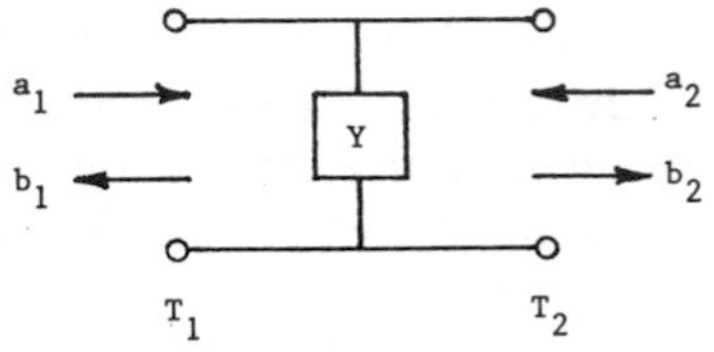

Figure 8-4 Shunt equivalent circuit element

In terms of reflection and transmission coefficients, S_{11} is the reflection coefficient seen by a wave incident at port 1 with port 2 having a matched termination, similarly S_{22} is the reflection coefficient seen from port 2, and S_{12} and S_{21} are the transmission coefficients from ports 2 to 1 and 1 to 2, respectively. For a shunt circuit with admittance Y, the scattering matrix is

$$[S] = \frac{1}{2+Y} \begin{bmatrix} -Y & 2 \\ 2 & -Y \end{bmatrix} \tag{8-25}$$

The scattering matrices necessary to define the other elements of a multilayer dielectric are the interface between dielectric layers and the phase shift due to a dielectric layer. A material interface is described in terms of the admittances on the "left" (Y^+) and "right" (Y^-) sides of the interface by

$$[S] = \frac{1}{Y^- + Y^+} \begin{bmatrix} Y^- - Y^+ & 2Y^+ \\ 2Y^- & Y^+ - Y^- \end{bmatrix} \tag{8-26}$$

where the admittances depend on the polarization and angle of incidence and are given by

$$Y_{TM} / Y_0 = \epsilon_r / \sqrt{\mu_r \, \epsilon_r - \sin^2 \theta_0} \qquad (8\text{-}27)$$

for the electric field parallel to the interface and

$$Y_{TE} / Y_0 = \sqrt{\mu_r \, \epsilon_r - \sin^2 \theta_0} / \mu_r \qquad (8\text{-}28)$$

for the magnetic field parallel to the interface, where θ_0 is the angle of incidence at the left side of the interface.

A slab of dielectric of thickness d simply introduces a phase shift, and its scattering matrix is given by

$$[S] = \begin{bmatrix} 0 & \exp(-ikd) \\ \exp(ikd) & 0 \end{bmatrix} \qquad (8\text{-}29)$$

where

$$k = k_0 \sqrt{\mu_r \, \epsilon_r - \sin^2 \theta_0} \qquad (8\text{-}30)$$

While the parameters of the scattering matrix are convenient because they naturally arise from many measurement techniques, they are not convenient for calculation of the reflection and transmission properties of a multilayer. For that purpose the cascade matrix, $[R]$, is normally used, because the properties of a cascade connection of shunt elements and spacers is given by a total cascade matrix $[R_T]$, which is simply the product of the component matrices:

$$[R] = \begin{bmatrix} R_{11} & R_{12} \\ R_{21} & R_{22} \end{bmatrix} = \frac{1}{S_{21}} \begin{bmatrix} (S_{12} \, S_{21} - S_{11} \, S_{22}) & S_{11} \\ -S_{22} & 1 \end{bmatrix} \qquad (8\text{-}32)$$

and

$$[S] = \frac{1}{R_{22}} \begin{bmatrix} R_{12} & (R_{11} \, R_{22} - R_{12} \, R_{21}) \\ 1 & -R_{21} \end{bmatrix} \qquad (8\text{-}33)$$

The wave matrix approach is particularly useful in the analysis of circuit analog designs, for which it is often desirable to use the measured properties of fabricated CA sheets to predict performance.

8.5 SCATTERING FROM CYLINDRICAL DIELECTRIC MULTI-LAYERS

The same concepts can be used to treat plane waves at normal incidence on cylindrical multilayers as on flat ones. However, in the place of a single pair of coefficients per layer, cylindrical geometries require an infinite number of pairs per layer because the formal representation of the layer fields requires an infinite sum of Bessel functions. In practice, of course, the infinite series must be truncated after a finite number of terms, but the resulting error can be made as small as desired. The following discussion is based on the formulation given by Richmond and Bussey [4], but includes the effect of resistive sheets between the dielectric layers.

The modal representation for the fields can be made in terms of independent solutions J_n and Y_n of Bessel's equation. Because the order n of these Bessel functions is the index of summation, the m, n notation (the layer numbers) of Fig. 8-1 are converted to l,m as shown in Fig. 8-5. Moreover, because the admittance symbol Y may be confused with the Bessel function of the second kind, Y_n, the layers on either side of an interface are characterized instead by impedances.

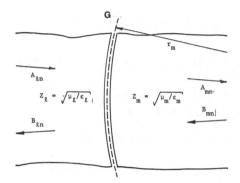

Figure 8-5 Layer interface in cylindrical geometry

The internal field structure in the mth layer is taken to have the following forms:

$$E_{zm} = \sum_{n=0}^{\infty} \left[A_{mn} J_n (k_m r_m) + B_{mn} Y_n (k_m r_m)\right] \cos n\phi \qquad (8\text{-}34)$$

$$H_{\phi m} = \sum_{n=0}^{\infty} \frac{i}{Z_m} \left[A_{mn} J_n' (k_m r_m) + B_{mn} Y_n' (k_m r_m)\right] \cos n\phi \qquad (8\text{-}35)$$

$$H_{zm} = \sum_{n=0}^{\infty} \left[A_{mn} J_n (k_m r_m) + B_{mn} Y_n (k_m r_m)\right] \cos n\phi \qquad (8\text{-}36)$$

$$E_{\phi m} = -\sum_{n=0}^{\infty} iZ_m \left[A_{mn} J_n' (k_m r_m) + B_{mn} Y_n' (k_m r_m)\right] \cos n\phi \qquad (8\text{-}37)$$

Equations (8-34) and (8-35) are for E-polarization (incident electric vector parallel to the cylinder axis) while Eq. (8-36) and (8-37) are for H-polarization (incident magnetic vector parallel to the cylinder axis). The primes on the

Bessel functions denote the derivative with respect to the entire argument of the function. Note that the arguments, in general, are complex for layer materials that include loss. The azimuth angle ϕ is measured from the forward scattering direction, hence the backscattering direction is $\phi = \pi$.

Application of the boundary conditions yields the following transformer equations for E-polarization:

$$A_{ln} = \frac{\pi}{2} k_l r_m \{[A_{mn} J_n (k_m r_m) + B_{mn} Y_n (k_m r_m)] [Y_n' (k_l r_m) + iGZ_l Y_n (k_l r_m]$$

$$- \frac{Z_l}{Z_m} Y_n (k_l r_m) [A_{mn} J_n' (k_m r_m) + B_{mn} Y_n' (k_m r_m)]\}$$ (8-38)

$$B_{ln} = \frac{\pi}{2} k_l r_m \frac{Z_l}{Z_m} J_n (k_l r_m) [A_{mn} J_n' (k_m r_m) + B_{mn} Y_n' (k_m r_m)]$$

$$- [A_{mn} J_n (k_m r_m) + B_{mn} Y_n (k_m r_m)]$$ (8-39)
$$[J_n' (k_l r_m) + iGZ_l J_n (k_l r_m)]$$

The corresponding pair for H-polarization is

$$A_{ln} = \frac{\pi}{2} k_l r_m \{Y_n' (k_l r_m) [A_{mn} J_n (k_m r_m) + B_{mn} Y_n (k_m r_m)]$$

$$- \frac{Z_m}{Z_l} [A_{mn} J_n' (k_m r_m)$$ (8-40)

$$+ B_{mn} Y_n' (k_m r_m)] [Y_n (k_l r_m) - iGZ_l Y_n' (k_l r_m)]\}$$

$$B_{ln} = \frac{\pi}{2} k_l r_m \left\{ \frac{Z_m}{Z_l} [A_{mn} J_n' (k_m r_m) + B_{mn} Y_n' (k_m r_m)] [J_n (k_l r_m) \right.$$

$$- iGZ_l J_n' (k_l r_m)]$$
$$\left. - J_n (k_l r_m) [A_{mn} J_n (k_m r_m) + B_{mn} Y_n (k_m r_m)] \right\}$$ (8-41)

The stepping procedure begins with A_{ln} and B_{ln} being assigned arbitrary values. If the innermost layer is non-metallic, the A_{ln}, evaluated at $r = 0$, are all assigned a value of one and the B_{in} are all set to zero. If the inner core is metallic, the A_{ln} remain unity, and the B_{ln}, evaluated at the outer surface of the core, are assigned the values

$$B_{1n} = - \frac{J_n (k_1 r_1)}{Y_n (k_1 r_1)} \text{ for } E\text{-polarization}$$ (8-42)

$$B_{1n} = - \frac{J_n' (k_1 r_1)}{Y_n' (k_1 r_1)} \text{ for } H\text{-polarization}$$ (8-43)

which satisfy the boundary conditions on Maxwell's equations at a cylindrical conducting surface. The transformer equations (8-38) through (8-41) are applied across successive interfaces until the last interface has been traversed. It should be noted that the equations for both polarizations have been given, but that only the appropriate pairs need be used if interest lies in the scattering for a single polarization.

As in the flat surface *TEM* case, the final values of A and B (i.e., $A_{N+1,n}$ and $B_{N+1,n}$) are not the correct ones, because the starting values were chosen arbitrarily. All that remains is to determine the normalization factor using the form of the plane wave incident on the multilayer structure. The scattering amplitude P is given by

$$P = - \sum_{n=0}^{\infty} \frac{\epsilon_n B_{N+1,n}}{B_{N+1,n} - i A_{N+1,n}} \cos n\phi \qquad (8\text{-}44)$$

where $\epsilon_n = 1$ for $n = 0$ and $\epsilon_n = 2$ otherwise. In the backward direction $\phi = \pi$, and substituting in Eq. (8-44) the backscattered amplitude is

$$P = - \sum_{n=0}^{\infty} \frac{(-1)^n \epsilon_n B_{N+1,n}}{B_{N+1,n} - i A_{N+1,n}} \qquad (8\text{-}45)$$

The radar cross section per unit length is

$$\sigma_{2d} = \frac{4}{k} |P|^2 \qquad (8\text{-}46)$$

The three-dimensional broadside radar cross section of a cylindrical structure of finite length l is related to the two-dimensional scattering of an infinite structure by the realation:

$$\sigma_{3d} = 2 l^2 \frac{\sigma_{2d}}{\lambda} \qquad (8\text{-}47)$$

8.6 AN APPROXIMATE REFLECTION ANALYSIS PROCEDURE

The analyses of the preceeding sections provide excellent tools for the evaluation of RAM performance. For use in the design stage the formulations presented can be integrated with optimization routines to provide maximum performance within a given set of constraints. However, while current RAM design efforts are usually based on computer optimization techniques, it is desirable to begin the optimization routine with a parameter set as close to optimum as practical. This generally implies that an analytical solution should be used to establish "first-cut" admittance parameters. Fortunately, RAM design practices are closely related to filter theory, and many of the tools developed for that arena can be utilized. This section briefly outlines the use of some of those tools in the design of broadband absorbers using resistive sheets or circuit analog sheets separated by low-loss dielectric spacers.

Figure 8-6 illustrates the typical model assumed for this analysis. As noted in sec. 8.4, each resistive or circuit analog sheet can be represented as an admittance shunted across the transmission line. A multi-element, broadband

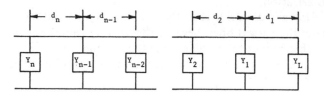

Figure 8-6 Circuit for approximate reflection analysis

design implies that the reflection coefficient at each shunt admittance will be small. A small reflection coefficient implies a small admittance for each shunt element. The small individual reflection coefficients allow us to assume that the total reflection coefficient is simply the sum of the reflection coefficients for each shunt element, modified by the appropriate phase shift due to line length. With this simple assumption, the characteristics of the reflection coefficient of an absorber design can be tailored to standard functions, such as a maximally flat or minimum ripple (Chebyshev) behavior with frequency. The analysis shown here is very similar to that of Collin [5], who considered the design of multilayer transformers using small reflection theory. In fact, an absorber can be considered a transformer between free space and a very small impedance. The only modification of Collin's analysis is to use shunt elements instead of transmission lines with varying characteristic impedances.

The reflection coefficient R of the absorber circuit is approximately

$$R \simeq R_n + R_{n-1} \exp{(i2kd_n)}$$
$$+ R_{n-2} \exp{[i2k\,(d_n + d_{n-1})]} + \ldots \tag{8-48}$$
$$+ R_L \exp{[i2k\,[d_n + d_{n-1} + \ldots + d_1)]}$$

where multiple reflections have been neglected. The power reflection is the square of the absolute value of Eq. (8-48); and for one shunt element and two shunt element circuits is given by

$$|R|^2 = \begin{cases} 1 - 2R_1 \cos 2\theta, & n=1 \\ 1 - 2R_1 \cos 2\theta - 2R_2 \cos 4\theta, & n=2 \end{cases} \tag{8-49}$$

where all lines are assumed to be of equal length ($\theta = kd$), the reflection coefficients are assumed to be real (to simplify the examples) and small, and the load reflection coefficient is -1 (i.e., a short circuit). The extension to $n > 2$ is straightforward. Equation (8-49) can be written as a polynomial in powers of $\cos\theta$ by using standard identities, resulting in

$$|R|^2 = (1 + 2R_1) - 4R_1 \cos^2\theta, \qquad n=1$$
$$|R|^2 = (1 + 2R_1 - 2R_2) + (-4R_1 + 16R_2) \cos^2\theta \qquad (8\text{-}50)$$
$$\qquad -16R_2 \cos^4\theta, \qquad\qquad n=2$$

A maximally flat design is achieved by setting the coefficients of all but the highest power terms to zero:

$$R_1 = 1/2, \qquad\qquad n=1$$
$$R_1 = 2/3, \ R_2 = 1/6, \quad n=2 \qquad\qquad (8\text{-}51)$$

Similarly, a Chebyshev design is achieved by setting the coefficients equal to the Chebyshev polynomial coefficients:

$$R_1 = -1/(2 \sin\theta_1), \qquad\qquad n=1$$
$$\left.\begin{array}{l} R_1 = 4R_2 \sin^2\theta_1 \\ R_2 = 1/(2 - 8\sin^2\theta_1 - 2\cos^4\theta_1), \end{array}\right\} \ n=2 \qquad (8\text{-}52)$$

where θ_1 is evaluated at the edge of the frequency band.

Within the small reflection approximation, the reflection coefficient is related to the normalized shunt conductance via

$$R_n \simeq G_n/2 \qquad\qquad (8\text{-}53)$$

It is interesting that this approximation provides the exact Salisbury screen result of a unity conductance for the single-layer maximally flat design. Similarly, the exact maximally flat result for the two-layer case is known to be

$$G_1 = \sqrt{2}, \quad G_2 = 1 - 1/\sqrt{2} \qquad\qquad (8\text{-}54)$$

which is very close to the approximate result of $G_1 = 4/3$ and $G_2 = 1/3$, found by using Eq. (8-51). The importance of this analysis is that it can easily be extended to complex admittances, additional layers, and other complications which render an exact analysis impossible. The approximate results can then be used as a starting point for an optimization program.

Figure 8-7 illustrates the exact solution for the two-layer design given in Eq. (8-54). Also shown is the approximate result given in Eq. (8-51). Figure 8-8 illustrates the two Chebyshev designs. Note that the Chebyshev results are similar to the maximally flat designs for this small number of layers and the approximate theory.

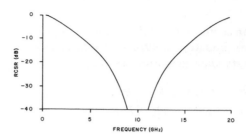

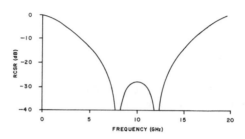

Figure 8-7 Two-layer maximally flat absorber for $l_1 = l_2 = 0.25 \lambda_0$, $f_o = 10$ GHz: (a) "exact" solution: $g_2 = 0.2929$, $g_1 = 1,4141$ (b) approximate solution: $g_2 = 0.333$, $g_1 = 1.333$

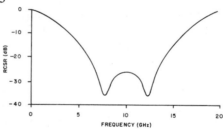

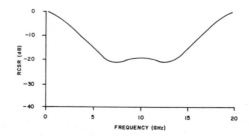

Figure 8-8 Two-layer Chebyshev absorber designs, $l_1 = l_2 = 0.25 \lambda_o$, $f_o = 10$ GHz: (a) $g_2 = 0.3806$, $g_1 = 1.3772$; (b) $g_2 = 0.5756$, $g_1 = 1.507$

8.7 SUMMARY

The intent of this chapter has been to provide the basic analytical tools necessary for RAM design and analysis. The approach has been to describe the reflection, transmission, and absorption of electromagnetic energy by multilayer dielectrics in terms of the material properties of the layers. For flat surfaces, the formulation is simple, and two different but equivalent methods of describing the process were presented. For most cases of specular RAM development, use of a normal incidence design program developed using either the stepping procedure or cascade matrix approach will provide sufficient design information. For geometries in which the RAM will have radii of curvature of the order of a few wavelengths, special purpose programs, such as that illustrated for cylindrical geometries, may be necessary to properly describe RAM behavior.

Because of the complexity of RAM designs which are generally of current interest, particularly within the military community, effective and efficient computer optimization is becoming mandatory in the design phase. For example, at the Air Force Avionics Laboratory, the analyses and computer programs contained in references [6, 7] have been combined with a general optimization program [8], and the combination has been used to optimize RAM design for *CA* and hybrid absorbers. Similar programs are in use by most other groups active in the design of RAM, and acquisition or development of computer performance prediction tools should be seriously considered by any engineer who is involved in absorber applications.

REFERENCES

1.   J.A. Stratton, *Electromagnetic Theory*, New York, McGraw-Hill, 1941, pp. 501-502.
2.   R.E. Collin, *Field Theory of Guided Waves*, New York, McGraw-Hill, 1960, pp. 79-87.
3.   D.M. Kerns, and R.W. Beatty, *Basic Theory of Waveguide Junctions and Introductory Microwave Network Analysis*, New York, Pergamon Press, 1967.
4.   H.E. Bussey, and J.H. Richmond "Scattering by a Lossy Dielectric Cylindrical Multilayer, Numerical Values," *IEEE Trans. Antennas Propag.*, Vol. AP-23, No. 5, September 1975, pp. 723-725.
5.   R.E. Collin, *Foundations for Microwave Engineering*, New York, McGraw-Hill, 1966, pp. 224-237.
6.   R.G. Wickliff, "Improved Analsyis of Circuit Analog Absorber Sheets (U)," AFAL-TR-72-64, Ohio State University, February 1972, (SECRET).

7.   C.H. Krueger, "A Computer Program for Determining the Reflection and Transmission Properties of Multilayer Plane Boundaries," AFAL-TR-67-191, Air Force Avionics Laboratory, September 1967.

8.   D.S. Hague, and C.R. Glatt, "An Introduction to Multivariable Search Techniques for Parameter Optimization," NASA-CR-73200, Boeing Company, April 1968.

CHAPTER 9

RADAR ABSORBERS

M. T. Tuley

9.1 INTRODUCTION

The ideal radar absorber should be thin, light, durable, easily applied, inexpensive, and have broadband frequency coverage. Alternatively, we might wish to have a structural RAM which is mechanically sound and has no size, weight, or cost penalty over standard structural materials. As might be expected, neither of these ideal RAM types has yet been formulated. Nevertheless, since the World War II era, many types of radar absorbers have been developed which are capable of significantly reducing the signature of radar targets. Two excellent classified compilations of what was then the state of the art in RCSR technology are found in references [1, 2], records of the 1975 and 1980 Radar Camouflage Symposia, which were sponsored by the Air Force Avionics Laboratory. An excellent unclassified treatment of RAM is provided in reference [3]. This chapter highlights some of the classes of absorbers available for radar applications, with emphasis on their design parameters and performance.

Radar absorbers may be loosely classified as "resonant" or "broadband." While the bandwidth of a resonant absorber may actually exceed that of some broadband absorbers, resonant absorbers derive their name from the fact that the conditions for reduction are satisfied, in general, only at one or more discrete frequencies. Broadband RAM, on the other hand, in principle provides absorption at all frequencies, but generally becomes ineffective outside a frequency band as a result of changes in material properties with changing frequency [3]. Practically speaking, current RAMs designed to cover a wide range of frequencies are generally made up of combinations of RAM elements. This chapter concentrates on the design of RAM for broadband applications, and it assumes that RAM volume is a constraint. Thus, little attention is given to the design of RAM for anechoic chambers and other similar applications, although those types of absorbers are briefly described as examples of geometric transition RAM.

It has already been pointed out in Ch. 8 that the design of RAM is essentially the design of a lossy matching network between free space and a conducting surface. Providing loss while minimizing reflection is the key idea in RAM application. Thus, a broadband RAM design basically confronts two questions. The first is: How do I get the incident wave to propagate into the material, rather than simply to reflect off the front surface? The second question is: How do I provide the required level of energy absorption, once the wave is interior to the RAM? As with many engineering problems, the two requirements are often conflicting, and compromises in bandwidth, level of performance, and RAM thickness must be made. The remainder of this chapter catalogs types of absorbers which have been designed over the years to provide practical solutions to the problems raised by the above design questions.

The first topics considered in the chapter are the simplest absorber types, the Salisbury screen and Dallenbach layer. Their analysis is followed by a consideration of Jaumann absorbers and graded dielectrics, absorbers composed of multiple Salisbury screens and Dallenbach layers, respectively. Magnetic RAM is then discussed, with an emphasis on the effect of material property variations with frequency on performance. Circuit analog materials are briefly touched on, although an entire chapter could easily be devoted to an analysis of circuit analog and tuned surface treatments. Hybrid materials and radar absorbing structures (RAS) are discussed, followed by a short description of non-specular RAM. Most of the RAM types described in this chapter are designed to act as specular RAM, and only a limited treatment of traveling wave absorber design is given.

9.2 SALISBURY SCREENS AND DALLENBACH LAYERS

Two of the oldest and simplest types of absorbers are represented by Salisbury screens and Dallenbach layers. The Salisbury screen [4] is a resonant absorber created by placing a resistive sheet on a low dielectric constant spacer in front of a metal plate. The Dallenbach layer [3] consists of a homogenous lossy layer backed by a metal plate. Each is analyzed below.

Figure 9-1 illustrates the geometry of the Salisbury screen. In the analysis of its performance, we assume that an infinitesimally thin resistive sheet of conductance G, normalized to free space, is placed a distance d from a metal plate. Typically, a foam or honeycomb spacer might be used, so spacer dielectric constants in the 1.03 to 1.1 range are normal. To simplify this analysis, the normalized permittivity of the spacer is assumed to be that of free space (i.e., $\epsilon_r = 1$).

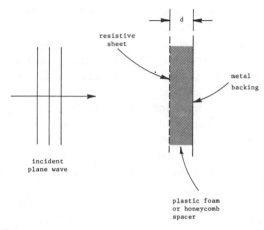

Figure 9-1 Salisbury screen

From Eq. (8-17), Ch. 8, the reflection coefficient for a dielectric multilayer will be zero if B_{N+1} is forced to zero. For the simple case of the Salisbury screen, substituting in Eq. (8-16):

$$B_{N+1} = B_2 = \frac{e^{-ikd}}{2} \left[-G\, e^{-ikd} - (2 - G)\, e^{ikd} \right] \qquad (9\text{-}1)$$

where $Y_1 = Y_2 = 1$ (free space), and $k_2 = k_0 = 2\pi / \lambda$ (the free space wavenumber). The magnitude of $B_2 = 0$ only if the quantity in brackets is zero. This requires that the magnitudes of the two exponentials in the brackets be equal and that their phase angles be opposite. The equal amplitude requirement forces G to equal one, or equivalently, the unnormalized resistance to be 377 ohms per square. In that case, Eq. (9-1) becomes

$$B_2 = -e^{-ikd}\, \frac{e^{ikd} + e^{-ikd}}{2} = -e^{-ikd} \cos \frac{2\pi d}{\lambda} \qquad (9\text{-}2)$$

The condition $B_2 = 0$ requires $\cos (2\pi d / \lambda) = 0$ which implies that

$$2\pi / \lambda = \frac{\pi}{2} + n\pi, \quad n = 0,1,2,\ldots,$$

or

$$d = \frac{\lambda}{4} + \frac{n\lambda}{2}$$

(9-3)

Thus, for zero reflectivity, a Salisbury screen requires a 377 ohm per square resistance sheet set at an odd multiple of an electrical quarter-wavelength in front of a perfectly reflective backing. Higher dielectric constant spacers may be used and still satisfy Eq. (9-2), but with a consequent reduction in band-width, because k for that case will be larger than k_0, and thus a given frequency change will cause a larger change in B_2 than for the $\epsilon_r = 1$ spacer.

Another way to think of the Salisbury screen is in transmission line terms. A quarter-wavelength transmission line transforms the short circuit at the metal plate into an open circuit ($G = 0$) at the resistive sheet. The sum of the sheet and open circuit admittances, which is the value seen by the impinging wave, is just that of the sheet, 1/377 mho, and thus a matched load is provided and no reflection occurs. By the same token, at multiple half-wavelength spacings, a short circuit is again seen and perfect reflection is obtained.

The screen performance for a 0.5in spacing is shown in Fig. 9-2. Note that the reflection coefficient reaches its minimum value at a frequency of 5.9 GHz ($\lambda = 2$in). The best performance is obtained for a resistivity of 377 ohms, but the performance is still a respectable –18 dB for a resistivity 20 percent lower (300 ohms). However, a resistivity of 200 ohms yields barely a –10 dB reflectivity level at the design spacing. The fractional bandwidth for the 377 ohm screen at a –20 dB reflectivity level is about 25 percent.

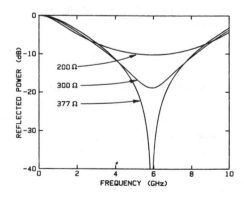

Figure 9-2 Theoretical performance of a Salisbury screen for a stand-off distance of 0.5 in.

To achieve similar performance at a lower frequency, the spacing must be increased because the wavelength becomes longer. That effect is shown in Fig. 9-3, and it will be observed that a pair of nulls now exist, one at three times the frequency of the other. The minimum reflectivity levels are the same as those in Fig. 9-2. The nulls will occur at odd integral multiples of the lowest frequency due to the fact that the design spacing can be any odd multiple of a quarter-wavelength.

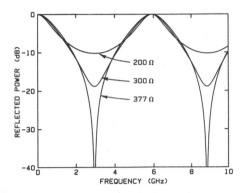

Figure 9-3 Theoretical performance of a Salisbury screen for a stand-off distance of 1 in.

The Salisbury screen has been used in varying degrees in commercial absorbing materials. However, the rapid oscillations for large spacings would render it ineffective over a wide frequency range. For increased mechanical rigidity, plastics, honeycomb, or higher density foams may be used as spacers. To maintain the electrical spacing, the resistive sheet would be mounted over a dielectric layer trimmed to be an electrical quarter-wavelength in thickness. As noted before, the gain in mechanical rigidity and decreased thickness obtained by using a higher dielectric constant spacer are paid for in reduced absorber bandwidth.

The above analysis assumed normal incidence of a plane wave on the absorber. It is interesting to explore the specular performance of the Salisbury screen at off-normal angles. It can be shown [3] that the magnitude of the reflection coefficients for both parallel and perpendicular polarizations are given approximately by

$$| R_\perp | = | R_\parallel | \simeq \frac{1 - \cos\theta}{1 + \cos\theta} \qquad (9\text{-}4)$$

where θ is the angle off-normal. Equation (9-4) is plotted in Figure 9-4. Note that performance is better than 20 dB (i.e., $|R| < 0.1$) for angles up to 35°. A more exact analysis of general RAM performance as a function of incidence angle can be found in [5], but the error in Eq. (9-4) is no worse than 5 dB, and thus it is useful as a rough estimate of Salisbury screen behavior with angle.

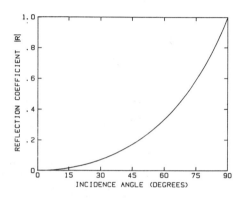

Figure 9-4 Reflection coefficient of a Salisbury screen as a function of the angle of incidence

Another simple resonant absorber, the Dallenbach layer, is constructed of a homogeneous lossy layer backed by a metallic plate. The reflection at the surface of a material is due to the impedance change seen by the wave at the interface between the two media. Thus, if a material can be found whose impedance relative to free space equals one (i.e., $\mu_r = \epsilon_r$), there will be no reflection at the surface. In this case the attenuation will depend on the loss properties of the material (ϵ''_r, μ''_r) and the electrical thickness.

Unfortunately, materials with the appropriate dielectric and magnetic properties to act as a matched RAM over any appreciable frequency range are difficult to find, and so the question becomes one of optimizing the loss at a given frequency using available materials. For a single material layer backed by a conducting plate, the reflection coefficient is given by substituting Eq. (8-7) into (8-8) to provide

$$R = \frac{\sqrt{\mu_r/\epsilon_r}\, \tanh\,(-ikd)-1}{\sqrt{\mu_r/\epsilon_r}\, \tanh\,(-ikd)+1} \tag{9-5}$$

where d is the thickness of the layer. Figures 9-5 and 9-6 from [6] provide curves of reflection as a function of material thickness in wavelengths for several hypothetical materials. The permittivity and permeability are written in polar form per Eq. (8-3).

Several things should be noted from the plots. First, for non-magnetic materials ($\mu_r = 1$), the best RCSR performance occurs when the material is near a quarter-wavelength thick electrically. The solid curve of Fig. 9-5 illustrates this point. However, adding magnetic loss shifts the optimum electrical

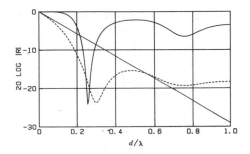

Figure 9-5 Reflectivity curves for dominantly electric materials. Solid trace is for $|\epsilon_r| = 16$, $|\mu_r| = 1$, $\delta_\epsilon = 20°$, and $\delta_\mu = 0°$. Dashed trace is for $|\epsilon_r| = 25$, $|\mu_r| = 16$, $\delta_\epsilon = 30°$, and $\delta_\mu = 20°$. Diagonal trace is for $|\epsilon_r| = |\mu_r| = 4$, and $\delta_\epsilon = \delta_\mu = 15°$

thickness to larger values (although it may actually reduce the physical thickness), with a pure magnetic absorber having an optimum thickness near an electrical half-wavelength, as illustrated in Fig. 9-6. Note on both figures that a hypothetical material with $\mu_r = \epsilon_r$, indicated by the diagonal traces, simply provides a linearly increasing loss in dB with increasing thickness in wavelengths.

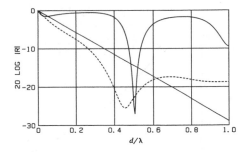

Figure 9-6 Reflectivity curves for dominantly magnetic materials. Solid trace is for $|\mu_r| = 16$, $|\epsilon_r| = 1$, $\delta_\mu = 10°$, and, $\delta_\epsilon = 0°$. Dashed trace is for $|\mu_r| = 25$, $|\epsilon_r| = 16$, $\delta_\mu = 20°$, and $\delta_\epsilon = 30°$. Diagonal trace is for $|\epsilon_r| = |\mu_r| = 4$, and $\delta_\epsilon = \delta_\mu = 15°$

For off-normal incidence, the behavior of a homogeneous layer is similar to that for Salisbury screens [3]. For the case where the index of refraction of the layer is much greater than one, Eq. (9-4) provides a much better approximation to the angular performance for the Dallenbach layer than it does for the Salisbury screen.

An additional question concerns the fractional bandwidth which can be achieved with the Dallenbach type absorber. Ruck [3] presents an analysis giving an approximate bandwidth for an ideal Dallenbach layer, assuming that the fractional bandwidth B is much less than one, in terms of the material properties, the material thickness, and the wavelength at maximum RCSR performance, λ_0, for a given reflection level R as

$$B = 2\left[\frac{f-f_0}{f_0}\right] \simeq \frac{2\,|R|}{\pi\,|\mu_r-\epsilon_r|\,(d/\lambda_0)} \tag{9-6}$$

Figure 9-7 provides plots of bandwidth for 20 dB RCSR *versus* thickness for single layers with only electric or magnetic losses. Note that a material with

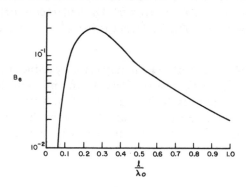

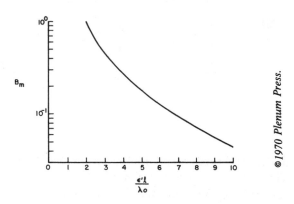

Figure 9-7 Bandwidths of thin homogeneous (a) electric, or (b) magnetic layers as a function of layer thickness (reprinted from Ruck *et al.* [3])

purely dielectric loss has a fractional bandwidth around 20 percent for a material thickness close to $\lambda/4$, which is somewhat less bandwidth than for the Salisbury screen. For magnetic materials, the bandwidth increases as the material becomes thinner. The values plotted are not accurate for small electrical thicknesses because a large bandwidth violates an initial assumption. However, in the limit, the infinitesimally thin magnetic lossy layer is equivalent to a magentic Salisbury screen which, in theory, has infinite bandwidth [3].

Another single layer absorber was recently postulated and analyzed by Gauss at the Ballistic Research Laboratory [7]. It is based on mixing filaments of *radar absorbing chaff* (RAC) in a solid binder of near unity dielectric constant. Attenuation of the incident wave is provided by resistive dissipation in the filaments, which are metallic strands with length to diameter ratios of about 1000, and diameters of about 500 Å. Filament separations in the binder are one-half to one-third the filament length.

An analysis is provided in [7] for two cases: the first is a regular array of filaments in the matrix and the second is for filaments with random orientation. Calculated RCSR for a two cm thick application using a regular array of filaments exceeds 30 dB from 10 GHz through 100 GHz. For the random filament orientation, RCSR is 13 dB at 10 GHz, and greater than 30 dB from 30 GHz through 100 GHz. It should be noted that the RCSR values quoted are theoretical and are not based on measurements of fabricated samples.

9.3 MULTILAYER DIELECTRIC ABSORBERS

As noted in sec. 9.2, it is difficult to achieve the bandwidths desired of microwave radar absorbers (typically 2-18 GHz) using a thin single-layer absorber. Thus, much work has been done in extending the bandwidth of absorbers through the use of multiple layers. The motivation behind this approach is the same as that for pyramidal and other geometric transition absorbers — slowly changing the effective impedance with distance into the material to minimize reflections. Two important types of multilayer absorbers will be discussed, Jaumann absorbers and graded dielectric absorbers.

The bandwidth of a Salisbury screen can be improved by adding additional resistive sheets and spacers to form a Jaumann absorber. To provide maximum performance, the resistivity of the sheets should vary from a high value for the front sheet to a low value for the back. The bandwidth is dependent on the number of sheets used, as illustrated in Fig. 9-8 and Table 9-1. For this illustration, the spacing between sheets was fixed at 0.295in (a quarter-wavelength at 10 GHz) and a quadratic resistance taper was used. The fractional bandwidth for slightly less than 20 dB performance is shown in the table. Note that a four sheet structure has about four times the fractional bandwidth of a single layer, but is four times as thick.

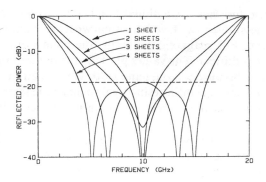

Figure 9-8 Performance of multiple resistive sheets

Table 9-1
Bandwidth of Jaumann Absorbers

Number of Sheets	Fractional Bandwidth	Total Thickness (in)
1	0.27	0.3
2	0.55	0.6
3	0.95	0.9
4	1.16	1.2

Even better performance is available for Jaumann absorbers with more sheets, as illustrated by a six-layer RAM in [8]. A 0.14in spacing between layers with a spacer $\epsilon_r = 1.03$ (probably styrofoam) was used. Table 9-2 provides resistivity values for the lossy sheets. Note the large change in resistivity from front to back provided by the approximate quadratic taper used. An average RCSR of 30 dB was measured for this design over the range of 7 GHz through 15 GHz, with a minimum of 27 dB at 8 GHz. One extremely important point brought out in [8] is the requirement for homogeneous and isotropic lossy layers if high levels of reduction are to be achieved in practice.

As with the Jaumann absorber, where sheet resistance values are tapered to reduce reflection, a graded dielectric can be used to help match the impedance between free space and a perfect conductor. The optimum method for design of such an absorber would be to determine analytically the μ and ϵ required as a function of distance into the material to limit $|R|$ over a given frequency range, subject to incidence angle and thickness constraints. Unfortunately, this general form of the problem has not yet been solved. [3]

Table 9-2
Resistive Sheet Values

Layer	Resistivity (ohms/sq)
Front	9425
2	2513
3	1508
4	943
5	471
Back	236

A more successful and useful approach has been to assume a model for ϵ and μ as a function of the distance z into the absorber, and then to solve for the resulting reflection coefficient. A number of models have been used for the taper including linear, exponential, and one (Jacobs) making the fractional rate of change in ϵ, per wavelength in the material a small constant. Table 9-3, extracted from [3], lists a half-dozen versions of tapers, along with the thickness required at the lowest frequency for 20 dB RCSR. Note that the minimum thickness is on the order of 0.3λ, implying that even in the ideal case, an absorber nearly 2in thick would be required for 20 dB performance down to 2 GHz.

Typically, practical graded dielectric RAMs are constructed of discrete layers, with properties changing from layer to layer. One commercial example is the AN series of graded dielectric absorbers made by Emerson and Cuming. AN-74, a three-layer foam absorber about 3 cm thick, is advertised to provide 20 dB RCSR down to 3.5 GHz. Dipped honeycombs, with successive dippings to lesser depths, have also been used to provide the conductivity gradient required for a graded dielectric absorber. Figure 9-9 provides measured reflectivity data for a commercial three-layer graded dielectric absorber about 1 cm thick.

Several other RAM types exist which are, in effect, graded dielectric absorbers. The first type appears to be a homogeneous single-layer absorber, but, due to its method of production, is actually a graded dielectric. The second type uses a geometric transition to provide an effective dielectric gradient.

A technique for reducing the reflection from the front face of a flat absorber is to produce a material whose intrinsic impedance is very close to unity. Two common examples of absorbers employing such a technique are the hair type and the carbon-loaded low density foam absorbers. However, both types employ a conductivity gradient to some degree.

Table 9-3
Several Graded Dielectric RAM Designs [3]

Type of Variation	$\mu_r'(z)$	$\mu_r''(z)$	$\epsilon_r'(z)$	$\epsilon_r''(z)$	Minimum l/λ for $R \leq 0.1$
Ideal Jacobs	1	0	$(1 - z/l)^{-2}$	Small Constant $\ll 1$	0.3
Finite Jacobs	1	0	$(1 - 0.95 z/l)^{-2}$	$1/2$	0.42
Linear	1	0	1	$3 z/l$	0.55
Exponential	1	0	1	$0.285 \exp(2.73 z/l)$	0.35
Exponential	1	0	$2^{z/l}$	$5^{z/l} - 1$	0.56
Three-layer discrete approximation to exponential	1	0	1	$\begin{cases} 0.58 \text{ for } 0.344l \\ 1.16 \text{ for } 0.359l \\ 3.48 \text{ for } 0.297l \end{cases}$	0.33

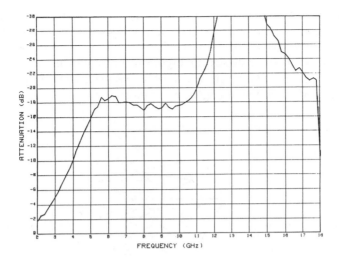

Figure 9-9 Measured reflectivity of a three-layer graded dielectric absorber

The original hair type absorber was developed by NRL in the late 1940s for an anechoic chamber used to cover from 2.5 to 25 GHz [9]. The material is constructed by impregnating mats of curled animal hair with a mixture of conducting carbon black in neoprene. Because the mats are normally laid flat to dry after dipping, gravity tends to provide a dielectric gradient, as more of the conductive mixture ends up toward the back of the mat. Currently available commercial versions of hair type absorbers require approximately a half-wavelength material thickness for 20 dB of RCSR [10]. Because of its poor structural properties and poor RCSR performance compared to pyramidal absorbers and graded dielectrics, hair type absorbers tend to be used less than they once were.

A more recent version of the "hair mat" absorber is a netting absorber produced by the Plessey Corporation. A 1.2 cm thick plastic netting is provided with a conductive coating. Again, there is a variation in the amount of conductive material from front to rear, providing a dielectric gradient. Advertised RCSR performance of the netting is better than 10 dB from 6 GHz through 100 GHz, with better than 15 dB performance over an 8 to 14 GHz band [11].

Another class of single layer absorber (and the one most commonly used in anechoic chambers) depends upon carbon-loaded foam to provide loss, but also uses a geometric transition from free space to the highly lossy medium to provide a dielectric gradient and thereby reduce reflections. The most common form used in anechoic chambers is the pyramidal absorber illustrated in

9.4 MAGNETIC RAM

Magnetic materials generally depend principally on magnetic rather than dielectric losses, although there is always dielectric loss as well. Compounds of iron are most often exploited for these losses, and ferrites and carbonyl iron, a form of pure iron powder, are common ingredients. Ceramic materials employing ferrites are useful for high temperature applications, provided the ambient temperature remains below the Curie point. When the temperature approaches or exceeds the Curie temperature (usually from 500° to 1000°F), the magnetic properties deteriorate, although the physical properties are usually retained to much higher temperatures. These ferrite materials are typically sintered in the form of small, rigid tiles, and application to a surface requires careful consideration of bonding techniques.

Another method of manufacturing involves embedding the magnetic materials in a flexible matrix of natural or synthetic rubber, which can then be glued to the surface to be shielded. Again the bonding method requires attention. Several firms have developed spray-on materials in which the magnetic "dust" is suspended in an epoxy vehicle. Because the solid particles are heavy, they tend to settle at the bottom of the container used for spraying, and constant agitation is required. The lossy coating is built up to the desired thickness by the deposition of several thin layers. Uniform thickness, and therefore uniform properties, are difficult to achieve unless skilled operators are available or can be trained for the task. The material can also be brushed on. Such materials are often used to reduce traveling and creeping waves, thus providing non-specular RCSR.

The spray-on RAM, also referred to as "iron paint," has the advantage that irregular surfaces can be covered more easily than with the flexible sheets, although singly curved surfaces (cylinders, cones, *et cetera*) are amenable to the use of the sheets. For both forms of material, adequate surface preparation is required or else the absorbing layer may peel off. Because these materials all contain iron in one form or an other, they tend to streak rust in an oxidizing environment. At high temperatures, oxidation of carbonyl iron to non-magnetic α Fe_2O_3 is a severe problem because the oxidation causes a permanent loss of absorbing properties.

Although magnetic RAM tends to be heavy, its virtue lies in the ability to provide extended low frequency performance with reasonable material thicknesses. Where it would require an ordinary dielectric absorber many inches thick to achieve low-frequency coverage down to 100 MHz, the magnetic materials can often be much less than a tenth as thick in order to achieve comparable performance. The reason for this is that the magnetic losses can be tailored for low frequencies, as sketched diagramatically in Fig. 9-12.

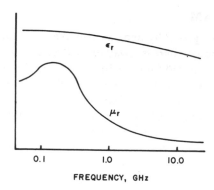

FREQUENCY, GHz

Figure 9-12 Schematic illustration of the frequency behavior of ferrites

Because the losses tend to increase for the lower frequencies via increasing μ_r, and the electrical thickness of the material tends to "keep in step" with the frequency, performance persists at lower frequencies. At the higher frequencies, the magnetic properties no longer contribute much to the performance, and the dielectric properties (ϵ_r) now account for whatever loss that occurs. Table 9-4 graphically illustrates the above traits for a nickel-zinc ferrite for which the electrical properties, index of refraction, and front-face reflection coefficient are listed. Note that at 100 MHz the ferrite's electrical thickness will be more than 50 times its physical thickness and, because $\mu_r \approx \epsilon_r$, the front-face reflection will be more than 30 dB below the incident level. Thus, most of the return will be that portion of the wave not attenuated during its two-way path through the ferrite. In contrast, at 10 GHz, the ferrite's electrical thickness is only 2.3 times its physical thickness, and there is a high front-face reflection which can only be cancelled by use of resonant techniques.

Besides the essentially "single-layer" magnetic materials, multiple-layer magnetic materials can also be used. These are designed to take advantage of the fact that different magnetic materials will have permeability curves which peak at different frequencies. The remainder of this section considers single-layer magnetic materials, and then briefly surveys some recent efforts in multilayer magnetic RAMs.

The analysis of an electric Salisbury screen showed that a resistive sheet should be placed at the maximum of the electric field, offset $\lambda/4$ from the surface. By analogy, a magnetic Salisbury screen would require a magnetic lossy layer at the peak of the magnetic field, which is immediately on the metal sheet [14]. With the assumption that $\mu_r'' \gg \mu_r'$ and ϵ_r, reference [3] provides an analysis for which $R = 0$ when

$$\omega \mu'' d = Z_0$$

(9-7)

Table 9.4
Electrical Properties
Sintered Nickel Zinc Ferrite

	Frequency (GHz)				
	0.1	0.5	1.0	3.0	10.0
ϵ_r'	27	24	20	18	15
ϵ_r''	54	24	9.0	6.3	6.3
μ_r'	15	9	1.2	0.9	0.1
μ_r''	45	45	12	6.3	.32
$\|\sqrt{\mu_r \epsilon_r}\|$	53.5	39.5	16.3	11.0	2.3
$\|R\|$	0.03	0.17	0.31	0.39	0.78
$\|R\|$ (dB)	−30.5	−15.4	−10.2	−8.2	−2.1

Note: μ_r and ϵ_r values were provided by William R. Cuming.

Note that meeting the requirements of Eq. (9-7) for practical peak values of μ_r'' encountered in magnetic materials would permit a very thin absorber to be used at microwave frequencies. However, finding materials which meet the conditions on μ'', μ' and ϵ required of a magnetic Salisbury screen is extremely difficult, although some of the ferrites have properties resembling those required over small frequency ranges.

The difficulty in finding materials with properties suitable for magnetic Salisbury screens means that most single-layer magnetic absorbers are not designed in accordance with Eq. (9-7). However, an attempt is often made to adjust material properties so that the intrinsic impedance is as near that of free space as can be obtained, over as wide a frequency range as possible. For iron or ferrite powder loaded into a dielectric matrix such as rubber or neoprene, this can be done by control of the percentage loading of the magnetic material. An additional technique useful with magnetic materials is to form a magnetic Dallenbach layer. The guidelines of [6] can be used in that case to determine optimum thickness as a function of material properties.

Magnetic absorbers, particularly those made of sintered ferrites, have the advantage of being useful at high temperature. However, while their physical properties may hold to above 1000°C, Curie temperatures, above which the magnetic properties essentially disappear, are typically much lower. For example, cubic spinel ferrites have Curie temperatures below 600°C, and hexagonal ferrites below 500°C. Ferromagnetic materials can be constructed with Curie temperatures in the 500° to 1000°C range, but they present serious problems of chemical stability [15].

Figure 9-13 from reference [16] provides plots of RCSR for four commercially available sintered ferrite absorbers. Note that they are relatively heavy
(about 7 lbs per sq ft), but they provide significant low frequency performance
in a treatment one-quarter to one-third of an inch thick.

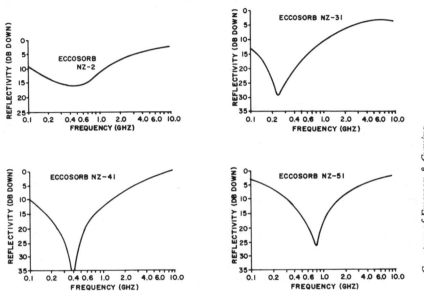

Courtesy of Emerson & Cuming

Figure 9-13 Typical sintered ferrite absorber properties (reprinted from reference [16])

As in the case of dielectric materials, magnetic materials may be layered to
take advantage of changes in properties between layers to enhance performance. Several methods can be used. For materials such as hexagonal ferrites in
which the frequency at which μ_r' peaks can be controlled by doping, matched
layers over different frequency ranges can be produced. Alternatively, different magnetic materials can be layered to optimize use of the properties of each.

The natural ferrimagnetic resonant frequency of a hexogonal ferrite can be
controlled by replacement of a portion of the Fe^{+3} ions in the lattice by divalent
and tetravalent metal ions such as Co^{+2} and Ti^{+4}. Thus, magnetic materials can
be constructed with properties similar to those of Fig. 9-14, with different
layers exhibiting different magnetic resonance frequencies [17]. Figure 9-15
illustrates one design of an optimized four-layer absorber, which required

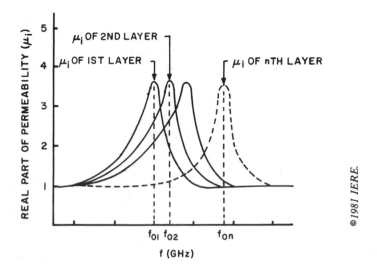

Figure 9-14 Idealized magnetic properties of the materials at different layers (reprinted from Amin and James [17])

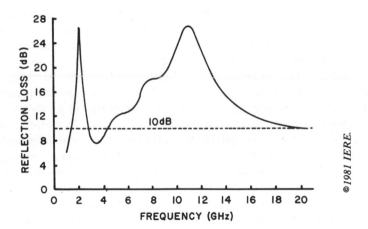

Figure 9-15 Reflection loss characteristics of an optimized absorber comprised of four layers of hexagonal ferrites specified in Table 9-5 (reprinted from Amin and James [17])

minimum reflectivity over a 1 to 15 GHz range, subject to a maximum absorber thickness of 7.5 mm. Table 9-5 provides the relative permeabilities at the resonant frequencies, where layer 1 is the outer layer and layer 4 is against a reflecting plate. Note that, on the average, better than 10 dB of RCSR is predicted from below 2 GHz up to 20 GHz, and this is obtained from a RAM less than 0.3in thick.

Table 9-5
Ferrite Properties for a Four-layer Optimized Absorber [17]

Layer Number	Resonant Frequency (GHz)	μ_r (peak)	Layer Thickness (mm)
1	10.35	2.21	0.85
2	7.56	2.34	1.43
3	5.23	2.76	2.22
4	3.50	4.00	3.00

9.5 CIRCUIT ANALOG ABSORBERS

As pointed out previously, the design of specular RAM is equivalent to a transmission line matching problem, where the goal is to limit the reflection seen at the input caused by a short-circuit termination. The Salisbury screen and Jaumann absorbers use resistive sheets, which have only a real part to their admittance, as the matching elements. Significant flexibility can be gained in the design process if the sheets can have a susceptance as well as a conductance. This imaginary part of the admittance can be obtained by replacing the continuous resistive sheet with one whose conducting material has been deposited in appropriate geometrical patterns (e.g., dipoles, crosses, triangles, *et cetera*), such as those shown in Fig. 9-16. The term "circuit analog" for such absorbers is derived from the fact that the geometrical patterns are often defined in terms of their effective resistance, capacitance, and inductance, and then equivalent circuit techniques are used in the subsequent analysis and design of the resulting absorber.

A design problem closely related to that of circuit analog sheets is that of band-stop or band-pass surfaces. However, in contrast to CA RAM, such *frequency selective surfaces* (FSS) do not absorb RF energy. Rather, an FSS is a frequency filter which might be employed, for example, as a band-pass radome in front of a radar antenna, or as a diplexer for a dual-frequency antenna. Figure 9-17 illustrates typical geometries used for band-pass applications. Duals of those geometries are often used as band-stop filters for antenna feed diplexers, *et cetera*. In FSS applications, a highly conductive pattern is

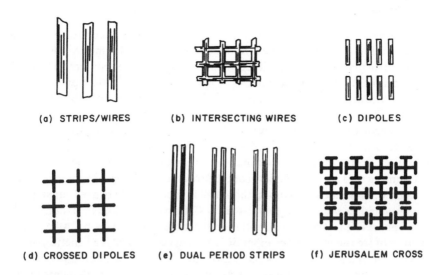

(a) STRIPS/WIRES (b) INTERSECTING WIRES (c) DIPOLES

(d) CROSSED DIPOLES (e) DUAL PERIOD STRIPS (f) JERUSALEM CROSS

Figure 9-16 Typical circuit analog element geometries

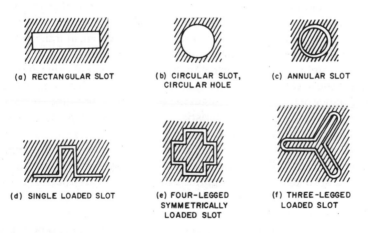

(a) RECTANGULAR SLOT (b) CIRCULAR SLOT, CIRCULAR HOLE (c) ANNULAR SLOT

(d) SINGLE LOADED SLOT (e) FOUR-LEGGED SYMMETRICALLY LOADED SLOT (f) THREE-LEGGED LOADED SLOT

Figure 9-17 Typical frequency selective surface element geometries

used because no absorption is desired. Thus, the impedance of the sheet is purely imaginary, and the design relies on changes in reactance with frequency to provide appropriate band-pass or band-stop characteristics.

Because so much more can be done to tailor the admittance properties of a circuit analog design than with Salisbury screens or Jaumann absorbers, better performance can be achieved within the same space constraints. However, optimization of the variables controlling the admittance properties is also more complicated. Thus, current CA design practice typically relies on rather sophisticated, and usually time consuming, computer programs, often with internal optimization routines. Nevertheless, an understanding of the design techniques used does not depend on the details of the computer implementation, and it is in that area that the following discussion focuses.

Assuming that a minimum level of performance over some frequency range is required, and that a maximum thickness is specified, there are four steps to be completed in the design process for a CA absorber. The first step of the design process is to arrive at admittance characteristics for each CA layer as a function of frequency. The number of CA sheets to be used is a function of the RCSR required and of the desired bandwidth, as with the Jaumann designs. As a rule of thumb, a broadband CA design can typically be implemented with one or two fewer sheets than would be possible using resistive sheets in a Jaumann absorber.

The second step of the design process is to find realizable geometry and conductance combinations which match as closely as possible the desired admittance characteristics for each sheet. Typically, the geometry to be used (e.g., dipoles, Jerusalem crosses, *et cetera*) will be specified, and the geometry variations will involve the size and spacing of the elements.

Because it is unlikely that, upon detailed analysis, any realizable design will exactly meet the desired admittance characteristics, the third step is to calculate performance of the design based on the achievable admittance properties. In this step, performance characteristics as a function of polarization and incidence angle may also be calculated.

The final step is to iterate the design until an achievable and acceptable combination is found. Given specifications and appropriate constraints, a computer program or series of programs will often be used to perform all four steps.

Nevertheless, because most optimization programs work much more efficiently if they are given a "nearly correct" answer as a starting point, it is very useful to be able to make approximate performance calculations. For an initial estimate of the optimum admittance characteristics, an extension of the approximate reflection analysis of sec. 8.6 to include complex admittance is helpful. Those initial admittance parameters can serve as input to an optimization program utilizing either the wave matrix or stepping procedure approach to calculate exact admittance parameters. The remainder of this section discusses equivalent circuit and integral equation techniques which can be used to translate admittance characteristics to CA geometries. Much of the following material has been drawn from notes prepared by Dr. J. P. Montgomery of Georgia Tech.

Once the desired admittances have been derived, either analytically or through numerical techniques, those values must be translated into a geometric design. Several methods may be used to perform this translation. The earliest technique, and one still useful in design, is to draw on equivalent circuit analyses originally intended for waveguide filter design. Marcuvitz[18] provides the equivalent circuits for many configurations which can be used for CA or FSS design. More complex geometries can often be modeled with combinations of circuit components for which circuits are known. An example is the analysis of the "Jerusalem cross" capacitively loaded dipole geometry [19, 20].

The alternative to a closed form solution using equivalent circuits is to seek a numerical solution. Because the equivalent circuit analyses often provide only a first-order solution, ignoring higher-order modes, numerical techniques are often used to refine equivalent circuit results for the final circuit realization. Below, a simple analysis of a round-wire geometry is used to illustrates equivalent circuit techniques. Then, numerical techniques currently used in CA and FSS design are discussed and evaluated.

Typically, CA and FSS designs consist of elements, such as dipoles or slots, which are on the order of a half-wavelength long. A periodic structure is used to ensure a uniform surface. In analysis of circuit analog sheets, the Floquet theorem [21] allows one to confine attention to a single cell in the periodic structure. In essence, it is assumed that the structure is infinite and illuminated by an infinite plane wave. This being the case, the result for a single cell applies to all cells in the surface. In practice, the structure is finite, and some differences must be expected because the cell-to-cell coupling for cells near the center of the array is different from that of those near its edges.

For circuit analog RAM which is planar and thin along the normal to the surface, as with resistive sheets, the tangential electric field will be continuous across the sheet, and the magnetic field will exhibit a discontinuity directly related to the current on the surface. It follows that any equivalent circuit chosen to represent the CA sheet must have these same properties. The analytical procedures in Ch. 8 recognized the transmission line analogy to RAM design, and Fig. 8-4 in that chapter pointed out that the equivalent circuit of a thin sheet can be modeled as an admittance shunted across the transmission line. The exact nature of the admittance depends on the geometry of the periodic surface. A common example is a dipole array, for which the simplest equivalent circuit is a series RLC network shunted across the line. However, that equivalent circuit is only valid near the first resonance. In general, the broadband equivalent circuit will take the form of multiple, parallel, shunt RLC circuits, each with different loss, resonant frequency, and bandwidth characteristics.

As a simple example of equivalent circuit techniques consider the periodic array of infinitely long round wires shown in Fig. 9-18, which will appear as an inductive or capacitive admittance, depending on whether the polarization is parallel or perpendicular (respectively) to the wires. For the inductive case,

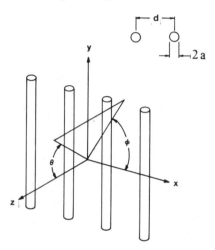

Figure 9-18 Wire geometry

Wait [22] gives the following shunt impedance for a wire spacing, d, and wire radius, a, as

$$Z = [dZ_{int} - (iZ_m\, d/\lambda_m)\, \ln{(d/2\pi a)}] \cos\theta \qquad (9\text{-}8)$$

where Z_m and λ_m are the impedance and wavelength for the medium surrounding the grid, θ is the incidence angle, and Z_{int} is the internal impedance of the wire. Ramo, Whinnery, and Van Duzer [23] give the general solution for the internal impedance of a round wire. Depending on the order of the ratio of the radius to the skin depth, δ, two important limiting forms of the internal impedance result:

$$Z_{int} = \begin{cases} \dfrac{1}{\pi a^2\, \sigma} - i\, \dfrac{\omega\mu}{8\pi}\;,\; a/\delta \ll 1 \\[2ex] \dfrac{R_s}{2\pi a}\,(1-i)\;,\; a/\delta \gg 1 \end{cases} \qquad (9\text{-}9)$$

where σ is the conductivity. The skin depth is given by

$$\delta = (\pi f\mu\sigma)^{-1/2} \qquad (9\text{-}10)$$

and the planar surface resistivity is

$$R_s = 1/(\sigma\delta) \tag{9-11}$$

Thus, when the wire radius is small compared to the skin depth, the current is essentially uniform within the wire, and the dc resistance and quasi-static filament inductance apply. When the wire radius is large compared to the skin depth, the current is concentrated at the surface, and the internal impedance is the planar surface resistivity divided by the wire circumference. Note in the low frequency limit that the resistance is not a function of the permeability and the reactance is not a function of the conductivity. In the high frequency limit, the resistance and reactance are equal, and both are functions of the ratio of permeability to conductivity through the skin depth and planar surface resistivity relations.

The impedance associated with a wire grid is that of a series RL circuit, and the wire radius, spacing, and conductivity can be adjusted to provide specific impedance behavior as a function of frequency. As an example of the use of Eq. (9-8), consider the simple case of the design of a Salisbury screen with lossy inductive wires. In this case, the spacing between the ground plane and the grid must be greater than a quarter-wavelength because of the inductance, while a capacitive grid would allow a spacing less than one-quarter of a wavelength. Figure 9-19 illustrates the return loss in dB between 5 and 15 GHz for three cases: (1) a perfect Salisbury screen (i.e., the real part of the shunt admittance exactly equals unity, and the imaginary part is equal to zero); (2) a good approximation of a Salisbury screen obtained by using closely spaced thin wires, so that the shunt reactive component is small; and (3) a poor approximation of a Salisbury screen, obtained with larger wires spaced further apart. The good approximation is indistinguishable from the perfect Salisbury screen, but, because of the large reactive component, the poor approximation has considerably less bandwidth.

The above discussion has been confined to the case of the shunt circuit immersed in a homogeneous medium. When the circuit is on the surface of a dielectric or magnetic medium, the intrinsic susceptance or reactance of the circuit changes. When the interface is with a non-magnetic dielectric, the inductive terms remain unchanged, while the capacitive terms are modified by the parallel combination of two admittances, each being half the susceptance of the circuit immersed in the appropriate homogeneous medium (see Fig. 9-20). A similar modification would apply for purely magnetic materials, except that the inductive reactance would be modified, while the capacitive terms would not change. When the medium interface has both dielectric and magnetic properties, both capacitive and inductive terms are modified as discussed.

It is important to note that the expressions given for the wire grid circuit admittances are only the *zero*th order approximations. These approximations

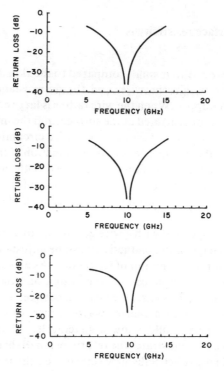

Figure 9-19 Inductive grid absorbers above a ground plane: (a) Salisbury screen; (b) a = 1 μ, d = 0.0137 in., σ = 3x10^5 $\mho$/m, l/λ_0 = 0.2579; (c) a = 4μ, d = 0.1172 in., σ = 3x10^5 $\mho$/m, l/λ_0 = 0.3707

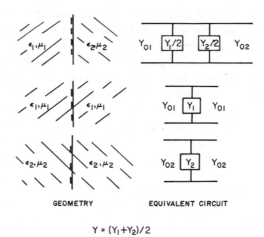

$$Y = (Y_1+Y_2)/2$$

Figure 9-20 Equivalent circuits for a circuit analog sheet inserted between various materials

are not a function of angle of incidence except for the characteristic admittance normalization given by the cosθ term. Often the variation of the normalized admittance with angle of incidence will degrade the performance to an unacceptable level. This can be compensated for by the use of material with a large propagation constant compared to that of free space. The near-normal propagation in such material ensures that the admittance will vary minimally with angle, and hence the performance will remain uniform. Several authors have used this compensation technique [24].

Numerical analysis of periodic planar surfaces has been the subject of many research papers in the 1970s and 1980s [24-38]. The majority of the techniques explored are based on an application of the method of moments to solve a vector integral equation using a finite matrix approximation, where either the transverse electric field or the surface current in the periodic cell is the unknown, depending on the geometry. If the periodic cell consists of apertures, the electric field is a natural unknown, because a convenient modal series may be available if the aperture is rectangular, circular, or some other elementary shape. If the periodic cell consists of conductors of simple shape (such as dipoles or intersecting wires), the current on these wires is a convenient unknown because of similar logic. Some authors have used a universal expansion such as the Floquet series, but have generally met with limited success because of the large number of unknowns. Recently, Tsao and Mittra [28] have introduced the spectral iteration technique which employs a Floquet expansion, but which avoids matrix inversion and its memory requirements. It is important to note the similarity of the above analysis techniques to phased array analysis because the conversion of many phased array analysis computer codes for use in circuit analog and FSS design is quite simple [39]. Alternate techniques include the analysis of finite periodic structures for dipole arrays [29]. Additionally, several computer codes are available for analysis of thin-wire structures [40].

The analysis of a single-layer circuit excited by a single plane wave is generally sufficient for design work when using scattering matrix techniques. Some authors have examined multilayer circuits by solving for the circuit parameters simultaneously [38, 41]. However, the single-layer technique can be extended to heavily coupled circuits by using the generalized scattering matrix which includes higher-order modes [42]. This, of course, requires the use of multiport scattering matrices. When excess coupling is suspected, the analysis of a two-layer circuit can be implemented by adding a magnetic and electric ground plane to the geometry. The admittance of the circuit with and without the adjacent circuit can then be examined.

The method of moments is quick and accurate for most shunt circuits. However, there are practical difficulties. The geometry must generally have some elementary shape for a modal solution. For example, for a thin linear dipole, a simple Fourier series whose terms are zero at the ends of the dipole will be complete. For a flat dipole, the dual of the rectangular waveguide modes can be used [25, 35]. Similar dual expansions can be used for circular elements [27]. Often, approximate modal expansions can be used if the wire is thin. As an example, consider a bent dipole geometry, for which an approximate modal expansion is a Fourier series in the local coordinate of the dipole. This technique has been used extensively [24, 37] to analyze the three- and four-legged slot geometry used by Pelton for metallic radomes [43]. More complex geometries can be modeled using a pulse basis expansion. However, the expansion must be complete to ensure an accurate solution. If the expansion does not include critical current terms, an inaccurate solution may result.

One technique which overcomes some of these difficulties is spectral iteration, recently introduced by Tsao [28]. This technique avoids matrix inversion by seeking an iterative solution for the tangential electric field and the surface current by using transform relationships. Spectral iteration begins with a Floquet mode expansion. The key to the procedure is the recognition that the Floquet summation represents a discrete Fourier transform (DFT), and that, in the transform domain, the convolution represented by the integral containing the unknown current becomes an algebraic product. The unknown current is then found by iteration, bypassing the memory and matrix inversion problems of MOM. Because the expansion of both the current and the electric field is in the Floquet series, the spectral iteration technique is complete and can provide accurate solutions for complex geometries. However, for practical CA and FSS structures, convergence of the spectral iteration is often slow, and much remains to be done to understand its limitations and define the conditions under which it is useful.

9.6 HYBRID RAM AND RADAR ABSORBING STRUCTURES (RAS)

Most of the absorbers discussed have been designed to be added to existing structures, and the assumption is that the RAM has no significant structural properties of its own. Much recent effort has been expended in developing RAM with load bearing properties. Concurrently, work has been done in combining RAM types (e.g., magnetic and CA, graded dielectric and CA, graded dielectric and magnetic) to provide broader bandwidths within thinner packages. While not all of the efforts have been correlated, so much of the ongoing RAS work involves hybrid techniques that the two subjects might logically be grouped together.

A continuing thrust in RAM research has been for a thin, broadband absorber. It was recognized early that a RAM employing purely dielectric loss was constrained to be physically thick for significant low frequency performance. Magnetic RAM, on the other hand, can provide significant loss at low frequencies with layers only 0.04 or 0.05in thick, and it operates best when adjacent to a conducting structure. A hybrid RAM might make use of CA sheets and honeycomb spacers in front of a thin layer of magnetic absorber. The magnetic material provides a loss mechanism at frequencies below those where the CA absorber is effective.

One advantage of CA designs is that they can also be built in honeycomb to provide the structural properties necessary for aircraft applications. Addition of magnetic material to improve low frequency performance is a mixed blessing because, along with the improvements in low frequency RCSR, come limitations or restrictions on structural integrity, temperature capability, weight, and cost [44].

9.7 NON-SPECULAR RAM

The previous discussions in this chapter have concentrated on the reduction of specular, or mirror-like radar return. Non-specular radar returns arise from edges, surface discontinuities, and changes in the surface curvature of a body. The sources of non-specular return include edge diffraction, traveling waves, and joints. Non-specular returns are second-order effects for bodies with strong specular returns, but when shaping and RAM reduce the specular return, the non-specular return may dominate.

Reference [45] provides an analytical method for predicting surface wave absorber performance, and for guiding material selection and configuration. The method is based on the dielectric slab guided wave analysis of Collin [46], where the surface impedance is the key variable in relating the performance to the material configuration.

The conclusions of the analysis are couched in terms of a normalized surface impedance which has the same form as the normalized impedance η in the previous chapter:

$$Z_s \simeq \sqrt{\mu_r/\epsilon_r} \tanh\left(-ik_o\, d\, \sqrt{\mu_r\,\epsilon_r}\right) = R_s - i\, X_s \tag{9-12}$$

where d is the thickness of the surface wave absorber. The conclusions drawn are that a good surface wave absorber should have sufficiently large values of X_s and R_s to achieve good attenuation, where for small values of Z_s the loss in dB per wavelength of absorber is approximately

$$L \simeq 54.6\, X_s\, R_s \tag{9-13}$$

Also, for good performance $X_s \simeq R_s$ to assure that the guided wavelength, λ_s, does not differ markedly from the free space wavelength λ_0. The condition for $\lambda_s = \lambda_0$ is given by

$$X_s = R_s \,/\sqrt{1 + R_s^2} \tag{9-14}$$

The surface reactance should be inductive (i.e., the imaginary term in the impedance should be negative) to assure that attenuation occurs, which requires an electrical layer thickness of less than one-quarter of a wavelength. A capacitive surface reactance will tend to shed the wave from the surface, rather than guiding and attenuating it.

9.8 SUMMARY

Satisfactory broadband RAM performance is predicated on getting the RF energy into the RAM and then providing sufficient loss to absorb the necessary energy within the allowed RAM thickness. These two requirements often conflict, because high loss materials often have intrinsic impedances much different from that of free space, and thus suffer high front-face reflections.

There are two basic solutions to this dilemma, but each has it drawbacks. The first solution is to taper the loss from the front to the back of the absorber; this method is employed in Jaumann, graded dielectric, and geometric transition absorbers. However, the increased performance obtained by tapering the admittance is accompanied by increased thickness. Typically, such absorbers will be near a half-wavelength in thickness at the lower end of the frequency band over which they provide a 20 dB or greater level of RCSR. Some reduction in thickness can be obtained through use of circuit analog sheets to replace resistive sheets, or through hybrid (e.g., CA and graded dielectric) techniques. Again, a penalty must be paid, this time in terms of design complexity and cost.

The second solution is to employ materials with both a high loss and an instrinsic impedance near that of free space, which implies a material with a high value for both μ and ϵ. Many practical magnetic materials come close to meeting those requirements, but only over a very limited frequency range because of the highly resonant nature of the permeability. Again, multilayer techniques utilizing different magnetic materials can be used to extend the bandwidth at the cost of RAM thickness and complexity.

When very wide bandwidths are required, hybrid techniques which take advantage of the low frequency performance of magnetic materials and the high frequency performance of circuit analog or graded dielectric are attractive. However, while such hybrids can provide significant performance in a reasonable thickness, problems in bonding, complexity in production and maintenance, and high cost are typical of such designs.

Integration of RAM into the structure of a vehicle is an obvious advantage from a design standpoint. Thus, RAS is a fruitful area of current research. For aerospace applications in which honeycomb is normally utilized, substitution of a circuit analog, Jaumann or graded dielectric absorber may be relatively straightforward. Problems arise, however, in designing a front face for the honeycomb which meets mechanical requirements with minimum reflection of the incoming RF energy. For primary structures which require significant strength, various laminates using internal resistive or CA sheets to provide loss have been studied. In general, such designs surrender RCSR performance level and bandwidth for added strength.

For applications in which a specific threat sector can be identified, every effort will normally be made to shape a vehicle to minimize the specular RCS.

In such cases, attention must be paid to the return caused by traveling waves and edge diffraction. The design of absorbers to suppress these effects is much different from that for reduction of specular reflection. Analyses are available which discuss the suppression of edge diffraction [47], and an analysis was presented which deals with traveling wave absorption.

This chapter has cataloged typical types of RAM and discussed their performance characteristics. A wide range of commercial material exists which can be obtained for RCSR application. Nevertheless, whether commercially available RAM is purchased or an effort is made to tailor a new design to the specific problem, RAM application will usually involve a trade-off between performance, cost, complexity, and ease of manufacture and maintenance. Only through knowledge of RAM performance, familiarity with the types of RAM available, and good engineering judgment will the best RCSR solution to a given problem be chosen.

REFERENCES

1. *Proceedings of the 1975 Radar Camouflage Symposium*, AFAL-TR-75-100, December 1975.
2. *1980 Radar Camouflage Symposium (U)*, AFWAL Avionics Laboratory and Martin Marietta Aerospace, Orlando, Florida, March 1981.
3. G.T. Ruck, editor, *Radar Cross-Section Handbook*, Vol. II, Ch. 8, New York, Plenum Press, 1970.
4. W.W. Salisbury, "Absorbent Body for Electromagnetic Waves," U.S. Patent No. 2, 599, 944, June 10, 1952.
5. R. Redheffer, "The Dependence of Reflection on Incidence Angle," *IRE Trans. Microwave Theory Tech.*, Vol. MTT-7, October 1959, pp. 423-429.
6. E.F. Knott, "The Thickness Criterion for Single-Layer Radar Absorbents," *IEEE Trans. Antennas Propag.*, Vol. AP-27, September 1979, pp. 698-701.
7. A. Gauss, "A New Type of EM Absorbing Coating," Ballistic Research Lab, June 1982, AD A117472.
8. T.M. Connolly and E.J. Luoma, "Microwave Absorbers," U.S. Patent No. 4, 038, 660, July 26, 1977.
9. A.J. Simmons and W.H. Emerson, "Anechoic Chambers for Microwaves," *Tele-Tech and Electronic Industries*, July 1953, pp. 47-49.
10. "Eccosorb H, Hair Type, Broadband Microwave Absorber," Technical Bulletin 8-21, Emerson and Cuming, January 1973.
11. "Advance Information, External Netting Absorber (ENA)," Plessey Materials, Plessey UK Limited, April 19, 1983.
12. W.H. Emerson, "Electromagnetic Wave Absorbers, Useful Tools for Engineers," Emerson and Cuming (n.d.).

13. Eccosorb VHP Very High Performance Absorber, "Technical Bulletin 8-23, Emerson and Cuming, July 1979.

14. H. Severin, "Non-Reflecting Absorbers for Microwave Radiation," *IRE Trans. Antennas Propag.*, Vol. AP-4, pp. 385-396, July 1956.

15. J.M. McGrath and H. Kirtchik, "Development and Performance of High Temperature Magnetic Radar Absorbing Material (RAM) (U)," *1980 Radar Camouflage Symposium (U)*, March 1981, pp. 79-93.

16. "Eccosorb NZ Thin Ferrite Absorber for 50 MHz to 15 GHz," Technical Bulletin 8-2-17, Emerson and Cuming, May 1979.

17. M.B. Amin, and J.R. James, "Techniques for Utilization of Hexagonal Ferrites in Radar Absorbers, Part 1 Broadband Planar Coatings," *The Radio and Electronic Engineer*, Vol. 51, No. 5, May 1981, pp. 209-218.

18. N. Marcuvitz, *Waveguide Handbook*, New York, Dover Publications, 1965.

19. I. Anderson, "On the Theory of Self-Resonant Grids," *Bell Syst. Tech. J.*, Vol. 54, Dec. 1975, pp. 1725-1731.

20. R.J. Langley, and A.J. Drinkwater, "Improved Empirical Model for the Jerusalem Cross," *IEEE Proc.*, Vol. 129, Pt. H., Feb. 1982, pp. 1-6.

21. E. Meyer, and H. Severin, "Absorption Devices for Electromagnetic Waves and Their Acoustic Analogs," *Z. Angew. Phys.*, Vol. 8, March 1956, pp. 105-114.

22. J.R. Wait, "Reflection at Arbitrary Incidence from a Parallel Wire Grid," *Applied Scientific Research*, Vol. 4, sec. B, 1955, pp. 393-400.

23. S. Ramo, J.R. Whinnery, and T. Van Duzer, *Fields and Waves in Communication Electronics*, New York, John Wiley and Sons, 1965, pp. 291-298.

24. R.J. Luebbers, "Analyses of Various Periodic Slot Array Geometries Using Modal Matching," Ohio State University, AFAL-TR-75-119, Feb. 1976.

25. J.P. Montgomery, "Scattering by an Infinite Array of Thin Conductors on a Dielectric Sheet," *IEEE Trans. Antennas Propag.*, Vol. AP-23, Jan. 1975, pp. 70-75.

26. C.C. Chen, "Transmission Through a Conducting Screen Perforated with Apertures," *IEEE Trans. Microwave Theory Tech.*, Vol. MTT-18, Jan. 1975, pp. 627-632.

27. J.P. Montgomery, "Scattering by an Infinite Periodic Array of Microstrip Elements," *IEEE Trans. Antennas Propag.*, Vol. AP-26, Nov. 1978, pp. 850-854.

28. C.H. Tsao, and R. Mittra, "A Spectral-Iteration Approach for Analyzing Scattering from Frequency Selective Surfaces," *IEEE Trans. Antennas Propag.*, Vol. Ap-30, March 1982, pp. 303-308.

29. E.L. Pelton, and B.A. Munk, "Scattering from Periodic Arrays of Crossed Dipoles," *IEEE Trans. Antennas Propag.*, Vol. AP-27, May 1979, pp. 323-330.

30. D.A. Hill, and J.R. Wait, "Electromagnetic Scattering of an Arbitrary Plane Wave by Two Nonintersecting Perpendicular Wire Grids," *Can. J. Phys.*, Vol, 52, 1974, pp. 227-237.

31. D.A. Hill, and J.R. Wait, "Electromagnetic Scattering of an Arbitrary Plane Wave by a Wire Mesh with Bonded Junctions," *Can. J. Phys.*, Vol. 54, 1976, pp. 353-361.

32. R.W.P. King, and B.H. Sandler, "Analysis of the Currents Induced in a General Wire Cross by a Plane Wave Incident at an Angle with Arbitrary Polarization," *IEEE Trans. Antennas Propag.*, Vol. AP-29, May 1981, pp. 512-520.

33. E.L. Pelton, and B.A. Munk, "Comments on 'Currents Induced in a Wire Cross by a Plane Wave Incident at an Angle'," *IEEE Trans. Antennas Propag.*, Vol. AP-29, May 1981, pp. 520-522.

34. R.W.P. King, "Authors Reply," *IEEE Trans. Antennas Propag.*, Vol. AP-29, May 1981, pp. 522-523.

35. C.C. Chen, "Scattering by a Two-Dimensional Periodic Array of Conducting Plates," *IEEE Trans. Antennas Propag.*, Vol. AP-18, Sept. 1970, pp. 660-666.

36. B.A. Munk, R.G. Kouyoumjian, and L. Peters, "Reflection Properties of Periodic Surfaces of Loaded Dipoles," *IEEE Trans. Antennas Propag.*, Vol. AP-19, Sept. 1971, pp. 612-617.

37. R.J. Luebbers, and B.A. Munk, "Cross Polarization Losses in Periodic Arrays of Loaded Slots," *IEEE Trans. Antennas Propag.*, Vol, AP-23, March 1975, pp. 159-164.

38. B.A. Munk, R.J. Luebbers, and R.D. Fulton, "Transmission Through a Two-Layer Array of Loaded Slots," *IEEE Trans. Antennas Propag.*, Vol. AP-22, Nov. 1974, pp. 804-809.

39. N. Amitay, V. Galindo, and C.P. Wu, *Theory and Analysis of Phased Array Antennas*, New York, Wiley-Interscience, 1975, pp. 310-313.

40. J. Richmond, "Thin Wire Computer Code," Ohio State University.

41. T.K. Wu, "Analysis and Applications of Multilayered Periodic Strips," *AEU*, Vol. 33, No. 4, 1979, pp. 144-148.

42. R. Mittra, and S.W. Lee, *Analytical Techniques in the Theory of Guided Waves*, New York, MacMillan, 1971.

43. E.L. Pelton, and B.A. Munk, "A Streamlined Metallic Radome," *IEEE Trans. Antennas Propag.*, Vol. AP-22, Nov. 1974, pp. 799-803.

44. L.E. Carter, "Hybrid RAM," *Proceedings of the 1975 Radar Camouflage Symposium*, December 1975, pp. 393-435.

45.   R.D. Stratton, "A Design Procedure for Surface Wave Absorbers (U)" *1980 Radar Camouflage Symposium*, March 1981, pp. 117-122.
46.   R.E. Collin, *Field Theory of Guided Waves*, New York, McGraw-Hill, 1960.
47.   E.F. Knott, V.V. Liepa, and T.B.A. Senior, "Non Specular Radar Cross Section Study," AFAL-TR-73-70, University of Michigan, April 1973.

CHAPTER 10

RADAR ABSORBER MEASUREMENT TECHNIQUES

E. F. Knott

10.1 OVERVIEW

As pointed out earlier, the principles involved in designing radar absorbing systems are relatively well known, provided that we have an adequate description of the electromagnetic properties of the materials used in design and fabrication. Because practical absorber systems must satisfy mechanical as well as electrical requirements, the end product represents a compromise between ideal mechanical and ideal electrical performance goals. Thus, in addition to catalogues of the mechanical properties of materials, the designer must have access to catalogues of the electrical properties of materials. If those properties are not available, they must be measured for the materials the designer feels may be useful in his design.

Once a product has been designed and fabricated, it must be evaluated. Large manufacturers of absorber products are interested in two kinds of materials measurements: one is for quality control of a product line and the other is the evaluation of products under development. The quality of the product line can usually be assured by ordinary reflectivity measurements of samples selected from the production line, whereas more detailed measurements may be necessary for products under development. Thus, we can identify two different kinds of material measurement requirements: the reflectivity of absorber panels for quality control and the electromagnetic properties of individual materials used in the fabrication process. The former is significantly easier to perform than the latter.

It was mentioned earlier that a very desirable property of radar absorbers is bandwidth and, hence, we are seldom interested in the reflectivity of a product or the electromagnetic properties of component materials at a single frequency, or even a handful of discrete frequencies. Because all the quantities of interest vary with frequency, we need instrumentation systems having some bandwidth, typically from 1 to 18 GHz. The actual selection of a measurement system hinges on other factors as well — cost, for one — and, consequently, the system finally selected is itself a compromise. This is illustrated in the selection process of Fig. 10-1.

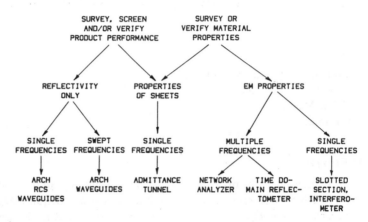

Figure 10-1 The objective of the measurement dictates the equipment and techniques used

The two basic objectives of the measurements are listed at the top of the chart, and there are three forms of material to be measured. The three are finished absorber panels or products, thin sheets used as components in the design and fabrication process, and samples of more or less uniform bulk materials. For product assessment and evaluation, it is almost always sufficient to simply measure the reflectivity of the absorber panels or sheets. Because this is usually a finished product ready for shipment or warehousing, the tests must be non-destructive and, the test fixtures are relatively large. The methods available are the the NRL arch technique, RCS measurements on outdoor ranges, interferometers, and within waveguides. The first two of these can be regarded as free space techniques, because the energy source and receiver are relatively remote from the test sample. This is discussed in sec. 10.3.

The only way to conduct waveguide measurements on finished panels is to use very large waveguides, and their use for this kind of testing is quite distinct from that used to measure the electromagnetic properties of materials. Interferometers are hybrid systems, in that the measurement system is not a closed one, yet the energy used in testing is confined essentially in the vinicity of the test panel. These are discussed in sec. 10.4.

The measurement of the electromagnetic properties of bulk materials is almost always performed within a closed (waveguide) system, although some interferometers are exceptions, as noted above. Although reflectivity measurements can certainly be performed within closed systems, there is a vast difference in the sample sizes required in the two cases. Typical waveguides have transverse dimensions measured in a few inches at most, and coaxial waveguides are even smaller. Consequently, the samples to be measured must be small, homogeneous, and fabricated to close tolerances. Depending on how much capital we are willing to invest in the equipment, the material properties can be measured quickly over a wide range of frequencies with swept-frequency systems, or more laboriously at discrete freqencies with less sophisticated instrumentation. In either case, the instrumentation systems are essentially bench set-ups that can be installed in a corner of the laboratory.

Whether or not measurements of small samples are conducted at single or swept frequencies, the results are based on the measurement of the reflection or transmission characteristics of a test sample inserted in the transmission line. For that reason, the fundamental theory of transmission lines as measurement devices is undertaken in some detail in sec. 10.2. Because the reflection and transmission of energy within the line may be influenced by the sample holder as well as the sample itself, it is important that the sample holder be carefully fabricated, and that the system calibration account as much as possible for residual line reflections and losses. Occasionally, the system losses are on the order of the losses within the sample, hence, it is important to be able to separate the two. Similarly, small reflections from minute discontinuities in the sample holder may lead to errors in the measurements.

Sheet materials such as those used in the production of Salisbury screens, Jaumann absorbers, and circuit analog materials cannot be evaluated easily in either enclosed or free space systems. The primary reason is that the sheet is too thin to be mounted in a waveguide (on the order of .005in or less), and it can never used as an absorber all by itself; it is always a component in an absorber design, even in its simplest form. The evaluation of sheet materials, therefore, is accomplished in what is known as an admittance tunnel, two forms of which exist. The tunnel is basically an absorber-lined box with a source antenna at one end and a test aperture at the other. The electromagnetic properties of the sheet may be obtained from measurements of the reflection and transmission characteristics of the sheet when placed across the aperture. Admittance tunnels are discussed in section 10.4.

10.2 TRANSMISSION LINE METHODS

10.2.1 Transmission Line Theory

The transmission line is a basic device used to measure the electromagnetic properties of materials because the field structure within waveguides is well known and the energy used in the measurements is confined within the system. A specimen of the material to be measured must be machined to fit snugly within the waveguide system and a signal is launched down the transmission line containing the sample. The sample holder itself is a short section of transmission line and the sample represents an abrupt change in the impedance characteristics of the line. No matter how thin the sample, this discontinuity gives rise to a reflection, and the amplitude and phase of the reflection are clues to the electromagnetic properties of the sample. As we shall see, some measurement systems make use of the characteristics of the signal transmitted through the sample as well as the reflected signal.

The term "waveguide" includes any structure, even non-metallic ones, that leads electromagnetic waves along a designated path with little or no loss in signal intensity. Examples of waveguides are the fibers of fiber optic lines, the hollow metal pipes used in radar and microwave systems, two-conductor lines such as coaxial lines and cables, and twin-conductor lines. For the purpose of our discussion, however, the term waveguide will denote hollow metal pipes, in which the structure of the propagating wave can be quite complex, while two-conductor systems (generally coaxial) will be called simple transmission lines.

If the waveguide is electrically too large (too many wavelengths in transverse dimension), it is possible to excite higher order propagation modes. This can be a serious limitation of measurements made with waveguide systems, because the insertion of a test sample within the waveguide increases its electrical size. In the analysis of the test results and the subsequent data reduction, we assume that it is the dominant (lowest order) propagation mode that exists, and if other modes are present, the wrong answer is obtained because the field structure within the waveguide and within the test sample are not those assumed. Nevertheless, useful measurements can be made using waveguide sample holders and test systems.

The propagation characteristics of waves within transmission lines can be established by solving the wave equation and invoking the appropriate boundary conditions on the conductors of the waveguide system. The procedures for accomplishing the solution are well known and will not be repeated here; the reader may consult standard texts for more detailed information [1, 2, 3]. The velocity of propagation along any two-conductor transmission line, such as the twin-line and coaxial line, is essentially the speed of light. If the space between the conductors is filled with a dielectric material, the speed is reduced in proportion to the index of refraction of the material. In hollow pipes, on the other hand, the wave propagates considerably less than the speed of light because of the nature of the solution of the wave equation within such structure. In essence, the wave traces out a zig-zag pattern as it propagates down the tube, bouncing first off one wall, then the other. This slows its propagation along the transmission line. If the transverse dimensions of the waveguide are too small, generally less than about one half-wavelength, the wave cannot propagate at all; the frequency at which this occurs is called the cut-off frequency. Thus, one disadvantage of using hollow waveguides for material measurements is the limit on the lowest frequency of operation.

Coaxial transmission lines are favored over waveguides for material measurements because of the bandwidth offered by coaxial lines: there is no lower cut-off frequency. However, coaxial transmission lines can support higher-order modes, just as hollow pipes do, and care must be used despite their inherently greater bandwidth. Higher-order modes can be generated if the mean circumference (measured at a radius halfway between the inner and outer conductors) exceeds a wavelength. As with waveguides, the insertion of material between the inner and outer conductors increases the internal electrical size of the line. Consequently, although a coaxial line may be unable to support higher-order modes when filled with air, it may certainly support them when loaded with a test sample.

10.2.2 Propagation Modes

The solution of the wave equation yields the propagation modes within the waveguide structure. For air-filled coaxial lines, the propagation factor is identically the free space wavenumber k. If x is a distance along the line measured with respect to some reference plane, solutions of the wave equation can be expressed in the forms:

$$e^{-j(kx - \omega t)} \qquad (10\text{-}1)$$

$$e^{i(kx - \omega t)} \qquad (10\text{-}2)$$

where $i = j = \sqrt{-1}$. Engineers prefer the notation in Eq. (10-1), while physicists are accustomed to the notation in Eq. (10-2). Throughout this chapter we shall use the notation in (10-2); if the reader prefers the other, he may convert any result to the notation of (10-1) merely be replacing i by $-j$ wherever it appears. The presence of the i (or $-j$) in the exponent implies that the waves in transmission lines have imaginary components. This is not true: the waves are all real, and the complex notation is merely a mathematical convenience in expressing phase relationships. The amplitudes of the field quantities can be obtained by multiplying any field component by its complex conjugate and extracting the square root of the result.

In Eq. (10-1) and (10-2), $\omega = 2\pi f$ is the radian frequency of the wave propagated down the line. The propagation constant k is a function of the electromagnetic properties of the material filling the space between the inner and outer conductors of the line:

$$k = \omega \sqrt{\mu\epsilon} = k_0 \sqrt{\mu_r \, \epsilon_r} \tag{10-3}$$

where ϵ and μ are the permittivity and permeability, respectively, of the material. As suggested in Eq. (10-3), these quantities may be normalized with respect to the free space values:

$$\mu_0 = 4\pi \times 10^{-7} \text{ henries per meter} \tag{10-4}$$

$$\epsilon_0 = 8.854 \times 10^{-12} \text{ farads per meter} \tag{10-5}$$

The subscript 0 indicates the free space values, while the subscript r indicates the normalized (relative) values.

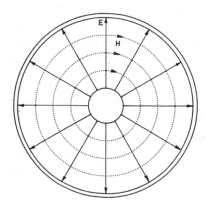

Figure 10-2 Electric field lines are radial and magnetic field lines are circumferential in a coaxial transmission line. Microwave energy does not radiate from this system

We have already seen (Ch. 3) that the geometric mean $\sqrt{\mu_r \epsilon_r}$ is the index of refraction of the material, and that both μ_r and ϵ_r can be complex numbers. As shown in Eq. (8-1), the real and imaginary parts are denoted by single and double primes, respectively. The signs of the imaginary components are consistent with the solutions in Eqs. (10-1) and (10-2), and they ensure that waves decay with increasing distance if any loss is present. Because the index of refraction is complex, the propagation constant k is also complex, and the imaginary part of k is responsible for the attenuation of the wave.

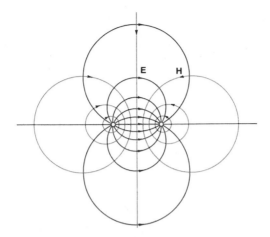

Figure 10-3 Electric and magnetic field lines lie on families of circles for the twin conductor line. Energy can radiate from this system

In the absence of higher-order modes of propagation, the electric and magnetic fields in any two-conductor transmission line are at right angles to each other and the direction of propagation. This is called a *transverse electromagnetic mode* (TEM). Higher-order modes in two-conductor lines, and all modes in hollow pipes, have longitudinal components of either the magnetic or electric field. These are called TE (transverse electric) and TM (transverse magnetic) modes. The TEM field structure within two-conductor lines is of great importance because it accurately simulates the nature of a plane wave striking a flat absorber panel at normal incidence. Figures 10-2 through 10-4 illustrate the transverse field configurations in three different TEM lines. Of the three, the coaxial line is by far the most common in microwave work. Microwave energy can radiate from the open structures of the twin-line and the parallel-plate line, leading to energy losses that might otherwise be attributed to the test sample. Because the coaxial line is a closed system, energy is not lost through radiation from the structure.

Every two-conductor transmission line can be characterized by an impedance called the surge impedance or the characteristic impedance. This impedance is governed by the size and shape of the conductors forming the line and by the electromagnetic parameters of the material within the line. For a coaxial line, the characteristic impedance is given by

$$Z_l = \frac{1}{2\pi} \sqrt{\frac{\mu}{\epsilon}} \ln \frac{r_0}{r_i} = \nu Z_0 \qquad (10\text{-}6)$$

where r_i and r_0 are the radii of the inner and outer conductors of the line, respectively, $\nu = \sqrt{\mu_r / \epsilon_r}$ is the intrinsic impedance of the material filling the line, and Z_0 is the impedance of an air-filled line of these dimensions,

$$Z_0 = \frac{1}{2\pi} \sqrt{\mu_0 / \epsilon_0} \ln \frac{r_0}{r_i} \qquad (10\text{-}7)$$

The dimensions of most air-filled coaxial lines are chosen for a line impedance of 50 or 100 ohms.

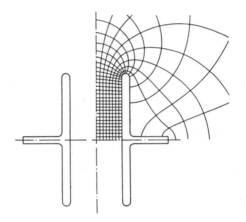

© *1960 John Wiley & Sons.*

Figure 10-4 The electromagnetic structure is very uniform in the region between the conductors of the parallel plate line. The fields are much weaker outside this region than inside (taken from reference [2], courtesy of John Wiley and Sons, New York and London)

10.2.3 Sample Holders

Figure 10-5 shows the shapes required of test samples for coaxial transmission lines and rectangular waveguides. Good measurement practice dictates that the sample be less than an eighth-wavelength thick, as measured in the materials. This implies that the material properties are known before the measurements are undertaken. So why measure the materials in the first

place? The answer is that we do not know the elctromagnetic material properties in advance, but we can usually make an estimate on the basis of the known
properties of similar materials or of the expected values.

The samples must be carefully machined to fit snugly within the inner
dimensions of the transmission line, and good electrical contact must be
maintained between the sample and the conductor walls. For this reason, it is
very difficult to fabricate and measure thin materials like resistive sheets in
transmission lines. The sample holder itself must be a precision device, or else
residual reflections from the sample holder may be interpreted as arising from
the test sample.

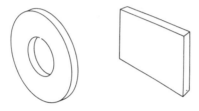

Figure 10-5 Test sample shapes for the coaxial line and rectangular waveguide are quite different

Standard coaxial line sections have nominal inner dimensions (of the outer
conductor) of 7 mm and 14 mm, and even the larger dimension does not allow
a very large sample. Special purpose coaxial sections have been built with an
inner diameter (of the outer conductor) of 1 in, and a tapered section must be
built to flare up to this diameter from the conventional size. The tapered
section must be carefully designed to minimize residual reflections, of course.
The samples must be homogenous to minimize the possibility of generating
higher-order propagation modes within the sample. If the material has inhomogeneities, such as the carbon-latex coated fibers of some materials or the
cell structure of carbon-loaded foams, it is best not to measure the material in
a transmission line. The sample holder itself must be a precision device, or else
residual reflections from the sample holder may be interpreted as arising from
the test sample.

10.2.4 Reflections in Transmission Lines

Shown schematically in Fig. 10-6 is a transmission line with forward and backward propagating waves. The backward traveling wave may be due to reflections from a sample placed within the line or from unintentional discontinuities in the line, or both. Whatever the source of the reflection, the backward wave combines with the forward wave to produce a standing wave pattern. The standing wave pattern represents the intensity of either the electric or the magnetic field strength, and because the two waves travel in opposite directions, the periodicity of the standing wave pattern is precisely $\lambda/2$. The ratio of the maximum field strength to the minimum is called the voltage standing wave ratio, or VSWR (often pronounced "*viswar*"). In terms of the amplitudes of the forward and backward traveling waves, the VSWR is

$$\text{VSWR} = \frac{E_{max}}{E_{min}} = \frac{|A| + |B|}{|A| - |B|} \tag{10-8}$$

where A and B may be complex numbers. The VSWR may be expressed in decibles, if desired.

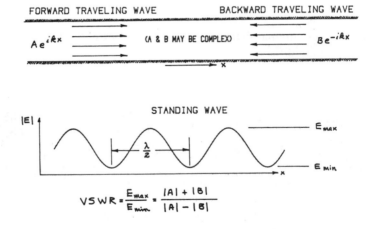

Figure 10-6 Two waves propagating in opposite directions create a standing wave pattern with a periodicity of a half-wavelength

Note that in theory the VSWR can rise to infinity, which is the case for perfect reflection: the amplitude of the reflected wave is exactly that of the incident wave. The VSWR may also be unity, for which case there is no reflection in the line at all. High quality radio-frequency lines have very low residual losses, hence, the peaks of the standing wave pattern several wavelengths toward the load or toward the source are essentially constant. This is not the case if there are losses in the line, as suggested in Fig. 10-7.

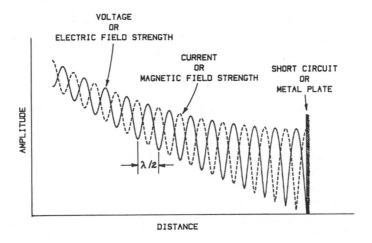

Figure 10-7 The standing wave pattern in a lossy medium does not have uniform peaks and nulls

This illustration was constructed to demonstrate the intensities of the fields in a transmission line several wavelengths long and filled with lossy material. The electric and magnetic field strengths may be expressed as

$$E = Ae^{ikx} + Be^{-ikx}$$
$$vZ_0H = Ae^{ikx} - Be^{-ikx}$$

(10-9)

where E represents the intensity of the electric field and H represents the intensity of the magnetic field. The reversal of the sign of the second term in the magnetic field expression assures that energy is indeed transmitted to the left by the backward wave. These are also the equations for the field intensities of two plane waves, one due to a reflection by a flat infinite obstacle presented normal to the direction of propagation of the incident plane wave.

There are direct parallels between voltage and electric field strength, and between current and magnetic field strength. The line in Fig. 10-7 is terminated by a short circuit, which forces the voltage to zero at the end of the line. This is a consequence of the fact that perfect conductors can support no tangential components of electrical field. On the other hand the magnetic field strength attains a maximum value at the short circuit, demonstrating that the effective current at the short circuit is a maximum.

The electric and magnetic field strengths both trace out standing wave patterns whose periodicities are $\lambda/2$, but the maximums of one coincide with the minimums of the other, and *vice versa*. Near the short circuit, the standing wave ratio is much higher than it is near the input of the transmission line because the intensities of the incident and reflected waves are nearly the same there. Near the input of the transmission line, however, the standing wave ratio is much smaller because the amplitude of the reflected wave is much smaller at the input than it is at the short circuit, and because the incident wave is similarly attenuated by line losses as it propagates from the input toward the load.

In practice, samples placed in sample holders for measurements are fractions of a wavelength long, not several, and only the small portion of the standing wave structure near the short circuit is present within the material. Nevertheless, even a thin sample presents a discontinuity to the incident wave, generating a reflection. The measurement of that reflection can be used to deduce the electromagnetic properties of the test samples.

Figure 10-8 illustrates an elementary measurement system for conducting measurements at discrete frequencies. Because the actual fields within the test sample cannot be probed, the sample properties must be deduced from measurements of the standing wave pattern in front of the sample. Consequently, the heart of the system is a slotted section, which is a precision section of transmission line with a longitudinal slot milled into the outer conductor. A fine probe is inserted into the interior of the slotted section so that a small sample of the voltage standing wave pattern may be extracted. The probe cannot be inserted too far, or else it will disturb the pattern being measured, and the slot must be narrow, or else energy will be lost by radiation from the system.

The probe is fitted to a carriage that may be slid along the length of the slotted section; not shown is a distance scale attached to the outside of the slotted section. This scale may even be fitted with a vernier or a precision dial gauge, so that the operator can record the physical position of the carriage. The carriage is moved back and forth manually, and in addition to readings of the carriage position by means of this scale, the operator must have an electronic device that allows him to read and record the amplitude of the probe signal. In the simple bench set-up shown in Fig. 10-8, this may be an ordinary crystal detector or bolometer detector, with a built-in audio amplifier to boost the detected signal to a level strong enough to display on a microammeter.

In order to make this simple detection system work, the RF source must be modulated at an audio rate, usually 1000 Hz. The source may be a klystron or other RF signal generator, and a few milliwatts of power (10 mW) is adequate for most measurements. The source is generally isolated from the slotted section by a fixed attenuator to minimize frequency changes because the measurement system connected to the output of the signal generator generally has a very low impedance, and this may "pull" the signal generator off the desired frequency. Attached to the output port of the slotted section is the sample holder, usually a precision device, followed by a section of transmission line of the same quality with a short circuiting plunger in it.

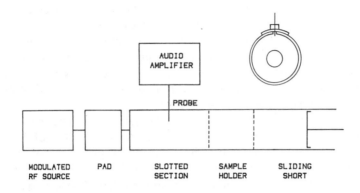

Figure 10-8 A simple bench set-up uses a slotted section to sample the standing wave pattern inside the line. A transverse section (inset) shows an electric field probe mounted on a carriage and inserted into the space between conductors

The purpose of the sliding short is to allow the operator to place any reactive load at the rear face of the test sample. Some operators prefer to work with fixed open and short circuits that may be connected to the output of the sample holder, but since the open circuits are designed for fixed frequencies, this may limit the range of frequencies at which the measurements may be made. It is also possible and permissible to use matched loads, although perfect matched loads are costly, and we must measure and account for the characteristics of the load.

In general, the standing wave pattern in the slotted section is characterized by only two numbers: the VSWR and the relative location of a null. These are sufficient to establish the amplitude and phase of the reflection coefficient associated with the test sample. If the test material has a relative permeability of unity, which is a very good approximation for non-magnetic materials, the measured reflection coefficient is sufficient to deduce the relative permittivity (the dielectric constant) of the sample. This is because the dielectric constant has two parts (real and imaginary), and the reflection coefficient is described by two numbers. However, if the relative permeability cannot be assumed to be unity, there are two more unknowns to be determined, and a second reflection coefficient must be measured for a different impedance placed at the rear face of the test sample. Thus, at least four measurements must be performed for a totally unknown material, but these may be reduced to two if it can be assured that the sample is non-magnetic. For improved accuracy, the measurements may be performed for a wide variety of impedances placed at the rear of the sample, but this takes more time, of course.

The two values most commonly used for the impedance placed at the rear face of the sample are zero and infinity, corresponding to short and open circuits, respectively. (An open circuit can be obtained by placing a short circuit an odd multiple of a quarter-wavelength from the rear face of the sample.) Better accuracy is obtained in measuring the losses of non-magnetic materials by the use of an open circuit than a short circuit, and completely unknown materials require the use of both.

Before measuring the test sample, we must first establish the characteristics of the measurement system. This involves measuring the losses of the system, including the sample holder, but without the sample present. There are methods for measuring the high VSWR of the line with no sample present, but we will not go into those in detail here. The reader is referred instead to the classic text by Ginzton [4]. In addition, because the only connection between the outside world and the sample holder is via the output port of the slotted section, we have only the standing wave pattern of the empty sample holder to gauge where the rear face of the sample is ordinarily located. Because this is the point along the line where we must place the open or short circuit, we must find out where it is. Thus, the sample holder must be designed so that a short circuit (an annulus or disk) may be physically placed inside such that the front face of the disk actually lies in the plane normally occupied by the rear face of the test sample. Then the position of one or more nulls in the standing wave pattern within the slotted section may be recorded. Further measurements of the standing wave pattern generated by the test sample must be calibrated against these initial measurements.

10.2.5 Transmission Line Equations

As we have seen, the electric and magnetic fields within a transmission line vary with the distance along the line. Because the ratio of the electric field strength to the magnetic field strength is an impedance, a wave impedance can be identified at any point along the line. Moreover, an effective reflection coefficient can be written in terms of the characteristic impedance of the line, the impedance terminating the line, and the physical or electrical distance from the terminating impedance to the point at which the line impedance is measured. These relationships are

$$Z = Z_0 (1 + \Gamma)/(1 - \Gamma) \tag{10-10}$$

$$\Gamma = \frac{Z_R - Z_0}{Z_R + Z_0} \, e^{i2kd} \tag{10-11}$$

where Z is the impedance at any point along the line, Γ is the voltage reflection coefficient, Z_R is the value of the impedance placed at the end of the transmission line, and d is the physical distance between the end of the line and the point at which Z is measured (see Fig. 10-9). In general, these are all complex numbers characterized by amplitude and phase, or, alternatively, by real and imaginary parts.

Figure 10-9 A load at the end of a transmission line causes a reflection if the load impedance is not equal to the characteristic impedance of the line

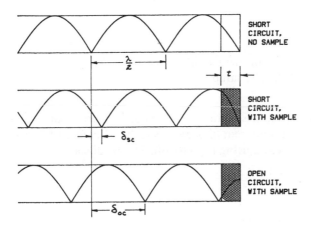

Figure 10-10 Measurement of the standing wave patterns in front of the test sample when backed by open and short circuits leads to a determination of the sample properties

Let us now make a set of measurements of the test sample. First, we attach the empty sample holder, with a shorting disk at its end, to the slotted section and establish a reference plane for the rear face of the sample, as shown in the upper diagram of Fig. 10-10. We assume for simplicity that there are no residual losses in the slotted section or the sample holder. We then place the sample in the sample holder backed by a short circuit and repeat the measurement. If the short circuit is provided by a sliding plunger, the plunger position must be carefully positioned at the proper distance. We measure the distance δ_{sc} by which the standing wave pattern has shifted toward the load and the VSWR of the standing wave R_{sc} as shown in the center diagram. Finally, the procedure is repeated with an open circuit behind the sample instead of a short circuit, and we measure the open circuit null shift and VSWR, δ_{oc} and R_{oc} respectively. Thus we emerge from the measurements with four measured quantities $(\delta_{sc}, \delta_{oc}, R_{sc}, R_{oc})$ from which we hope to determine four unknown quantities $(\epsilon_n', \epsilon_n'', \mu_n', \mu_n'')$.

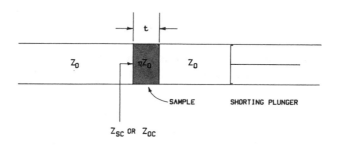

Figure 10-11 A sliding metal plunger may be used to place any reactive impedance behind the test sample

The data reduction is as follows: from the measured null shifts and VSWRs, first compute the apparent impedances at the front face of the sample, Z_{sc} and Z_{oc}, for the two cases using Eq. (10-10). The results are

$$Z_{sc} = Z_0 \, \frac{1 - R_{sc} \, e^{i2k\delta_{sc}}}{1 + R_{sc} \, e^{i2k\delta_{sc}}} \tag{10-12}$$

$$Z_{oc} = Z_0 \, \frac{1 - R_{oc} \, e^{i2k\delta_{oc}}}{1 + R_{oc} \, e^{i2k\delta_{oc}}} \tag{10-13}$$

where R_{sc} and R_{oc} are the measured values of the VSWR under short circuit and open circuit conditions, respectively. Equations (10-12) and (10-13) are the measured impedances at the front face of the sample when the sample is backed by open and short circuits. Consider now the diagram of Fig. 10-11, for which that part of the line containing the sample may be represented as a section of transmission line of length t. We may use Eq. (10-10), with the right-hand side multiplied by ν, to calculate the front face impedance. The reflection coefficient for the short circuit termination is $\Gamma = -1$, and $\Gamma = 1$ for the open circuit termination. Inserting these values in Eq. (10-10), we obtain for the front-face impedances:

$$Z_{sc} = \nu Z_0 \; \frac{1 - e^{i2kt}}{1 + e^{i2kt}} \tag{10-14}$$

$$Z_{oc} = \nu Z_0 \; \frac{1 + e^{i2kt}}{1 - e^{i2kt}} \tag{10-15}$$

where t is the thickness of the sample and k is the wavenumber characteristic of the material, as in Eq. (10-3). Because the unknowns to be determined appear in the intrinsic impedance ν in Eqs. (10-14) and (10-15), as well as in the exponents of the numerators and denominators, the solution is somewhat complicated.

One way to extract the desired values is to form the product and the ratio of Eqs. (10-14) and (10-15). The product yields

$$\frac{\mu_r}{\epsilon_r} = (Z_{sc} Z_{oc})^{1/2} / Z_0 \tag{10-16}$$

while the ratio yields

$$\mu_r \epsilon_r = \frac{1}{i2k_0 t} \ln \frac{1 - (Z_{sc}/Z_{oc})^{1/2}}{1 + (Z_{sc}/Z_{oc})^{1/2}} \tag{10-17}$$

The relative permeability and permittivity are obtained by forming the product and the ratio of Eqs. (10-16) and (10-17), with the result

$$\mu_r = \frac{(Z_{sc}/Z_{oc})^{1/2}}{i2k_0 t Z_0} \ln \frac{1 - (Z_{sc}/Z_{oc})^{1/2}}{1 + (Z_{sc}/Z_{oc})^{1/2}} \tag{10-18}$$

$$\epsilon_r = \frac{Z_0}{i2k_0 t (Z_{sc}/Z_{oc})^{1/2}} \ln \frac{1 - (Z_{sc}/Z_{oc})^{1/2}}{1 + (Z_{sc}/Z_{oc})^{1/2}} \tag{10-19}$$

Because the exponential function $\exp[i\,2kt]$ is multivalued, there is an uncertainty in the measurement. This is because the periodicity of the standing wave pattern in the slotted section is $\lambda/2$, and null shifts greater than that cannot be measured, as suggested in Fig. 10-12. There are several ways to resolve the ambiguity. One is to measure samples so thin that it would be impossible for the sample to be more than $\lambda/2$ in thickness, as measured within the material. However, it becomes progressively more difficult to accurately measure the losses of the sample as it becomes thinner. Another method is to repeat the measurements with a slightly different frequency. The comparison of two or more test results usually resolves the measurement ambiguity.

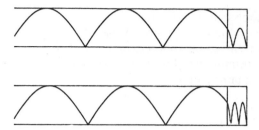

Figure 10-12 Because of the periodicity in the standing wave pattern, a single set of measurements cannot resolve the actual electrical thickness of the sample. The measurements should be repeated at a slightly different frequency or on a sample with a slightly different thickness

10.2.6 Multifrequency Measurements

The simple slotted section measurements described above require only a modest amount of laboratory equipment, but the price paid for this simplicity is that only a few samples may be measured in one day, or that it may take a day to measure a single sample for only a handful of frequencies. Because the goal of most absorber designs is the achievement of broadband performance, we must characterize the electromagnetic material properties for dozens of closely spaced frequencies, the smaller the frequency increment and the greater the frequency coverage, the better. There are two ways which this can be done: one is the implementation of swept frequency measurements and the other is time-domain reflectometry. Let us consider swept frequency measurements first.

At any frequency, a two-port device, such as a section of transmission line, can be characterized by the so-called scattering matrix. The scattering matrix describes the reflection of an incident wave from either of the two ports and the transmission of energy from one port of the other. The two-port device may be a passive device, or it may be an amplifier containing active sources, and each of the elements of the scattering matrix is a complex number.

Consider the circuit of Fig. 10-13, in which a device is inserted in a transmission line and excited by incident waves from both directions. The backward traveling wave to the left of port 1 arises from a reflection of the incident wave E_{i1} from port 1 and the transmission of the incident wave E_{i2} to port 1 from port 2. The forward traveling wave to the right of port 2 arises from a reflection of the incident wave E_{i2} from port 2 and the transmission of the incident wave E_{i1} to port 2 from port 1. The interaction may be expressed by the following pair of equations:

$$E_{r1} = S_{11} E_{i1} + S_{21} E_{i2}$$
$$E_{r2} = S_{12} E_{i1} + S_{22} E_{i2}$$
                                                                    (10-20)

In these expressions, the first subscript on the S-parameters signifies which port is being excited and the second subscript denotes the exit port at which the emerging wave will be measured.

Note that each of the four parameters may be measured independently by turning off one or the other incident wave. For example, S_{11} and S_{12} may be measured with $E_{i1} = 0$:

$$S_{11} = E_{r1} / E_{i1}$$
                 for $E_{i2} = 0$ (10-21)
$$S_{12} = E_{r2} / E_{i1}$$

Similarly, S_{21} and S_{22} may be measured with $E_{i1} = 0$:

$$S_{21} = E_{r1} / E_{i2}$$
                 for $E_{i1} = 0$ (10-22)
$$S_{22} = E_{r2} / E_{i2}$$

For most linear passive devices, such as uniform transmission line segments, $S_{12} = S_{21}$ represents the transmission from port 1 to port 2. The coefficients S_{11} and S_{22} are reflection coefficients associated with ports 1 and 2, respectively. It should be remembered that each of the S-parameters is a complex number characterized by phase as well as amplitude. Thus, in general, it requires no fewer than eight numbers to characterize the two-port device at any one frequency. This may be reduced to six for linear passive devices.

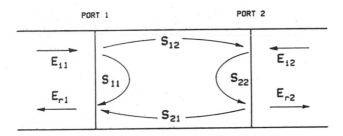

Figure 10-13 Four complex numbers comprise the scattering matrix for a two-port network or device

With the appropriate equipment, we can contrive to measure and display any of the four parameters over a band of frequencies. We must first have an RF source whose frequency can be swept or stepped across the desired band and a sensitive receiver that can track the source frequency as the frequency changes. As suggested by Eqs. (10-21) and (10-22), provisions must be made to inject the signal into either port of the device and to measure the signals emerging from either port. Secondly, because we must be able to measure the phase of the signals as well as their amplitude, the receiver must be furnished with a reference signal to operate a phase comparator. Typically, the phase comparison is made at IF, the intermediate frequency, after superheterodyne detection. Thirdly, a ratio-forming circuit must be included in the receiver so that the ratios in Eq. (10-21) and (10-22) can be formed and displayed. As with the phase information, this too is performed at the intermediate frequency.

Finally, appropriate display or output devices must be included in the equipment so that the operator can observe the results of the measurement. CRT displays are available that plot any of several parameters in any of several formats. The device's gain and phase shift may be plotted in rectangular or polar coordinates. These may all be done in real time, and if the results of the measurement must be recorded, the output can be diverted to an XY plotter. The entire system makes it very convenient for device development work and to evaluate devices off the production line. With appropriate interface components, the system may even be driven by a computer system.

This, in fact, is essential if we are to extract the electromagnetic properties of test samples from measurements of reflection or transmission coefficients because, as shown in sec. 10.2.4, considerable manipulation of the measured quantities is necessary. The Hewlett-Packard Corporation markets a complete system that performs these kinds of measurements over multi-octave bands from 110 MHz to 18 GHz, and over 2 GHz windows from 18 to 40 GHz, for less than $100,000. If the system is used for material sample measurements, waveguide sample holders must be fabricated for frequencies above 18 GHz, and even for frequencies below 18 GHz, it may be necessary to fabricate special coaxial sample holders.

In practice, the system must be calibrated, of course, usually by terminating the output ports of the test equipment and the test device with short circuits or matched loads. In making material measurements, the operator must ensure that the equipment is properly interfaced with the sample holder, and the operator must account for additional line lengths needed to route the signal to the sample holder. Although the automated equipment makes it possible to accurately measure the characteristics of material samples rapidly over a wide range of frequencies, the accuracy of the simple system described in sec. 10.2.4 is probably better. Nevertheless, the advantages of automated swept frequency measurements are far more important than sheer accuracy for most work.

Another way of rapidly obtaining the electromagnetic properties of materials over a broad range of frequencies is through the use of time-domain reflectometry, as discussed by Nicolson and Ross [5]. In this method, narrow pulse waveforms are transformed to the frequency domain by means of the fast Fourier transform (FFT). The heart of the measurement system is a computer-controlled broadband sampling oscilloscope which scans the wave-forms, digitizes them, and the performs the FFT and the subsequent computations.

A schematic diagram of the measurement system is shown in Fig. 10-14. The physical lengths of the transmission line segments between the pulse generator and the sampling head, between the sampling head and the test sample, and between the test sample and the short circuit, must be selected such that the time interval over which the transmitted and reflected wave-forms are sampled are not contaminated by other reflections in the system. This uncontaminated time window has a width of $2l/c$, where l is indicated in Fig. 10-14 and c is the speed of light. The ideal pulse for the measurement is a unit step function with zero rise time, but this theoretical pulse cannot be produced in practical systems. The generator in the system described by Nicolson and Ross developed pulses with a rise time of 30 picoseconds ($30 \cdot 10^{-12}$ sec) and an amplitude of about 0.12 V.

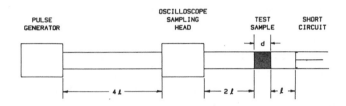

Figure 10-14 A time-domain reflectometer measures wave-forms over critical time windows. The heart of this system is a sampling oscilloscope. (The computer associated with the system is not shown)

The pulse propagates down the line and through the sampling head of the oscilloscope and strikes the sample, where part of the pulse is reflected and part is transmitted through the sample. The component transmitted through the sample is reflected from the short circuit a short time later, and is transmitted through the sample again, this time in the reverse direction, and then passes through the sampling head. Thus a pulse appears at the sampling head at three different times: first the incident pulse, then the reflected pulse from the sample, then the pulse reflected from the short circuit and which passes through the sample twice. A comparison of the second waveform with the first yields the scattering matrix element S_{11}, and a comparison of the third wave-form with the first yields the scattering matrix element $(S_{21})^2$, the square appearing because of the double transmission of the pulse through the sample.

In the practical implementation of the method, four waveforms are measured and digitized by the system, two with and two without a sample in place.

The first two measurements are performed with a metallic short circuit placed first in the empty line at precisely the location normally occupied by the front face of the test sample, then some distance *l* behind the rear face of the sample. In actual practice, because the transmission line is uniform and the short circuit is assumed perfect, both empty line measurements may be performed with the shorting plug placed at the location normally occpied by the front face of the sample. The first of these measurements records the waveform at the front face of the sample due to the incident pulse, and the second records the waveform at the same location, but at an additional time delay corresponding to a double traversal of the distance from the front face of the sample down to the short circuit and back again. These two waveforms are digitized by the sampling oscilloscope and saved by the computer (not shown).

The sample is then placed in the sample holder and installed in the system, and the measurements are repeated at precisely the same time delays. The four data sets are then transformed to the frequency domain by means of the FFT and the transformed sample data are normalized with respect to the transformed empty line data. The first set of normalized transformed data is the frequency dependent function $S_{11}(\omega)$. The phase of each datum of the second normalized data set must be shifted by $2kd$ (twice the sample electrical thickness in free-space wavelengths). When this is done, the second set of normalized transformed data is the frequency dependent function $S_{21}^{2}(\omega)$. These are complex numbers, of course, and the measurements yield four frequency domain functions from which the four frequency dependent electromagnetic parameters (ϵ_r', ϵ_r'', μ_r', μ_r'') may be obtained. Nicolson and Ross give the mathematical expressions for the computations in their paper, and Figs. 10-15 and 10-16 illustrate typical results for a sample of magnetic absorbing material. The rather large oscillations in the reported values for the relative permittivity in the range from 6 to 10 GHz are probably not real, rather are due to system errors.

Some of the measurement errors are due to small shifts in the timing of the windows used in the measurements. The effective window width reported in [5] was 2.5 ns, and timing errors as small as 1 ps can cause 4 percent errors in the measured values, much smaller than suggested by Figs. 10-15 and 10-16. Nicolson and Ross offer suggestions for improving the accuracy, among them more stable pulse generators with shorter rise times and longer line lengths to expand the time windows. Whatever the case, a complete set of measurements can be collected for one sample covering the range from 100 MHz to 10 GHz in as little as 30 minutes. It would take at least a month of concentrated effort to duplicate the measurements using the simple slotted line method discussed earlier, hence the advantages of time domain reflectometry for the measurement of material samples are clear.

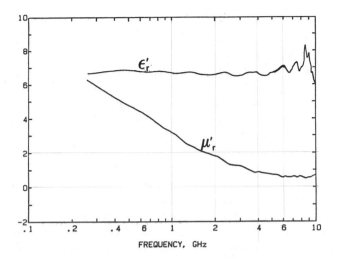

Figure 10-15 A display of the measured results for a commercial magnetic material. Shown are the real parts of the relative permeability and permittivity

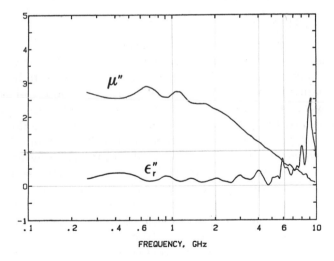

Figure 10-16 A display of the measured results for a commercial magnetic material. Shown are the imaginary parts of the relative permeability and permittivity

10.3 FREE SPACE METHODS

The evaluation of absorbers, either for quality control or product development, does not require such elaborate test equipment and data processing as the measurement of the intrinsic properties of materials. Moreover, nondestructive testing is required so that test panels may be shipped to distributors or warehoused until needed. Even if panels could be sacrificed to supply small test specimens for transmission line measurements, the physical construction of most finished absorber products is ill-suited for insertion into small transmission lines. Thus, the simplicity offered by free space methods is attractive.

It should be emphasized at the outset that the term "free space" refers only to the fact that the test panel is not installed within a waveguide or transmission line. In one of the free space methods (the NRL arch technique), the test panel is often placed within a few feet of a set of illuminating and receiving antennas, and the framework of the arch itself is no less than that distance from the test panel. Although the data collected by means of this technique is extremely useful and can be quite accurate with adequate controls, the arch hardly simulates a free space environment. The other of the two methods discussed below, the RCS method, does simulate free space conditions more closely, but is essentially limited to one frequency at a time.

10.3.1 The NRL Arch

The arch concept was displayed at the Naval Research Laboratory some 40 years ago as a way of evaluating absorber panels. The arch itself was simply a vertical wooden framework that allowed a pair of antennas to be fixed at a constant radius from the center of a test panel for a variety of subtended angles, as suggested in Fig. 10-17. Each antenna, usually a small horn, was mounted on a carriage that could be clamped anywhere along the arch. The carriage was designed to keep the horn pointed at the center of the test panel no matter where the carriage was positioned. The transmitting and receiving antennas could be brought no closer together than the width of one antenna aperture and, consequently, they subtended a small angle at the test panel. However, the angle was small enough to be negligible in the measurement of the monostatic performance of the test materials. Sometimes it was necessary to slip a small layer of absorbing material (not associated with the material to be tested) edgewise between the two horns to reduce the direct coupling energy from one horn to the other.

Early arches were no more than about six feet tall and were sawed out of plywood sheets. The test panel was simply laid on a metallic plate attached to a short pedestal placed at the center of the arch. This plate was pre-cut to the dimensions of the test panel, usually 1×1, 1×3, or 2×2ft in size. The metal

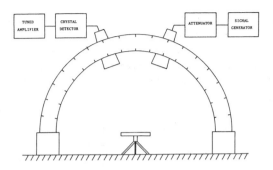

Figure 10-17 The classic NRL arch provides a framework along which a pair of antennas may be moved. The test sample is mounted on the small pedestal at the center of curvature of the arch

plate provided the proper backing for the material for the tests and was also used for calibration of the measurements.

Depending on the preference of the user of the system, the metal plate on the pedestal could be used for calibration, or another plate of identical size could be placed on a spacer above it. In the latter case, the surface of the reference plate was deliberately elevated to coincide with that of the top of the test panel to be installed. This was thought by some to increase the accuracy of the measurements, but it is doubtful that it made much difference, especially for angles of incidence and reflection close to the normal.

As suggested in Fig. 10-17, the equipment needed for measurements at a single frequency was quite modest. The amplitude-modulated output of a signal generator was passed through a precision attenuator to the transmitting antenna, and the reflected signal captured by the receiving antenna was delivered to a simple crystal detector and audio amplifier, where the amplified output was displayed on a meter calibrated in decibels. Two measurement options were available, depending on the linearity of the detection system.

If the linearity was not good enough over a 30 dB range, as might be the case for some detectors and some absorbers, the precision attenuator could be used in a substitution method. First, the metal plate was exposed and the attenuator set for some nominal deflection on the output indication device. Then the test sample was installed on the plate and the attenuator was adjusted to yield the same output indication. The performance of the test sample was then simply the difference in the two attenuator settings. If the linearity of the detector was adequate, the attenuator was left fixed and the material performeance was taken to be the difference in the two output indications, with and without the sample in place. If performance data were required at other frequencies, the frequency of the signal generator had to be changed and the measurements repeated. If the measurements had to be performed for different microwave bands, the signal generator had to be exchanged for another and the horns had to be appropriately replaced.

The availability of frequency swept sources, especially those controlled by computer, vastly expanded the utility of the arch. The single-band horn antennas must be replaced by broadband antennas, of course, but the system remains relatively simple because the signals to be measured are relatively strong. The size of the arch itself may be expanded simply by laying the arch over on its side, and other capabilities may be implemented. For example, Georgia Tech recently built a pair of horizontal arches with radii of approximately 8ft, fitted with sprocket drives that can move the antennas continuously along the arch. With swept frequency instrumentation, this makes it possible to measure the performance of a test sample over a broad range of frequencies as functions of the angles of incidence or scattering or both.

Figure 10-18 is an example of a swept frequency measurement of a Salisbury screen made on a horizontal arch. The test was made at near normal incidence, and the screen was designed for optimum performance at 6 GHz.

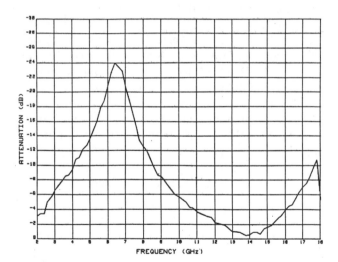

Figure 10-18 This measurement of a Salisbury screen was performed on a horizontal arch operated by the McDonnell Aircraft Company in St. Louis. In this inverted display, better performance is indicated by a higher position on the curve

10.3.2 The RCS Method

Because the horns in the arch method are only a few feet from the test panel, the incident phase fronts are spherical. If we seek to evaluate absorbers under the more realistic conditions in which they may be used, we might prefer the flatter phase fronts available in compact ranges or conventional RCS ranges. Whichever the case, the RCS method is very simple, requiring only that the absorber panel be mounted on a sturdy metal plate and the RCS pattern of the absorber-covered plate be recorded. The patterns of a bare plate and one covered with a test sample are shown in Figs. 10-19 and 10-20.

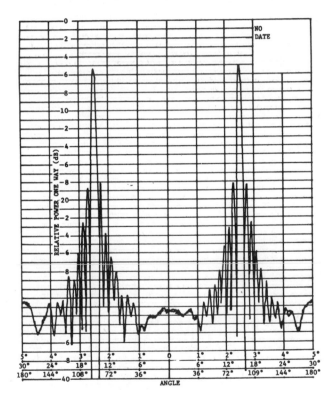

Figure 10-19 RCS pattern of a six-inch square plate measured at 10 GHz

These patterns were recorded at a frequency of 10 GHz, and the plate was 6in square, or about 5.1 wavelengths along a side. The specular (normal incidence) returns from the plate are prominently displayed. When the test panel is mounted on the plate, patterns like the one in Fig. 10-20 will be recorded. It can be seen that the specular lobe to the right has been attenuated by about 20 dB compared to its previous level; this number represents the performance of the material sample at this particular frequency. Although the bare plate pattern of Fig. 10-19 is included here for illustration, it is not necessary to make two separate measurements (one pattern with the test panel installed and one without). The left specular lobe (from the bare side of the plate) is clearly unchanged by the absorber mounted on the reverse side, hence we only need to make one measurement and compare the amplitudes from the bare and absorber-covered sides. Thus, it is a self-calibrating measurement.

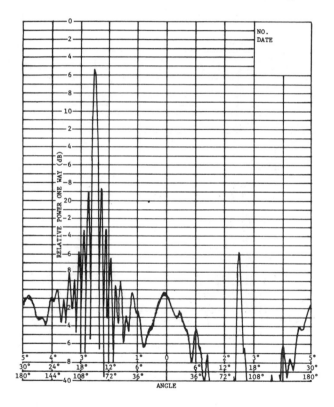

Figure 10-20 The RCS pattern of the same plate with a panel of AN-75 absorber (produced by Emerson and Cuming) attached to one side

There are some precautions that we should follow in making these kinds of measurements, however. First, the plate cannot be too small, or else the edge contributions will contaminate the measurements. Generally, the plate should be at least three wavelengths along an edge, and preferably five or more, and this requirement may even have to be extended to larger plate size if the material sample has very high performance characteristics. This is because the surface contribution to the echo may be reduced to the order of the edge contributions by very good materials, and the edge effects are not good indications of material performance.

On the other hand, a very large plate has a narrow specular lobe, and if this lobe is too narrow, it may be difficult to properly align the plate on its support fixture. The null-to-null width of the main lobe is approximately equal to λ/l radians, where l is the length of that edge of the plate perpendicular to the axis of rotation. From this relationship, we may make some estimates of the size of the plate required for accurate alignment. Assume, for example, that the typical alignment in the vertical plane cannot be held much closer than 0.5 degree, and that it is acceptable to be off the main lobe by as much as 1 dB. These requirements suggest that the plate be no more than about 15 wavelengths in size. Thus, it is a rule of thumb for the evaluation of material samples mounted on flat plates that the plates be from five to 15 or 20 wavelengths in size. These are not very strict requirements for most samples at most frequencies. Larger samples may in fact be used if the test plate can be mounted on an azimuth-over-elevation positioner, for which more control over the plate tilt angle is possible.

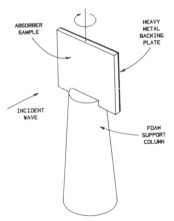

Figure 10-21 The absorber-covered plate may be mounted on a slotted foam support column for measurement

The plate support fixture may be designed as shown in Fig. 10-21. The support column is typically made of a low density foamed plastic material (see Ch. 11) with a slot cut into the top to accomodate the test plate and the material fastened to it. Low reflectivity support struts may have to be designed to stabilize large plates. Because the performance of some absorbing material designs, notably magnetic absorbers, depend on intimate contact with the backing plate, care must be taken to establish that contact. If adhesives are used, the adhesive layer should be the one recommended by the manufacturer or designer.

10.4 OTHER METHODS

10.4.1 Large Waveguides

Large waveguides can be used to evaluate production line samples of the pyramidal urethane foams used to line indoor anechoic chambers at low frequencies. The first such system employed this way was fabricated by the B.F. Goodrich Company Sponge Products Division, which was a large manufacturer of carbon-loaded foam absorbers until it discontinued production in the mid-1970s. This measurement system was quite distinct from the waveguide or transmission line measurements discussed in sec. 10.2.3 because only the reflectivity of the finished product was desired and no attempt was made to measure the intrinsic properties of the material. A larger and more modern version of the system was designed and fabricated by the Emerson & Cuming in 1981, as suggested in Fig. 10-22. This newer system is 90ft long, and is comprised of a broadband wave launcher at one end, a large exit aperture at the other, and a section of transition waveguide that flares out from the smaller dimension to the larger. The test panels must be mounted on a large plate and inserted into the open end of this structure, which measures 2×12ft. Thus, six material samples can be tested simultaneously.

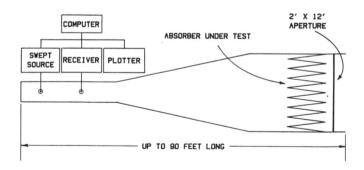

Figure 10-22 Schematic diagram of a large waveguide system for testing production line absorber products. Emerson and Cuming Corporation operates a system like this over a frequency range of 100 to 400 MHz. This system is intended primarily for evaluation of absorbers installed in commercial anechoic chambers

In the earlier B.F. Goodrich system, a probe was installed in the waveguide between the source and the flared section to sample the standing wave pattern inside this huge waveguide. The probe was not movable, as in the slotted section used in the bench set-up discussed in sec. 10.2.3, but instead was fixed. The standing wave pattern was actually measured by moving the test samples. A large metal plate was mounted on a rolling carriage that could be slid in and out of the open end of the flared waveguide. The absorber test panels were mounted on this plate for evaluation.

The test "panels" are actually large pyramids of varying sizes, depending on the production samples involved, but they all have bases measuring 2×2ft. Some production absorbers have pyramids as long as 15ft, and several panels must be installed side-by-side on the movable plate and inserted into the waveguide. The materials are evaluated by recording or noting the output of the probe section as the shorting plunger, with its test panels attached, is moved to and fro in the waveguide. In effect, the standing wave pattern due to the small reflection from the test samples was slipped back and forth past the probe, rather than sliding the probe along a stationary standing wave pattern, as described in sec. 10.2.3. The measurements could only be made at one frequency with this arrangement.

In the newer Emerson & Cuming system, the test panels are inserted into the waveguide and held fixed while the frequency is stepped over the frequency range from 100 to 450 MHz. The signal reflected by the test samples is recorded as a function of frequency, and the data are then transformed to the time domain with the FFT, essentially the reverse of the process used in time domain reflectometry. A display of the resulting reflected signal shows that reflections emanate from small discontinuities in the system as well as from the test samples, but the undesired reflections may be eliminated by filtering out all returns except those from the absorber samples. After the data have been filtered, they may be transformed back to the frequency domain, and we then have a display of the reflection characteristics of the material as a function of frequency. Emerson & Cuming claims sensitivities on the order of –60 dB with this system [6].

10.4.2 Interferometers

As mentioned earlier, a standing wave pattern is created when the wave reflected by a test slab of material combines with the incident wave impinging on the sample. The standing wave pattern can be used to deduce the nature of the reflection of waves and, for unknown test samples, the nature of the reflection can be used to deduce the electromagnetic properties of the test sample. When the standing wave pattern is measured outside the confines of a closed system, the measurement device can be called an interferometer. Interferometers were invented to investigate the nature of light and the term has carried over to microwave work.

A very simple interferometer is shown in Fig. 10-23, and it will be noted that much of the equipment, from the source up to and including the slotted section, is identical with that of Fig. 10-8. However, instead of a sample holder and a sliding short circuit, the interferometer of Fig. 10-23 uses a small horn and a tuner. The test sample is mounted on a metal support plate relatively close to the horn (no more than a few feet), hence the incident phase fronts are spherical. This is a disadvantage of the system, but the test results are not usually seriously in error because of the wavefront curvature.

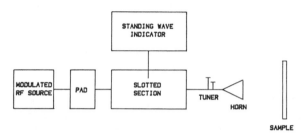

Figure 10-23 A one-horn interfermoeter uses some of the elements of Fig. 10-8 but this set-up is used for much larger samples

Before the test sample is measured, residual system reflections, primarily due to impedance mismatches associated with the small horn, must be minimized, and this may be accomplished by adjusting the tuner. Residual reflections from objects in the laboratory must also be minimized by isolating the test set-up, or by covering the objects with absorbing material (not part of the test sample), or both. Once the residual reflections have been eliminated or minimized, a metallic plate is placed some distance in front of the horn to verify that the sample will be adequately illuminated. Obviously, if the metal plate is too far away, the signal reflected back into the system will be small and the measurements will be meaningless.

After the plate has been positioned to give a reasonably high VSWR, the test panel is attached to the plate and the VSWR measured. The reflectivity of the test sample can be deduced from a comparison of the two readings. The one-horn interferometer is best used for measurements of the reflectivity of sample materials, rather than the measurement of the intrinsic material properties, and the simple equipment shown in Fig. 10-23 is best used for measurements one frequency at a time. This sample system is a crude one at best, but it can yield acceptable results if care is taken in setting it up and using it.

At very high frequencies, e.g., 35 GHz and up, it is no longer feasible to use coaxial transmission lines, or even waveguides, for the measurement of the intrinsic properties of materials because the transverse dimensions of the line become very small. This imposes sample fabrication tolerances that become impractical to meet and another approach is required. One such approach, the Fabry-Perot interferometer, is based on the design of a device used to measure the wavelength of light.

The Fabry-Perot interferometer is basically a microwave cavity without sidewalls [7]. A microwave cavity is a hollow metal box that can be excited by means of a small aperture or loop. Depending on the size and shape of the box, and the frequency of the electromagnetic excitation, a three-dimensional standing wave structure can be created inside the box. These modes exist for only certain frequencies, which can be related to the internal dimensions of the cavity. For elementary shapes, the modes are well known and the resonant cavity can be used as a frequency measuring device (see Ch. 11 of reference [2]).

A cavity can be characterized by its Q, the ratio of the energy stored in its internal electromagnetic fields to the energy lost, primarily due to ohmic conduction in its imperfectly conducting inner walls. The intrinsic properties of very small test samples can be determined by noting the change in the resonant frequency and Q of a cavity when a sample is inserted. For the purposes of our discussion, however, we will consider the Fabry-Perot interferometer in which the side walls of the cavity are eliminated, and in which slabs of material are inserted instead of very small samples.

The classical Fabry-Perot interferometer used in optical work is a kind of open resonator with two partially silvered mirrors facing each other. The plates may be placed from 0.1 to 10 cm apart, which can be several thousand wavelengths at optical frequencies, and the source is usually a large aperture, uniformly illuminated by a monochromatic light source, as suggested in Fig. 10-24. One of the plates is mounted on a precision traversing mechanism for reasons that will become clear in moment. A collimating lens is installed behind the second plate to project the transmitted rays onto a screen for observation.

As shown in the diagram, a ray due to some point on the extended source is partly reflected from, and partly transmitted through, the first plate (the reflected ray is of no interest and will be ignored). Because the transmitted ray is passed through the first plate at a slight angle to the axis of the system into the region between the two plates, it suffers several reflections back and forth along a zig-zag path. A portion of each zig-zag ray impinging on the second plate is transmitted through the plate and through the lens, and finally reaches the observation screen. It can be seen that a collection of rays impinging on the first plate at the angle θ is brought together at the screen. This generates a circular fringe pattern on the screen, centered on the axis of the collimating lens.

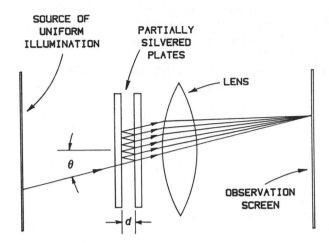

Figure 10-24 The optical Fabry-Perot interferometer creates a fringe pattern from which the wavelength of the light source can be measured. One of the partially silvered plates must be mounted on a precision adjustable carriage, (not shown)

The fringe pattern is a collection of alternating bright and dark circles, with the bright circles satisfying the condition

$$2d \cos\theta = m\lambda \tag{10-23}$$

where m is an integer and d is the spacing between the inner plate surfaces. We can deduce the wavelength of the light by counting the number of fringes on the screen and then moving one plate a very short distance until a different number of fringes appear. The change in the plate spacing can then be used to calculate the wavelength.

When applied to the measurement of test materials at millimeter wavelengths, the classic configuration of the Fabry-Perot interferometer may be changed somewhat [8,9]. For one thing, microwave energy is not projected on an extended screen and a small antenna is used instead to capture the energy. Similarly, a small antenna or slot is used to inject energy into the space between a pair of highly conductive plates or mirrors. Depending on the particular design employed, the mirrors may be flat or they may be curved in what is called a *confocal* configuration. The mirrors form a kind of cavity because a standing wave pattern is generated by the reflections back and forth between the two surfaces, and the device may even be characterized by its Q. However, the energy loss is now due primarily to waves diffracted by the edges of the mirrors and other apertures in the system instead of conduction losses in the "walls" of the cavity.

In operation, the equipment is carefully set up and aligned, and it must be remembered that one of the mirrors must be movable. The sample holder may be a frame that is large enough so as not to interfere with the standing wave structure generated between the plates when there is no sample present. The position of the movable plate is adjusted for a maximum indication of detected signal, and because the spacing between the plates is large compared to the wavelength, there may be several such positions. These positions may be measured by means of fine dial gauges or precision screws in the traversing mechanism, hence allowing the microwave signal wavelength to be measured as well. Then the sample, usually a slab of the material of interest, is placed in the sample holder, and the movable plate is again adjusted for maximum received signal. Because the insertion of the sample slab increases the electrical distance between the reflecting plates, the distance the plate must be moved to restore the resonance condition is an indication of the electrical thickness of the sample. If no adjustment of the plate is necessary after the sample slab has been inserted, the sample must be an integral number of half-wavelengths in thickness. This seldom occurs for unknown materials, however, and if it does, it is simply by coincidence. But just as with the measurement of samples in transmission lines, a single measurement is not sufficient to determine the material properties unambiguously. The measurement should be repeated with a slightly different sample thickness, and preferably two additional thicknesses, to obtain unambiguous results.

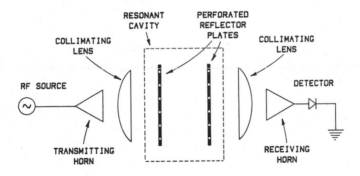

Figure 10-25 This microwave version of the Fabry-Perot interferometer depends on measurements, with and without a sample inserted between the perforated plates. The perforated plates play the same role as the partially silvered plates in the optical interferometer of Fig. 10-24

An interesting version of the Fabry-Perot interferometer is shown in Fig. 10-25, taken from reference [10]. Collimating lenses are placed near the transmitting and receiving antennas to flatten out the incident and received phase fronts, and the reflecting plates on either side of the sample are perforated with small holes. The size and shape of these small perforations are chosen to optimize the reflection of energy between the plates while allowing energy to be injected into and sampled from the test region. In every sense of the word, the perforated plates are the microwave counterparts of the half-silvered mirrors used in the optical version of the Fabry-Perot interferometer.

10.4.3 The Admittance Tunnel

To the best of our knowledge, the admittance tunnel was developed by the Rockwell International Corporation (Tulsa Division) for evaluating thin sheet-like materials used in the production of circuit analog absorbers. We have already mentioned that it is very difficult to mount thin sheets in transmission lines for the evaluation of intrinsic material properties. Moreover, the geometric patterns deposited on circuit analog sheets are often on the order of the transverse dimensions of the waveguide or transmission line, hence they represent large scale inhomogeneities that would lead to serious errors in transmission line measurements. The admittance tunnel is a solution to these problems.

The admittance tunnel is essentially a long absorber-lined box illuminated by a small antenna at one end and fitted with a test aperture at the other. The test sample to be measured is fastened across the aperture, and the box itself is intended to isolate the sample from unintentional reflections from objects in the laboratory, and to provide a reasonably planar incident wave on the sample. The admittance tunnel may have transverse dimensions of perhaps 2ft and may be 12ft long, depending on the operating frequency.

In one form of the admittance tunnel, the sample may be backed by a large metal plate, as shown in Fig. 10-26. The plate plays identically the same role in these measurements, as does the short circuit in transmission line measurements. In the particular version shown, a hybrid *tee* (T-section) separates the signal reflected by the sample (and the shorting plate behind it) from the transmitted signal. The reflected signal is routed to a phase-amplitude receiver which, for proper operation, requires a small reference signal sampled from the microwave source. With this particular type of receiver, the source is a continuous wave (CW) device that needs no amplitude modulation.

Another sample of the transmitted signal is coupled, routed through a waveguide attenuator and phase shifter, and combined with the received signal so that residual reflections from the empty box and the antenna itself may be minimized. This is a version of the classic cancellation technique often used in indoor RCS ranges, discussed in a later chapter. To "tune out" these residual reflections, the aperture at the far end of the admittance tunnel is plugged with high quality absorbing material, and the phase shifter and attenuator are adjusted for a minimum signal indication at the receiver. The system is calibrated by placing the metal plate against the aperture, then pulling the plate away with the automatic drive system. In the system shown in Fig. 10-26, a synchronizing signal proportional to the plate position is delivered to the chart drive of a recorder, making it possible to record the amplitude and phase of the reflected signal as a function of the position of the plate. When referenced to the amplitude of the incident wave, the amplitude and phase of the recorded signal is essentially the complex reflection coefficient of the metal plate. In an error-free system, the normalized amplitude should be precisely unity and the phase angle should be proportional to $2kd$, where d is the distance between the aperture and the shorting plate.

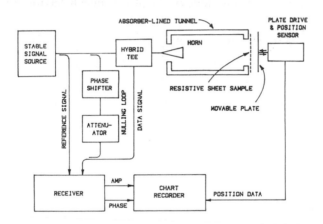

Figure 10-26 The admittance tunnel is an absorber-lined box with an aperture at one end. In this particular version, a movable plate is positioned behind the sheet to be measured

The purpose of the measurement is to determine the admittance of the test sheet, which may be characterized in terms of a conductance G and a susceptance B,

$$Y = G + iB \tag{10-24}$$

It is convenient to normalize these values with respect to the admittance Y_0 of free space,

$$y = Y/Y_0 = g + ib \tag{10-25}$$

for which g and b are the normalized conductance and susceptance of the sheet.

After the system has been calibrated, the test sheet is attached to the aperture, and the measurement is repeated. The amplitude and phase of the reflection coefficient associated with the sheet backed by the variable short circuit will trace out curves similar to those shown in Fig. 10-27. Note that the amplitude oscillates between maximum and minimum values, with the maximum being unity (0 dB), and that the phase of the reflection coefficient does not increase uniformly. In fact, for sheet conductances greater than the admittance of free space, the relative phase angle will attain a maximum value, and then decrease. The null in the amplitude trace does not occur midway between the two maxima, but is shifted somewhat to the left or right of the mid-point, depending on the value of the susceptance. The complex admittance of the sheet may be calculated from the depth of the null in the amplitude curve, the position of the null, and the slope of the phase angle behavior at the null.

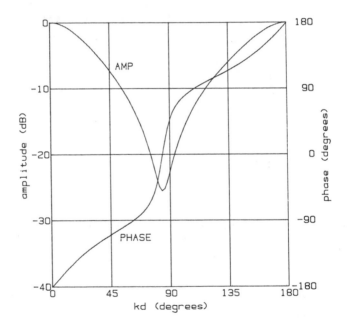

Figure 10-27 Theoretical traces of the amplitude and phase of the reflection coefficient for a sample with g = 0.9 and b = 0.1. The sheet conductance is obtained from the minimum value at the null and the susceptance is obtained from the position of the null. The slope of the phase behavior establishes whether the conductance is less than or greater than that of free space

Specifically, let Γ_m denote the minimum value of the recorded reflection coefficient at the null, and let d_0 be the position of the null relative to the left maximum in the amplitude trace. Then the normalized conductance is

$$g = \begin{cases} \dfrac{1 + |\Gamma_m|}{1 - |\Gamma_m|} & \text{for negative phase slope} \\[2ex] \dfrac{1 - |\Gamma_m|}{1 + |\Gamma_m|} & \text{for positive phase slope} \end{cases} \tag{10-26}$$

Note that without the slope of the phase behavior, we do not know if the normalized sheet conductance is greater than or less than unity (i.e., greater or less than the admittance of free space). If we prefer to characterize the sheet in terms of its resistance per square, this is simply $R = 377/g$. The normalized susceptance is

$$b = -\cot(kd_0) \tag{10-27}$$

Note that, depending whether kd_0 is greater or less than $\pi/2$ radians, the susceptance will be either positive or negative, respectively.

In a simpler version of the measurement, the shorting plate may be omitted and a large panel of high quality absorber placed a few feet beyond the aperture. In this case, the test sheet is not backed by a reflection coefficient of unit amplitude and variable phase, as furnished by the metal plate, but by a reflection coefficient that is nearly zero. The low reflection is provided by the absorbing material placed beyond the test aperture. As before, however, calibration is effected by the placement of a metal plate at the aperture, and provisions must be made to measure the phase as well as the amplitude of the signal reflected by the plate. Then the sample is installed and the amplitude and phase of the signal reflected by the sample is recorded. The two measurements serve to establish the reflection coefficient of the sample, which we shall call Γ.

The reflection coefficient of the thin sheet backed by the admittance of free space (which is the admittance provided by the absorber placed beyond the aperture) is

$$\Gamma = -\frac{y}{y + 2} \tag{10-28}$$

which may be solved for y,

$$y = -2\Gamma/(1 + \Gamma) \tag{10-29}$$

Because Γ is the quantity resulting from the measurement, the normalized conductance is the real part of Eq. (10-29) and the susceptance is the imaginary part. As with many of the material evaluation techniques discussed in this chapter, the admittance tunnel may be converted to swept frequency operation, thereby allowing the measurement of sheet properties over a wide range of frequencies in a short time.

10.5 SUMMARY

As we have seen in this chapter, there are several ways to evaluate absorber materials and products, and the kind of information desired dictates the particular kind of measurements that should be made. The measurement of electromagnetic material properties generally requires more complicated procedures because there are no fewer than four uknown quantities that must be determined from the measured data. The evaluation of absorber performance, on the other hand, requires only the measurement of the reflectivity of selected test panels or samples.

Either kind of measurement may be performed with relatively simple laboratory equipment, but the simpler the equipment, the more labor is required to extract the desired information. In particular, we are almost always interested in the frequency dependence of product performance or electromagnetic properties, and manual methods are simply too time consuming for this kind of information. Automated swept frequency and stepped frequency equipment make it possible to collect large amounts of data in relatively short times, but require a significant capital investment. However, the investment can be justified in most cases.

The electromagnetic properties of materials can be accurately measured only within closed (waveguide) systems, while absorber products or panels are best evaluated in open systems. Examples of the latter are the NRL arch method and free-space RCS testing performed in an anechoic chamber or on an outdoor range. The arch is quite suitable for swept frequency testing, whereas RCS measurements are best used for measurements at a limited number of discrete frequencies. In one instance, a manufacturer of commercial absorber products uses a large waveguide system to evaluate product performance, but this is for low frequencies and is not used for the shorter wavelengths. Small samples are required for the waveguide measurements used to determine electromagnetic properties, and material samples must be uniform and homogeneous to reduce the possibility of exciting higher order modes within the sample. The effects of inhomogenieties are largely averaged out when we measure the reflectivity of standard test panels.

At very short wavelengths, it is no longer feasible to machine small waveguide samples to the tolerances required, and interferometric methods become attractive. Based on techniques used at optical wavelengths, these methods and equipment allow the use of much larger test samples than demand by the waveguide methods. Nevertheless, the samples still must be homogeneous, and the interferometric methods require very precise sample holders and positioning carriages.

REFERENCES

1. Walter C. Johnson, *Transmission Lines and Networks*, New York, McGraw-Hill, 1950.
2. Simon Ramo and John R. Whinnery, *Fields and Waves in Modern Radio*, New York, John Wiley and Sons, 1960 (second edition).
3. Robert E. Collin, *Field Theory of Guided Waves*, New York, McGraw-Hill, 1960.
4. Edward L. Ginzton, *Microwave Measurements*, New York, McGraw-Hill, 1957.
5. A.M. Nicolson and G.F. Ross, "Measurement of the Intrinsic Properties of Materials by Time-Domain Techniques," *IEEE Trans. Instrum. Meas.*, Vol. IM-19, No. 4, pp. 377-382, November 1970.
6. Private communication with Mr. William H. Emerson, Emerson & Cuming, 5 November 1984. Mr. Emerson was responsible for the design of the waveguide system currently in use at Emerson & Cuming as well as the older version used by the B.F. Goodrich Company Sponge Products Division.
7. K.H. Breeden and J.B. Langley, "Fabry-Perot Cavity for Dielectric Measurements," *Review of Scientific Instruments,* Vol. 40, No. 9, pp. 1162-1163, September 1969.
8. W. Culshaw and M.V. Anderson, "Measurement of Dielectric Constants and Losses with a Millimeter Wave Fabry-Perot Interferometer," National Bureau of Standards, Report 6786, 19 July 1961.
9. J.E. Degenford and P.D. Coleman, "A Quasi-Optics Perturbation Technique for Measuring Dielectric Constants," *Proc. IEEE*, Vol. 54, No. 4, pp. 520-522, April 1966.
10. Constantine A. Balanis, "Investigation of a Proposed Technique for Measurement of Dielectric Constants and Losses at V-Band Using the Fabry-Perot Principle," Master's Thesis, School of Engineering and Applied Science, University of Virginia, August 1966.

CHAPTER 11

RCS MEASUREMENT REQUIREMENTS

E. F. Knott

11.1 MEASUREMENT OBJECTIVES

There are five basic reasons for conducting RCS measurements, and each has a different influence on the way the measurements are carried out. The five reasons are:

- Acquire understanding of basic scattering phenomena
- Acquire diagnostic data
- Verify system performance
- Build a data base
- Satisfy a contractual requirement

Some of these may seem to overlap, but the overlapping depends on the particular objective of the measurement program. This will become more clear in a moment.

The most basic of the five reasons is the pursuit of an understanding of fundamental scattering phenomena. Despite the completeness of electromagnetic theory, there remain phenomena that are incompletely described, such as the scattering from the base of a cone. In the early 1960s, for example, Keller applied his new geometric theory of diffraction to the scattering from a cone [1,2]. Remarkable for its "cookbook" simplicity, this theory accurately predicted the radar return from a cone, except over certain angular sectors.

Bechtel noted errors in Keller's 1961 paper, and gave the corrected results in 1965 [3]. Nevertheless, comparisons between measurements and predictions made with the correction formula still showed disagreement, a problem reiterated by Bechtel in 1969 [4]. Blore conducted an experimental study of blunted cones [5], and noted the failure of the theory, but offered no explanation of the failure nor of a correction that might be applied.

Using an equivalent current approach, Burnside and Peters sought to improve the predictions by accounting for multiple diffraction across the base of the cone [6]. This was done for axial incidence only, and Knott and Senior extended the analysis to additional aspect angles [7]. Although this resulted in an improvement, it appears that even higher-order terms must be included. Knott and Senior conducted a careful series of measurements of right circular cones, measured with and without a pad of absorber cemented to the base, demonstrating the importance of the multiple cross-base diffraction effects [8].

Not all the researchers mentioned above carried out measurements, but the evaluation of the theory depended on comparisons between measurements and predictions, and during the 12-year span between Keller's comparisons and the measurements by Knott and Senior, it was the pursuit of a single idea that motivated the measurements: *a better way to explain diffraction by the base of a cone*. This is merely one example of measurements conducted to understand a basic scattering phenomena.

A second reason for making RCS measurements is to acquire diagnostic data. Although many of the cone measurements were used to gauge the validity of a theory being applied, it is doubtful that any of the data were used to assess the elusive cross-base interaction terms. For that purpose, it would have been necessary to isolate only those interactions, a difficult task considering the physical sizes of the targets being tested.

However, more often than not, diagnostic testing currently means the isolation of "flare spots" on major weapons systems. One way to accomplish this is to somehow conceal or cover up a suspected flare spot, and to compare measurements made with and without such treatment. The difference in the two patterns is an indication of the intensity of the flare spot, provided its contribution to the total return is not masked by contributions from the remainder of the target. Typically, all suspected flare spots must be treated, and measurements made selectively, with each flare spot exposed, one at a time.

The identification of flare spots can be made considerably easier with the use of specialized instrumentation, as mentioned in Ch. 13. Certain radars are designed to emit chirped pulses, and when the data are processed, a radar "image" of the target can be produced. The image is a plot of the intensity of the scattering as a function of range and cross range, and the resolution of some instrumentation systems is as good as 6in or better. Images can be produced for virtually any aspect angle, and their diagnostic value lies in identifying the major scattering centers as functions of aspect angle.

The third reason for RCS measurements is to verify that the RCS performance of a system, a new aircraft, for example, meets specified levels. Although, as we have seen, there may be several computer programs capable of predicting that performance, no prediction technique is all-encompassing and all codes have limitations. Therefore, it is often necessary to test the final, integrated weapons system, although portions of the system already may have been treated or investigated individually.

Verification of system performance must often be carried out at several threat frequencies, and the measured performance must be compared with specifications. This can be done informally at a contractor's facility, but the government may demand that the measurements be performed at a national facility, such as RATSCAT.

The fourth reason for conducting RCS measurements is to build a data base. Although this is an admirable objective, the data base seldom satisfies the needs of a *true* data base. This is because the users of the data often reduce an entire pattern to a single number used to develop evaluation criteria or to assess a sensor's target response. Moreover, the comparison of patterns from one target with those of another often becomes an excruciating task because of the sheer human effort required to sift through a multitude of patterns. As a result, the patterns are often reduced to a handful of numbers compared by computer algorithms, and the very necessary injection of human judgement is bypassed. The patterns comprising a large and significant data base are usually examined but once, if that, and then collect dust in a repository until the documentation itself crumbles to dust.

The fifth and final reason for RCS measurements is for contractual purposes. Although this may be necessary to ensure compliance with specifications, for example, some program managers order the measurements simply as insurance against criticism. RCS measurements are time-consuming and costly, and they should never be ordered off handedly.

Thus, the five basic reasons for conducting RCS measurements range from a purely academic endeavor to a purely contractual requirement. Whatever the reason, it influences (or should influence) the design of the experiment and the way in which the experiment is carried out. Measurements made for scientific purposes are best conducted when a single parameter is carefully varied from one test to another. The next test in the sequence is often based on a running analysis of previous tests, and, hence, the direction taken by experimental work often cannot be predicted in advance. By contrast, data-base measurements usually follow a rigid test plan, and, as such, there is little opportunity to explore unusual or inexplicable results. Thus, the purpose of the measurements should always govern the design of the experiments, and thereby precisely determine what information is to be collected.

11.2 TYPES OF RCS MEASUREMENTS

The purpose of a radar cross section measurement range is to collect radar target scattering data. Usually — but not always — the range user requires far-field data, corresponding to the case where the target is located far enough from the instrumentation radar that the incident phase fronts are acceptably flat. Many times this requirement can be satisfied only at an outdoor range. On the other hand, there are many research and development programs that can and should be conducted indoors in anechoic chambers.

Whether indoors or outdoors, an RCS measurement facility must have at least five features:

1.   An instrumentation radar capable of launching and receiving a microwave signal of sufficient intensity.
2.   Recording instruments, either analog or digital or both, for saving the information.
3.   A controllable target rotator or turntable.
4.   A low background signal environment, including "invisible" target support structures, to minimize contamination of the desired signals.
5.   A test target suitable for the measurements.

These five basic elements can, of course, be embellished and augmented by other equipment and instrumentation. The instrumentation used indoors is different from that used for outdoor measurements, as shown later in this chapter.

Once the decision has been made to pursue a measurement program, a suitable facility must be found. As we shall see, several large corporations have their own facilities, and the selection of a test range is automatically bypassed. Occasionally, however, corporate facilities may not be suitable, and negotiations will have to be conducted with outside agencies. Occasionally, a contractor must use the RATSCAT range in New Mexico as a contractual requirement.

Negotiations usually involve the specification of a set of test conditions and a test matrix, and the prospective range will submit a bid. Unfortunately, some customers are not sure of their own test requirements, and they may ask for more test conditions than they actually need. In many cases, the prospective range offering its services can suggest alternate conditions that satisfy the customer's needs at lower cost.

There are four basic kinds of RCS measurements, classified according to pulse width (long or short) and coherence (coherent or noncoherent). In long-pulse measurements, the transmitted pulse is wide enough to bracket the target by a comfortable margin. As such, the measured RCS characteristics are identical to those that might be measured using a CW radar. In fact, most measurements made in indoor chambers are CW measurements because chamber reflections complicate the design of pulsed radar receivers.

The pulse width for short-pulse measurements is usually narrow enough that range resolutions on the order of 10 to 20cm can be achieved, which is typically a fraction of the total target length. Narrow pulses are useful for diagnostic measurements, but they impose system bandwidth requirements that are not always easy to meet. The sensitivities of short-pulse systems are less than those of long-pulse systems because of the additional bandwidth required for the system.

Long- and short-pulse measurements are easiest to make if the radar system is noncoherent. Noncoherent systems sense only the amplitude of the return signal, whereas coherent systems sense the in-phase and quadrature components (I and Q) of the received signal. The phase angle of the return signal can be computed from the I and Q components. However, coherent radar instrumentation is more costly than noncoherent instrumentation.

Most RCS measurements of interest are for the monostatic case, in which the radar transmitter and receiver are sensibly at the same point in space. For the bistatic case, the transmitter and receiver are separated, sometimes at angles as high as 180 degrees (see Fig. 11-1). In fact, some radars employ separate transmitter and receiver antennas, even for monostatic measurements, and hence a non-zero bistatic angle is subtended at the target. However, the antennas are usually so close to each other that the measurement is indistinguishable from the monostatic case.

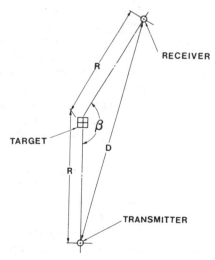

Figure 11-1 Bistatic measurement layout

Fig. 9-10. Other common shapes include an aggregate sine wave (convoluted) front, conical shapes, and off-normal angle wedges. These types of absorbers can provide reflectivity reductions in excess of 50 dB, but may require thicknesses in excess of 10λ to do so [12]. Figure 9-11 provides an indication of the RAM thickness needed for a given level of RCSR *versus* frequency, and of the performance of the RAM at angles far off-normal [13].

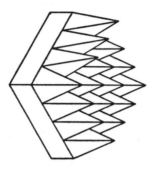

Figure 9-10 Geometric transition absorber

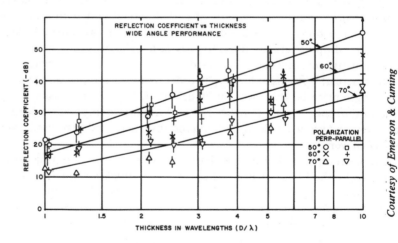

Figure 9-11 Reflection coefficient as a function of absorber thickness for pyramidal absorbers at various incidence angles (reprinted from reference [13])

Modern radio frequency generation technology makes it possible to synthesize high resolution measurements using frequency-stepping techniques. The RF source is typically a low power, voltage-controlled oscillator that is phase-locked to a highly stable quartz crystal oscillator. Thus, a controller or computer can step the radar frequency across a specified band at specified intervals. The output of the low power source usually must be amplified (with a traveling wave tube amplifier, for example) to achieve the necessary transmitted signal level. The received data are typically recorded on magnetic tape for subsequent processing. Software processing is then used to create radar imagery.

The following list includes almost all the types of radar signature measurements of interest:
- Conventional (amplitude only)
    - monostatic
    - bistatic
- High resolution
- Coherent (amplitude and phase)
    - glint
    - Doppler
- Frequency stepped, coherent
    - radar imagery
    - high resolution

Examples are shown in Fig. 11-2 through 11-7.

Figure 6-24, discussed earlier, is an example of a conventional long-pulse, amplitude-only measurement. The ordinate is the radar cross section of the model in dB relative to a square meter, and the abscissa is the yaw angle as measured between the longitudinal aircraft axis and the radar line of sight. The rectangular plot is a very common form of displaying measured patterns.

No bistatic patterns are available for display here. In practice, the transmitter and receiver are set up at the appropriate bistatic angle and their positions relative to the target are held fixed. The target is rotated for this fixed transmitter-receiver orientation, generating a pattern similar to the monostatic RCS pattern. However, the greater the bistatic angle, the fewer the pattern scintillations.

Figure 11-2 is an example of high resolution RCS measurements and how they may be displayed. This particular form of presentation is often called a "waterfall plot," and the information could easily be displayed as a contour chart. It shows the amplitude of the return as a function of range and target aspect angle, and, as judged from the plot, the range resolution for this particular chart was on the order of 15 cm. The target itself appears to have been about 10ft long. If the chart is examined very carefully, it can be seen that the returns from the near end of the body moved away from the radar, and those from the far end moved toward the radar, as the aspect angle increased. At the higher aspect angles, the returns become larger. This form of display helps isolate the scattering centers on the target, but the display itself cannot be generated unless the instrumentation radar is designed for short-pulse generation and reception.

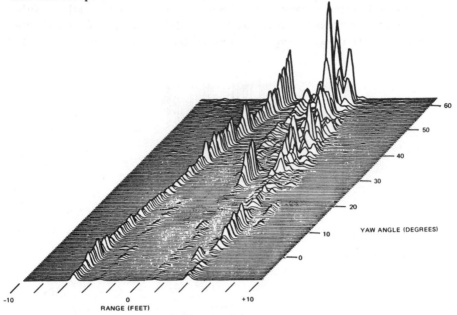

Figure 11-2 High resolution RCS data

A coherent radar system allows the measurement of the relative phase of the signal return in addition to the amplitude. Figure 11-3 is a typical phase plot, showing the phase of the return as a function of target aspect angle [9]. Because the phase angle is constrained to lie between zero and 2π radians (or between $-\pi$ and $+\pi$ radians), the recorded phase variation has a "sawtooth" behavior, as shown in the upper diagram. Under many circumstances, the phase plot can be "unwound" to account for the number of complete 2π cycles undertaken by the total phase excursion. This is shown in the lower diagram of Fig. 11-3. Interestingly enough, the lower diagram is skewed, suggesting that this particular target had not been placed symmetrically on the axis of rotation, or that the phase behavior has not been adequately "unwound." In many cases, especially over aspect angles where the phase changes rapidly, it cannot be established whether the change in the phase angle from one aspect to the next was positive or negative. The pattern of Fig. 11-3 suggests that the target apparently had a pair of echo sources, one of which was dominant over two-thirds of the pattern.

Phase information, as plotted in Fig. 11-3 is useful only for diagnostic, interpretive reasons. By processing the coherent data collected by a suitably designed radar, we can calculate the target glint. Glint is the cross-range location of the effective center of target scattering, and is due to the way the individual contributions of all the scatterers add together. It is possible for the apparent center of scattering to lie beyond the physical constraints of the target.

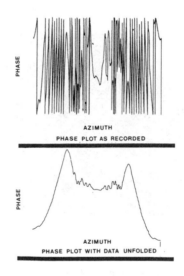

Figure 11-3 Phase plots [9]

A glint pattern can be generated by plotting a quantity proportional to the rate of phase charge with respect to the angle of rotation. Such a pattern is shown in Fig. 11-4 (also taken from reference [9]), and the data can be collected using a single antenna. However, it is not essential that a coherent radar be used. Glint patterns can also be generated by measuring the phase difference between the signals received by a pair of antennas stationed at some known distance apart.

A coherent radar system is essential for recording the Doppler characteristics of targets that have exposed moving parts. After coherent detection in the receiver, the received IF signal must be compared with a coherent reference IF signal so as to separate the negative Doppler shifts from the positive ones. The process generates data such as shown in Fig. 11-5, recorded for a two-engine, propeller-driven aircraft [10]. Note that the dominant Doppler frequency

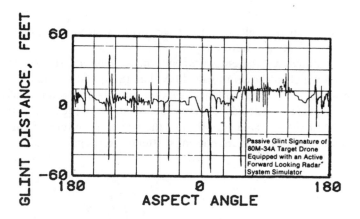

Figure 11-4 Target glint pattern [9]

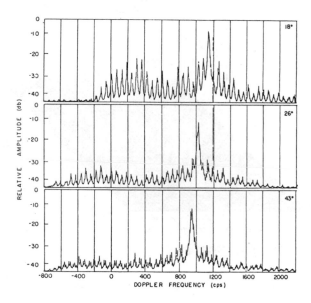

Figure 11-5 Measured doppler spectrum of a propeller-driven aircraft (from Gardner [10])

changes with aspect angle, and the spectral lines are not symmetrical with respect to the dominant frequency.

It also requires a coherent radar system to generate radar imagery, an example of which is shown in Fig. 11-6. The radar must transmit a chirped pulse in which the radar frequency is swept linearly across an interval in the radar band, preferably at least 1 GHz wide. A sample of the transmitted signal is mixed with the received signal, and the spectral content of the detected output serves to locate the individual scatterers in range.

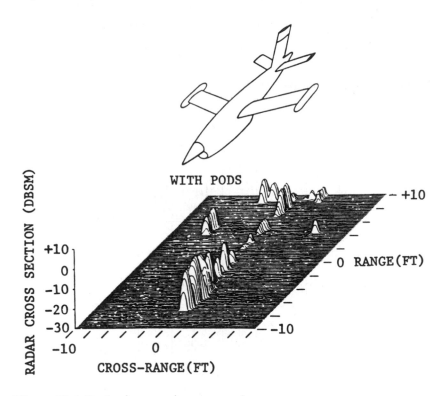

Figure 11-6 Radar imagery in range and cross range

The received data must be recorded and processed to generate the plot shown in Fig. 11-6. This involves a two-dimensional Fourier transformation, as was accomplished in the early days with optical processing techniques. However, the availability of frequency stepping technology makes it possible to accomplish the same effect using digital processing techniques. The resolution in range depends on the frequency intervals used, and the resolution in cross-range depends on the angular width of the aspect angle sector over which the data are processed.

As suggested by the display of Fig. 11-6, the engine of the drone is a large scatterer. Other important scatterers are the pods on the wing tips and the empennage at the rear of the fuselage. This plot was generated using data collected for a narrow angular sector centered on the nose-on aspect angle. Images can be created for any other aspect angle, of course, but the relative strengths of the scatterers will be different.

11.3 THE FAR FIELD REQUIREMENT

Equation (3-1), the formal definition of radar cross section, states that the distance r between the radar and target must become infinite. The reason for this is to eliminate any distance dependence in the RCS characteristics. The limiting process essentially requires that the target be illuminated by a plane wave, yet the wave incident on a target in practical measurement situations is nearly always a spherical wave, due to the finite separation between the target and the instrumentation radar. As a matter of practical interest, we may wonder how "spherical" the incident wavefronts can be while still being reasonably good approximations of a plane wave.

One way to resolve the question is to assume the radar to be a point source, and to examine the deviation of the incident phase fronts from perfect uniformity over an aperture having the same width as the target. From the geometry of Fig. 11-7, we can find the distance h in terms of the range r and the transverse width d,

$$h = r \left\{ 1 - \left[1 - \frac{d}{2r} \right]^2 \right\}^{1/2} \tag{11-1}$$

which, assuming $d \ll 2r$,

$$h \simeq \frac{d^2}{8r} \tag{11-2}$$

Thus, the phase of the incident wave at the center of the target is different from the phase at the extremes of the target by the amount kh. If we arbitrarily require that this phase deviation be less than $\pi/8$ radians (22.5 degrees), then the familiar far field condition is obtained,

$$r \geq 2 \, \frac{d^2}{\lambda} \tag{11-3}$$

Note that the requirement $kh \leq \pi/8$ is quite arbitrary, and does not consider any accuracy statements about the measurement of the far scattered field. Despite its arbitrary nature, Eq. (11-3) has become a widely used yardstick for "ensuring" the accuracy of far-field RCS measurements.

Kouyoumjian and Peters considered the range requirement in far greater detail [11], including the axial decay in incident field strength as well as the

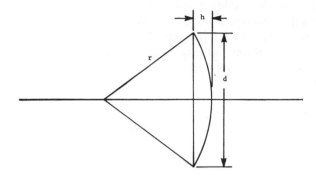

Figure 11-7 Phase deviation over a transverse aperture due to spherical incident wave-fronts

transverse phase deviation. They concluded that a measurement can "be optimized with respect to minimum range and maximum sensitivity by a proper
choice of antenna size," although no general statements could be made on the
accuracy of the measurements. Knott and Senior later conducted a numerical
study of errors on very simple targets [12] and assigned values to the errors,
and they concluded that even the standard far-field criterion, Eq. (11-3), might
not be sufficient for certain classes of targets. Some of the large errors noted
were due to near-field cancellation of the returns from various scattering
centers on the targets that would not, in fact, occur for true plane wave
incidence. Nevertheless, Eq. (11-3) remains a good rule of thumb for establishing the minimum range at which targets should be measured.

   A plot of the far field distance prescribed by Eq. (11-3) is given in Fig. 11-8
for a variety of frequencies and target size. The far-field requirement is not too
hard to meet at low frequencies, but typical radar sensitivities make it hard to
meet at high frequencies. For example, a 10ft target should be measured at a
range of not less than 200ft at 1 GHz, but not less than 2000ft at 10 GHz.

   In some cases, the radar sensitivity is not good enough for the target to be
measured at the far-field distance, and a considerably smaller range may have
to be selected to ensure adequate received signal strengths. For very simple
targets, such as cylinders and flat plates, the first symptom of near-field effects
is the filling in of nulls and the reduction of the amplitudes of the sidelobes
[13]. As the range is further shortened, the nulls become entirely filled in, the
sidelobes become shoulders in the pattern, and the main lobe amplitude decreases significantly.

   The effects of measuring more complex targets at less than the standard
far-field distance are less obvious. At high frequencies, each feature of the
target scatters energy more or less independently of other target features, and

each feature is usually significantly smaller than the overall target. Such features, which could be tail fins, engine intakes, nose tips, or other target surfaces, could each be in the far field with respect to its own characteristic size, while the composite target may not be. Thus, the amplitude of the scattering from each feature, as well as the locations of peaks and nulls in its own pattern, are less sensitive to the measurement range. The primary effect of a near-field measurement in this case is the slight shifting of the lobes and nulls of the composite pattern as compared with the true far-field pattern. Therefore, measurements performed at less than the standard far-field distance can often be defended.

Moreover, high accuracy in RCS measurements is often unnecessary unless the data are to be used as the basis for detailed analytical work or the development of scattering predictions. Often, users of test data require only median values, which are statistical representations of the return over an aspect angle window that is moved across the RCS pattern. The many other approximations of physical processes are made that could be worse than the errors in the RCS data. Thus, when determining how important the near-field effects may be, the end use of the data should be considered.

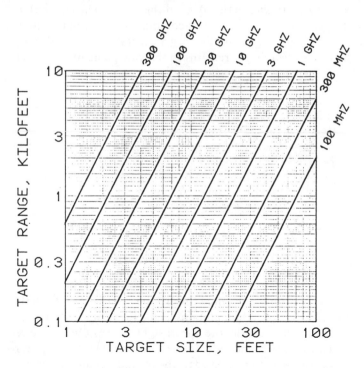

Figure 11-8 The far-field distance

Finally, it should be pointed out that it is the transverse target dimension which figures predominately in calculating the far-field range. In the nose-on or end-on aspect angle region, the transverse target dimension of missiles and slender bodies allows the target to be brought much closer to the radar. Because the end-on returns are usually much smaller than the broadside return, this helps to improve the RCS sensitivity of the system. RCS patterns in the broadside aspect angle region can usually be measured comfortably at a greater range. Although it requires two separate measurements and the "splicing together" of separate RCS patterns valid over separate aspect angle regions, this is often the price that must be paid for high quality data.

11.4 GREAT CIRCLE VS. CONICAL CUTS

One of the decisions faced by a program manager is whether to specify great circle or conical "cuts." A *cut* refers to an RCS pattern recorded for a complete revolution of the target in azimuth. Whether or not that cut is a great circle trajectory or a cone trajectory depends on the tilt angle of the axis of rotation, as shown in Fig. 11-9.

In Fig. 11-9, the radar is off to the left, out of view of the diagram, and the target axis of symmetry lies in the plane of the figure. The left-most diagram shows the target mounted on a support column in a level flight attitude and in the nose-on viewing position. As the target turntable is rotated through 360 degrees, the radar line of sight remains in the yaw plane of the target.

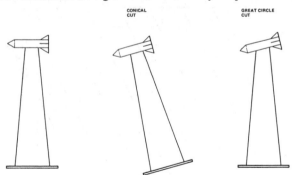

Figure 11-9 Axis of rotation is tilted for conical cuts

If the axis of rotation is now tilted toward the radar, as shown in the center digram, the radar line of sight maintains a constant angle with respect to the axis of rotation. Consequently, the line of sight sweeps out a cone center on the yaw axis (which is perpendicular to the yaw plane) as the target is rotated. This generates the conical cut referred to above, and is indicated schematically in Fig. 11-10. In Fig. 11-10, α is the angle by which the axis of rotation is tilted toward the radar.

If the axis of rotation is brought back to perpendicular to the line of sight, as shown in the left-most diagram of Fig. 11-9, the cone angle flares out to a disk, which is a special case of the conical cut. This represents a great circle cut, and it remains a great circle cut, even if the target is pitched atop its support column. As shown in the right-most diagram of Fig. 11-9, the axis of rotation remains at right angles to the line of sight, even if the target is pitched.

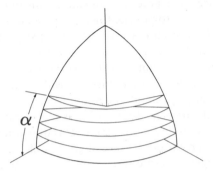

Figure 11-10 Conical cuts; α is the angle by which the rotator is tilted toward the radar

However, the plane of the great circle cut is inclined to the yaw plane, as suggested in Fig. 11-11. The angle of inclination is the pitch angle α, not to be confused with the tilt angle α used for conical cuts. The difference between a collection of conical cuts and a collection of great circle cuts can be grasped from a comparison of Figs. 11-10 and 11-11. Whereas all the great circle cuts intersect at the broadside aspect in the yaw plane, conical cuts never intersect. Thus, conical cuts provide more coverage of the spatial pattern of a target than great circle cuts.

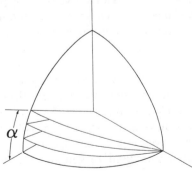

Figure 11-11 Great circle cuts; α is the pitch angle of the target as mounted on the support column

Great circle cuts can provide the same kind of coverage, however, if combinations of pitch and roll are used. If the target is rolled instead of pitched, the great circle cuts intersect along the roll axis (longitudinal target axis), instead of along the pitch axis. Clearly, the line of intersection between the yaw plane and a great circle cut can be "walked" around the yaw plane using a sequential selection of roll and pitch angles.

However, there are more practical considerations in the selection of great circle or conical cuts. The top of the foam support column must be custom-built to fit the surface contour of the target. This distributes the target weight over the contact area between the target and the support column, and it minimizes distortion of the column at the load bearing surfaces. The fabrication of these support "saddles" represents a considerable cost investment in an RCS test program and thus should be minimized.

The use of conical cuts helps to do this. Using a single support saddle, all that needs to be done to index from one conical cut to another is to change the tilt angle of the axis of rotation. The use of great circle cuts, on the other hand, requires that the target be dismounted from the column, the support saddle exchanged for one designed for a different pitch and roll angle, and the target remounted on the new saddle. In addition to the fabrication of a costly collection of support saddles, great circle cuts entail a great deal of target handling. This represents a significant portion of the test program cost, especially for large targets.

Nonetheless, there are limitations to the otherwise advantageous use of the conical cut measurement configuration. Target support columns cannot support large bending movements required when the rotation axis is tilted, although they can support considerable vertical loads in the upright position, and there is the risk of the target sliding off the support column. Thus, the cone angle is limited in practice to about 30 degrees at best.

Despite the disadvantages of the great circle cuts, it is possible to make measurements at much higher angles above the yaw plane than with conical cuts. Indeed, a target can be mounted on its nose or tail with its axis along the vertical, whereupon a great circle cut includes RCS data, which corresponds to looking directly down on, and directly up to, the target. This is impossible with conical cuts.

Thus, the conical cut is the favored method of target rotation for RCS measurements because much more data can be collected in much less time at less cost than a great circle cut can provide. Nevertheless, the great circle cut can provide high elevation viewing angles not possible with the conical cut. In the end it is the test objectives, test requirements, and cost that dictate whether great circle or conical cuts are used, or whether combinations of the two are necessary.

11.5 TARGET SUPPORT STRUCTURES

Several target support techniques are available. These include:
- Low density plastic foam columns
- High strength non-metallic support lines
- Absorber-covered metal pylons
- Hollow solid plastic support columns

Of these four, the first has been the most widely and successfully used, while the last has been the least used. Each has its advantages and limitations. Plastic foam columns offer relatively low background return when properly designed and used, but have limited weight handling qualities. Suspending targets from filamentary lines is attractive because support lines can have smaller returns than foam plastic columns, but target attitude control is hard to maintain. Metal pylons have excellent load bearing properties, but require special attachments to mate with the target, and the target may have to be disfigured at the mounting point. Moreover, an absorber-covered metal pylon increases the electromagnetic target-to-support coupling and is very expensive. Hollow plastic columns are probably useful only over a narrow frequency range where the front and rear column returns can be "tuned" to cancel each other; but, outside these narrow limits, the returns can be nearly as high as those of a bare metal column.

Freeney [14] discusses the four methods, but offers no data by which to assess the echo properties of metal or solid plastic support columns. He presents data for dielectric support lines and some data for foamed plastic columns. However, more detailed information on foamed plastic support is available in other references.

As pointed out by Plonus [15], and Senior, *et al.* [16], the radar echo of a plastic foam column can be ascribed to two mechanisms. One is a reflection from the surface of the column, and the other is a "noncoherent" volume return due to millions of tiny internal scatterers (i.e., the cell structure). The volume contribution is conceptually independent of the shape of the column, and therefore is irreducible; that is, column tuning can have no effect. Plonus' result for the incoherent return is

$$\sigma = \frac{\pi}{2} \, t^2 \, k^4 \, a \, |\epsilon_r - 1|^2 \, V \tag{11.4}$$

where t is the mean cell wall thickness, a is the mean cell radius, V is the volume of the column, and ϵ_r is the dielectric constant of the base polymer from which the foam is made. Note that the theoretical noncoherent return rises with the fourth power of the radar frequency.

Extensive experiments conducted at the University of Michigan [17] did not confirm this prediction. Because of the very small returns from test blocks and ogives, the measured data showed large deviations. Nevertheless, a value of about −58 dBsm per cubic foot seems typical of the averaged data for several classes of plastic foams, as suggested in Table 11-1. This table summarizes the properties of five foam materials examined in Reference [17]. It should be appreciated that the properties listed in Table 11-1 were obtained by measurement of a few small samples. The densities of foams are controllable, and hence foams can be made with properties different from those listed in the table.

The expandable bead foam and the three types of styrofoam are made of a polystyrene polymer. The expandable bead foam is easily molded to complex shapes, and is found in ice chests and as a packaging cushion for fragile products of all kinds. The effective dielectric constant of the material depends on the material density; whereby, the lower the density, the lower the dielectric constant.

The styrofoams listed in Table 11-1 are used in the construction industry as insulation materials. They are heavier and much stronger than the expandable bead foams, and one of the materials (styrofoam FR) has a very small cell structure. Were it not for the fact that foams are unavailable in large monolithic blocks, they would be candidates for target support structures.

The polyurethane foam is nearly twice as heavy as the expandable bead foam, but has about the same strength. The polyurethanes are useful because of their fine cell structure — despite their reduced strength, pound-for-pound, compared with styrofoam. Polyurethane columns are attractive because they can be molded in one piece, eliminating the need to bond several smaller pieces together in order to build a single column.

Figure 11-12 is a theoretical prediction of the radar return from a conical support column 12ft tall having a dielectric constant of 1.04, which is typical of a styrofoam or polyurethane foam. The column diameter at the top was taken to be 1ft and the base diameter was assumed to be 2ft, and hence the sides of the column were slanted at 2.4 degrees with respect to the column axis.

To obtain the plot of Fig. 11-12, the radar was assumed to be in a horizontal plane intersecting the column at its midpoint, and the column was tilted away from the vertical either toward or away from the radar by the amount shown on the abscissa. The two large peaks of the pattern are due to specular reflections from the front and rear slanted column surfaces, which occur when those surfaces are perpendicular to the radar line of sight. The rear surface is visible through the front surface because the foam is nearly — but not perfectly — transparent. The return from the rear surface attains its peak value when the column is tilted away from the radar. These peaks have amplitudes of about −12 dBsm, an objectionable value for some targets.

Figure 11-13 is a measured pattern for a polyurethane column of the same size used to generate Fig. 11-12. This pattern was recorded at the McDonnell-

Table 11-1
Properties of Some Plastic Foams

Material		Expandable Bead Foam	Styrofoam* FB	Styrofoam* DB	Styrofoam* FR	Polyurethane
Base Polymer	Type	Polystyrene	Polystyrene	Polystyrene	Polystyrene	Polyurethane
	Density, lb/cu ft	66.50	66.50	66.50	66.70	70.00
	Dielectric constant	2.55	2.55	2.55	2.55	2.06
Foamed Product	Call Size, in	0.125	0.025	0.057	0.011	0.019
	Density, lb/cu ft	1.150	1.760	1.800	1.970	2.040
	Elastic Modulus, psi	733	2061	1692	3000	710
Noncoherent Returns, dBsm/cu ft	Theoretical	−25.8	−47.5	−35.8	−54.6	−50.4
	Measured	−58.0	−56.7	−57.4	−60.1	−60.2

*Registered trade name, Dow Chemical Company, Midland, Michigan.

Douglas Microwave Test Facility near Palmdale, California. Differences between the measured and predicted patterns are due to the fact that the actual column was slightly bent and its base was shielded from the incident radar wave because it was mounted on a turntable that was below the ground level.

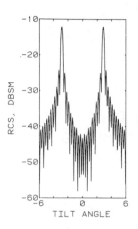

Figure 11-12 Theoretical return from a 12-foot conical support column at a radar frequency of 10 GHz

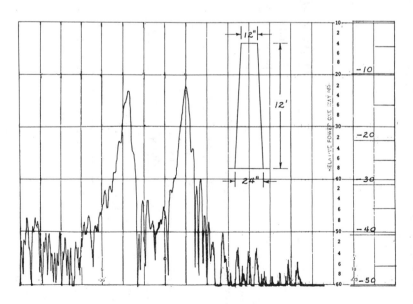

Figure 11-13 Measured pattern of a 12-foot polyurethane support column

The specular reflections from the sides of the column occur for any azimuth angle because the conical column is circularly symmetric. However, if the column is fabricated with a diamond cross section, as shown in Fig. 11-14, the specular reflections can be confined to four narrow azimuth regions. This is particularly useful when measurements are to be made of targets whose returns are very small in the nose-on region. If the support column of Fig. 11-14 were to be oriented with an edge facing the radar, and tilted either toward or away from the radar, only the noncoherent return would be recorded. The two peaks shown in Figs. 11-12 and 11-13 for the conical column would be absent. The diamond shaped design has been used routinely by the Boeing Company.

An effect in reflectivity measurements which is not always appreciated is the amount of contact area between the target and its support column. As shown by Knott and Senior [17], this area should be minimized so as to reduce distortion in the incident field in the vicinity of the target. The wave propagates slower within the foam material than it does in the surrounding air, but the electromagnetic boundary conditions require that the phase fronts be continuous across the boundary. After propagating over a considerable length of the interface, the phase fronts just inside the surface will lag behind those just outside, and they will become distorted in order to maintain continuity. Although the distortion is due to the phase front lag, the amplitude of the wave will also become distorted because energy must propagate in a direction normal to the phase fronts.

Minimizing the contact area is not always possible because of conflicting strength and safety requirements, but experimenters should be aware of the effects of extended target-to-support interfaces. It is possible to design "finger cradles" that will support light objects, much as an experienced waiter or waitress carries a large tray with one hand. Unfortunately, such fragile support cradles can seldom be used by heavy targets or for guyed targets.

A recent improvement (mid-1970s) in target support technology is the absorber covered metal pylon. The pylon is shaped much like the wing of a commercial jetliner, as shown in Fig. 11-15, and is mounted on a heavy base fixture embedded in the ground. The pylon has sharp leading and trailing edges, and is inclined toward the radar. The return from an inclined edge is very small (theoretically zero), and the inclination angle is selected so as to minimize interactions between the pylon and the ground.

The pylon is stationary to preserve the favorable edge-on orientation toward the radar, and the target rotator is mounted at the top of the pylon. In practice, the target must be modified so that the rotator fits inside the target. Thus, the rotator and the top of the pylon are concealed by the target itself. Unfortunately, this means that the target now has a hole in it which would not exist except for this mounting requirement.

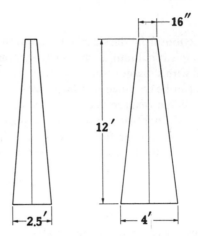

Figure 11-14 Support column with diamond cross section

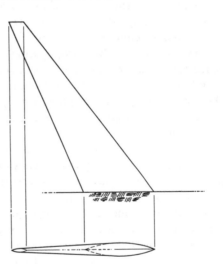

Figure 11-15 Absorber-covered metal support pylon

The hole itself can alter the radar signature of the target, hence some means must be devised for concealing the hole or suppressing the effect. Absorbers and special shields can be designed for this purpose. The metal pylon cannot be used to support operational missiles and aircraft, unless a military agency can be found which is willing to sacrifice a production vehicle.

RATSCAT has installed a 75ft pylon (at pit 7) and an even taller one (95-120ft) is planned for an advanced measurement site (see Ch. 12). The load bearing capabilities of these structures are not known, but they are probably limited to a few thousand pounds. Light developmental model targets will probably be used almost exclusively, while operational vehicles will continue to be measured using conventional foam support column technology.

The relatively large returns from foam support columns can also be circumvented by suspending the target from overhead with strong non-metallic lines. When implemented outdoors, this requires the erection or availability of a pair of towers straddling the target region. The towers should be placed outside the range gate, and well outside the main lobe of the radar antennas, as in Fig. 11-16. A sling or harness must be made to accomodate the target, which must be guyed to a rotator in order to control the aspect angle of the target (see Fig. 11-17). The support lines must be strong enough to bear the tension of these guy lines in addition to the weight of the target. The line tension is usually such that the target will oscillate in aspect angle even with the rotator moving at constant speed. It requires calm conditions, very careful attention to target alignment, and a great deal of patience to obtain RCS patterns using this suspension technique, but the effort may often be well worth it. The experimenter should be aware that all the guy lines to the rotator will be normal to the incident wave at two points in a 360 degree azimuth cut, and their returns, although small, will likely be visible in the recorded data.

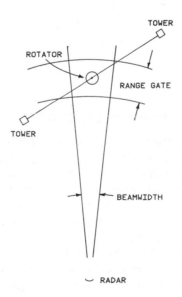

Figure 11-16 Tower placement for target suspension of the target rotator

The suspension line support technique has been considerably refined by the Boeing Military Airplane Company in the fabrication of its millimeter wave modeling facility [18]. The model is suspended in a harness of fine lines between a pair of turntables, one in the floor and one in the ceiling of an indoor anechoic chamber. The turntables are "slaved" together, and the support lines are wound on small reels driven by stepping motors. Under computer control, the turntables can be rotated and the stepping motors activated to simulate dynamic target motion. A television monitor in the control room gives the range operator a continuous view of the target motion.

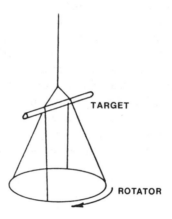

Figure 11-17 A method of target suspension

11.6 TARGET-GROUND INTERACTIONS

Interactions between the target and the rotator or ground are possible, and for some classes of target the effect can be surprisingly large. Examples are spherical shapes and cylindrical shapes near broadside incidence. The scattering for these cases is nearly omnidirectional, and multiple scattering can occur, as suggested in Fig. 11-18. The net effect, as far as the instrumentation radar is concerned, is that there is more than one target within the beam and within the range gate, and, depending on the actual geometry, they seem to be aligned one behind the other. The intensities of the reflections from these "ghost" targets is not as high as those of the real target, but they can be large enough to add constructively or destructively (depending on the target height), thus introducing measurement errors. Because metallic spheres are in very common use as calibration standards, the potential for error is more prevalent than might be supposed.

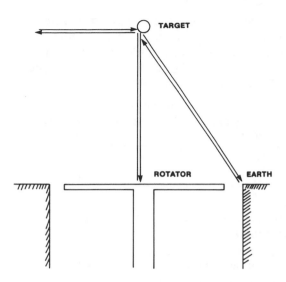

Figure 11-18 Target/rotator interactions

Obviously, the most expedient remedy is to cover the rotator and the ground in the vicinity of the target with radar absorbing material. This is not as easily accomplished as it seems because the target support column (if one is used) must somehow be mounted on the absorber. Moreover, especially on outdoor ranges, the rotator and the absorber coating are exposed to the weather. Consequently, the material must be rigid, able to withstand the environment, and rugged enough to support the target as well as personnel who may have to stand on the rotator during target mounting procedures. Despite these harsh requirements, radar absorbing coatings having the necessary qualities can be designed and manufactured.

Another way to remove interactions between the target and the ground is to use short radar pulses. This can be done with actual short pulses, or with the aid of modern frequency-stepping techniques, which allow short pulses to be synthesized. With either an actual short pulse or one that is synthesized, the target must be mounted at a height above the ground that is significantly greater than the target length. An example serves to illustrate the effect.

Figure 11-19 shows a hypothetical target mounted on foam support columns, and a single interaction is shown between the engine nacelle and the target rotator. In addition to this interaction, the net target return is comprised of direct contributions from other portions of the target. For the purpose of numerical illustration, various parts of the target have been assumed to contribute, and arbitrary amplitudes have been assigned, as indicated in Table 11-2. In addition to these amplitudes, the physical locations of the scatterers have been located along the target axis. Although it is unlikely that a target like the one shown in Fig. 11-19 could be mounted 40ft above the turntable, this has been assumed as well.

Table 11-2
Amplitudes and Location of Scatterers
(for numerical simulation purposes only)

Scatterer Number	Scatterer	$\sqrt{\sigma}$, Amplitude, m	Relative Range, ft
1	nose	0.5	20.0
2	canopy, leading edge	3.0	27.3
3	engine intake	5.0	33.3
4	canopy, trailing edge	2.0	38.0
5	wing root, leading edge	4.0	41.7
6	wing, trailing edge	2.0	53.0
7	vertical stabilizer, root 1	1.5	55.0
8	vertical stabilizer, root 2	0.5	62.4
9	vertical stabilizer, tip	0.5	66.6
10	horizontal fin, leading edge	1.5	57.4
11	horizontal fin, trailing edge	1.0	63.9
12	interaction with turntable	3.0	73.3

The numerical procedure is as follows:

1. Using frequency stepping techniques, transmit a synthesized short pulse; the effective pulse width must be short enough (and, therefore, the bandwidth large enough) to separate the target return from the interaction return.
2. Transform the received signals to the time domain defined earlier using FFT processing.
3. Gate out the interaction return by zeroing all signals beyond a predetermined time slot.
4. Transform the gated return back to the frequency domain.

The total return from the target can be represented as the coherent sum of the individual contributions:

$$\sqrt{\sigma} = \sum_{p=1}^{12} \sqrt{\sigma_p} \, e^{i2kr_p} \tag{11.5}$$

where σ_p is the RCS of the pth scatterer and r_p is its relative location along the radar line of sight. Assume now that a short pulse will be synthesized by stepping the radar frequency from 8.73 to 11.29 GHz in 256 equal intervals of

10 MHz each. At each of these frequencies, the total return can be computed using Eq. (11-5) and the net return plotted (as in Fig. 11-20) as a function of frequency. It is not apparent from this figure precisely how many scatterers there are, nor what their distribution along the line of sight might be.

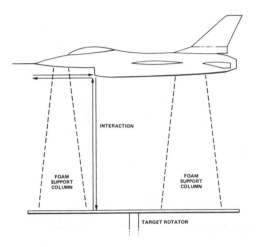

Figure 11-19 Illustrating an interaction between the target and the rotator

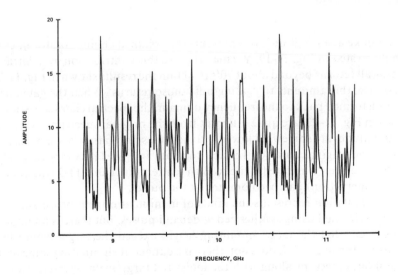

Figure 11-20 Total simulated target return over 2.56 GHz bandwidth

We may now transform the data in Fig. 11-20 to the time domain using the FFT, generating the results shown in Fig. 11-21. A sketch of the target has been superimposed for comparison. Note that all the scatterers can be resolved because the bandwidth of the synthesized short pulse was adequate.

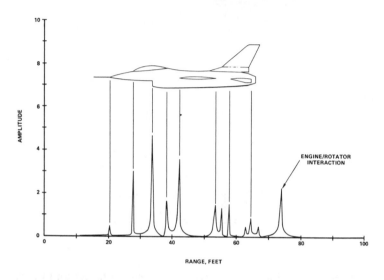

Figure 11-21 Results of transforming the data to the time domain using the FFT algorithm

The spike at a range of 72ft represents the assumed engine-rotator interaction illustrated in Fig. 11-19. We may remove this contribution by arbitrarily zeroing all returns beyond about 70ft to obtain the results shown in Fig. 11-22. In essence, this amounts to "gating out" those returns. When the gated data are transformed back to the frequency domain, the spectral characteristics, as shown in Fig. 11-23, are quite different from the original spectrum. Note that the amplitudes are somewhat lower and the total scintillation is somewhat less than in the original data set. In fact, the data in Fig. 11-23 represent what would have been simulated if the twelfth scatterer in Table 11-2 (the engine-rotator interaction) had been omitted at the outset.

This simulation represents the kinds of manipulation and processing that can be performed using synthesized wideband pulses, but there is an implicit requirement that the target mounting height exceed the target length by a comfortable margin. If it does not, there is a danger of eliminating some of the natural target returns along with the undesired target-rotator interactions.

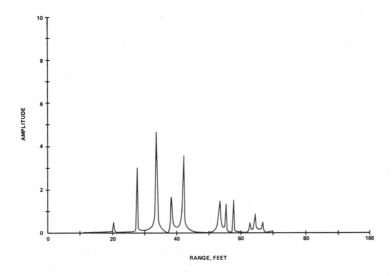

Figure 11-22 Time-domain data after engine/rotator interaction has been gated out

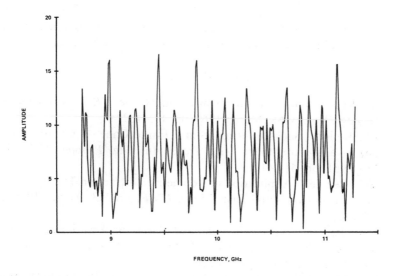

Figure 11-23 Frequency-domain data after the engine/rotator return has been gated out

11.7 SUMMARY

In this introductory chapter to RCS measurement ranges, we have covered some general considerations. Most of these considerations apply to indoor and outdoor ranges, but in a few cases, they are applicable only to one or the other. The measurement of large targets (for example, the one shown in Fig. 11-19) can be performed only on an outdoor range, while the use of upper and lower turntables for target suspension techniques is strictly an indoor operation.

We would like to support targets with "invisible" support structures, but truly invisible supports do not exist. Plastic foam columns can be quite serviceable in some cases and, in other cases, absorber-covered pylons are better. The far-field condition cannot always be realized, but, in many cases, measurements taken at less than this distance are quite acceptable.

Examples of typical coherent and noncoherent RCS measurements have been given. In general, coherent radar signature data yield better diagnostic information than noncoherent data. However, the instrumentation required for coherent data is more complex. Modern frequency-stepping techniques make it possible to synthesize short pulses, thereby generating high resolution data.

The peculiarities, advantages, and disadvantages of indoor and outdoor ranges can now be addressed within the context of the RCS measurement considerations in this chapter. These are contained in Ch. 12 and Ch. 13.

REFERENCES

1. J.B. Keller, "Backscattering from a Finite Cone," *IRE Trans. Antennas Propag.*, Vol. AP-8, March 1960, pp. 175-182.
2. J.B. Keller, "Backscattering from a Finite Cone — Comparison of Theory and Experiment," *IRE Trans. Antennas Propag.*, Vol. AP-9, July 1961, pp. 411-412.
3. M.E. Bechtel, "Application of Geometric Diffraction Theory to Scattering from Cones and Disk," *Proc. IEEE*, Vol. 53, August 1965, pp. 877-882.
4. M.E. Bechtel, "Vertically Polarized Radar Backscattering from the Rear of a Cone or Cylinder," *IEEE Trans. Antennas Propag.*, Vol. AP-17, March 1969, pp. 244-246.
5. W.E. Blore, "The Radar Cross Section of Spherically Blunted 8° Right Circular Cones," *IEEE Proc. Antennas Propag.*, Vol. AP-12, March 1973, pp. 252-253.
6. W.D. Burnside and L. Peters, Jr., "Radar Cross Section of Finite Cones by the Equivalent Current Concept with Higher Order Diffraction," *Radio Science*, Vol. 7, No. 10, October 1972, pp. 943-948.

7. E.F. Knott and T.B.A. Senior, "Comparison of Three High Frequency Diffraction Techniques," *Proc. IEEE*, Vol. 62, November 1974, pp. 1468-1474.

8. E.F. Knott and T.B.A. Senior, "Second Order Diffraction by a Ring Discontinuity," Report No. AFOSR-TR-73-1237, University of Michigan, Radiation Laboratory, July 1973.

9. "RATSCAT Facilities and Capabilities," brochure published by the 6585th Test Group, Air Force Systems Command, Holloman AFB, NM. (*Note*: Brochure is not dated, but was probably issued in 1977 or 1978.)

10. R.E. Gardner, "Doppler Spectral Characteristics of Aircraft Radar Targets at S-Band," NRL Report 5656, US Naval Research Laboratory, 3 August 1961, p. 17.

11. R.G. Kouyoumjian and L. Peters, Jr., "Range Requirements in Radar Cross Section Measurements," *Proc. IEEE*, Vol. 53, August 1965, pp. 920-928.

12. E.F. Knott and T.B.A. Senior, "How Far is Far?" *IEEE Trans. Antennas Propag.*, Vol. AP-22, September 1974, pp. 732-734.

13. Richard B. Mack, "Basic Design Principles of Electromagnetic Scattering Measurement Facilities," Report RADC-TR-81-40, Rome Air Development Center, Griffiss AFB, NY, March 1981.

14. C.C. Freeny, "Target Support Parameters Associated with Radar Reflectivity Measurements," *Proc. IEEE*, Vol. 53, August 1965, pp. 929-936.

15. M.A. Plonus, "Theoretical Investigation of Scattering from Plastic Foams," *IEEE Trans. Antennas Propag.*, Vol. Ap-13, January 1965, pp. 88-93.

16. T.B.A. Senior, M.A. Plonus, and E.F. Knott, "Designing Foamed-Plastic Materials," *Microwaves*, December 1964, pp. 38-43.

17. E.F. Knott and T.B.A. Senior, "Studies of Scattering by Cellular Plastic Materials," Report No. 5849-1-F, University of Michigan, Radiation Laboratory, April 1964.

18. H.S. Burke, T.G. Dalby, W.P. Hansen, Jr., and M.C. Vincent, "A Millimeter-Wave Scattering Facility," presented at the 1980 Radar Camouflage Symposium, Orlando, Florida, 18-20 November 1980, Report No. AFWAL-TR-81-1015, Air Force Wright Aeronautical Laboratories, Wright-Patterson AFB, March 1981, pp. 327-336.

CHAPTER 12

OUTDOOR RCS RANGES

E. F. Knott

12.1 OVERVIEW

We have examined some general considerations attending the measurement of radar target characteristics and now we will consider outdoor ranges in particular. There are two main reasons for conducting RCS measurements outdoors. First, it is necessary for large targets because of the far-field range requirement, even if that requirement cannot always be satisfied outdoors. Second, many indoor facilities cannot accomodate large targets, aside from the far-field requirement.

A major decision in the construction and operation of an outdoor RCS range is whether to utilize the ground plane effect. In such an environment, the ground itself can increase overall system sensitivity by concentrating more energy on the target. However, there is a price to be paid for using the ground plane, and it includes rigid constraints on where the target and the radar antennas are located.

The instrumentation for an outdoor range is relatively simple for conventional RCS measurements, but can become considerably more complex if coherent data are required. For diagnostic isolation of flare spots, for example, we must use a chirped pulse, as mentioned in Ch. 13, whether that pulse is an actual (analog) pulse or synthesized using frequency-stepping techniques. Radar and recording instrumentation for outdoor ranges are described in this chapter.

Several new outdoor ranges have been built since 1978. Information about some of the newer ones is harder to obtain than information about older ones, but a summary of known ranges is presented at the end of this chapter.

12.2 INSTRUMENTATION

The quality, quantity, and complexity of radar instrumentation varies from range to range, and we cannot hope to cover this very broad subject in so little space. Instead, the elements of a simple, amplitude-only RCS instrumentation radar will be described. Interested readers may find more detailed information in a great number of books and papers; perhaps the most widely known being Skolnik's *Radar Handbook* [1].

The elements of a basic radar system are shown in the block diagram of Fig. 12-1. This system has separate transmitting and receiving antennas, although separate antennas are not required. A single antenna can be used to perform both functions, in which case a receiver protector (not shown) must be installed. When a single antenna is used for transmission and reception, the receiver and transmitter must both be connected to the antenna. The receiver protector prevents the transmitted energy from reaching the receiver. Otherwise, the receiver will be destroyed the first time the radar is turned over.

The transmitted energy originates from a radio frequency source, which is generally tunable. These sources are low power devices that generate a few hundred milliwatts of power. Because most measurements performed on outdoor ranges require thousands of watts for adequate sensitivity, this low level signal is amplified by an RF power amplifier, typically a traveling wave tube amplifier (TWTA). The TWTA is a grid tube that is turned on for a short time by a trigger pulse applied to the grid. This trigger is provided by a pulse generator whose output waveform is a square wave.

The pulse generator usually has a selectable pulse width and, depending on the particular design, can have a built-in audio oscillator whose frequency can be controlled. The frequency of the pulse train fed to the TWTA is called the PRF (pulse repetition frequency). The TWTA amplifies the incoming RF signal from the tunable RF source only for the duration of the trigger pulse. Typical pulse widths used in outdoor measurements are 0.1 to 0.5 μs. The duty cycle of these amplifiers (ratio of the time on to the time between pulses) varies, but is on the order of 0.5%.

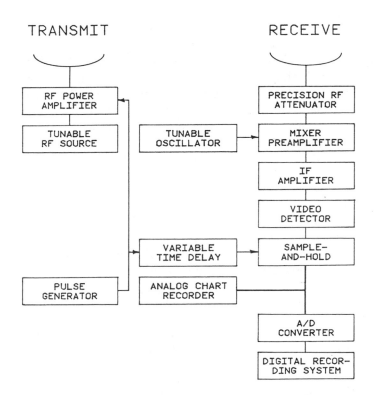

Figure 12-1 A simple instrumentation radar

There are some RF sources which oscillate at high enough power levels that further RF amplification is not necessary. These are called magnetrons and they have rated output levels ranging from kilowatts to megawatts. They are totally unrelated to TWTAs, however, and they generate power in an altogether different way. Occasionally, a high power TWTA is required and two stages of RF amplification may be necessary. When it is, the first stage uses a low level TWTA (output of about hundred watts) while the second stage uses a high level TWTA, generating hundreds of kilowatts.

The signal returned from the target is captured by the receiving antenna. A precision RF attenuator is shown in the diagram, and this can be used to adjust the signal level into the receiver to take best advantage of the sensitivity and dynamic range of the receiver characteristics. In some instrumentation systems, a low level RF signal source may be switched into the system and the precision attenuator may be used to measure the receiver transfer (gain) function.

A tunable local oscillator is adjusted to a frequency slightly above or below the transmitted frequency so that the difference frequency (or intermediate frequency, the IF) between the two is close to the center design frequency of the IF amplifier. Thus, the target signal and the local oscillator signal are both fed to the mixer-preamplifier, and the resulting IF is delivered to the IF amplifier. The local oscillator, it might be noted, is a low power device, typically developing 10 mW of output power at most.

In the diagram of Fig. 12-1, the local oscillator and the transmitter signal source are independent of each other and are "free running." For coherent radar systems (not shown), the two must keep in step and must be locked together. There are several ways to accomplish this, one of which is to phase-lock the pair to a low frequency crystal oscillator. A third source must also be locked into the system to provide a reference IF signal with which the signal IF can be compared to extract phase information from the target signal.

The IF amplifier can be of the linear or logarithmic variety. The output of a linear IF amplifier varies linearly with the input signal, whereas the output of a log IF varies as the logarithm of the input signal. These amplifiers typically have a 60 or 70 dB dynamic range over which the output varies faithfully with the logarithm of the input signal.

In some instrumentation systems (not shown), an automatic gain control (AGC) circuit is included. This circuit develops a slowly varying, low level voltage that is fed back to the IF amplifier and acts to keep the IF output signal at a constant level. In those kinds of systems, it is the AGC voltage that is recorded rather than the output of the IF amplification circuits.

The video detector is often an integral part of the IF amplifier package. It provides a slowly varying, low level dc voltage that can be recorded. However, such a signal contains contributions from clutter and unwanted targets in addition to the desired target signal. Consequently, a trigger pulse from the pulse generator instructs a sample-and-hold circuit to sample the video output at "target time." Because the received target signal is delayed by the time it takes for the transmitted pulse to reach the target, plus the time it takes for the reflected pulse to return to the radar, this trigger must also be delayed to ensure the video signal is sampled at the proper time. The delayed trigger is called a "range gate."

The sample-and-hold circuit maintains a voltage corresponding to the target signal until instructed by the next trigger pulse to sample again. Thus, the output of the sample-and-hold is a slowly varying DC voltage that can be recorded. With suitable amplification and processing, the output can be routed directly to an analog chart recorder. The position of the chart paper is slaved to a synchronizing device mounted on the shaft of the target rotator, so that the recorder chart position is directly related to the target aspect angle. Thus, the operator of the system can watch the pattern being traced as the target is rotated.

Modern radar instrumentation systems are equipped with digital recording systems. Thus, the analog output of the sample-and-hold must be converted to digital format. A variety of analog-to-digital (A/D) converters is available, with resolutions ranging from a few bits to as many as 20 bits or more. Selection of the actual A/D converter used depends on several factors, among them the resolution and the PRF. This is because it takes longer to provide more bits, and in extreme cases, there simply may not be enough time between pulses to perform the conversion.

Actual radar instrumentation can be much more complex, but the conduct of a basic measurement program is relatively insensitive to the actual equipment. Once a test matrix has been established and the requirements of the instrumentation have been identified, a typical outdoor measurement program has most of the following features:

1. Design and fabricate target support systems if none are available from previous test programs.
2. Select the range geometry, including antenna height, target height, range, and other parameters, if necessary.
3. Install or set up the instrumentation radar and data acquisition system.
4. Probe the incident field structure in the vicinity of the target to ensure acceptable field taper.
5. Depending on the results of the field probing, adjust the antennas and other systems as necessary.
6. Install the target support system.
7. Measure the background return, most of which is due to the target support system.
8. Adjust support columns or other systems, if feasible.
9. Install the primary calibration standard, preferably at or near the centroid of the target had it been in place on the support structure.
10. Calibrate the system based on the known (theoretical) return of the primary calibration standard.
11. Install a secondary calibration standard on the range outside the target range gate and establish its strength relative to that of the primary calibration standard.
12. Remove primary standard and install target.
13. Rotate target and record data as a function of aspect angle.
14. Remove target.
15. Re-check calibration.
16. Check background signals.
17. Examine "quick-look" data.
18. Verify digital data.
19. Obtain real-time plots of processed data, if available and if desired.
20. Next target configuration.

The above sequence is typical of that followed at RATSCAT and several other outdoor ranges, but may be somewhat abbreviated at other installations, depending on the test requirements.

12.3 THE GROUND PLANE EFFECT

The proximity of the ground to the antenna and the target is sometimes hard to avoid, and one solution is to exploit the reflection from the ground. This is done at RATSCAT and several other outdoor ranges. The exploitation of the ground plane effect requires maintenance in the form of grading and leveling the soil, unless an asphalt or concrete ground plane has been constructed.

The ground plane effect is illustrated in Fig. 12-2. Its performance depends on the reflection of an incident wave at the specular point P. For a flat, smooth earth, the location of this point can be found by placing an image antenna directly below the actual antenna just as far below the ground surface as the real antenna is above it. Transmitted energy and received target returns propagate along four distinct paths:

Direct path   ATA
"Image" path   APTPA
Diplane path # 1   ATPA
Diplane path # 2   APTA

Of the four, the last two are actually bistatic returns because the angles of incidence and scattering at the target are different. For most measurement situations, the bistatic angle is insignificant, however.

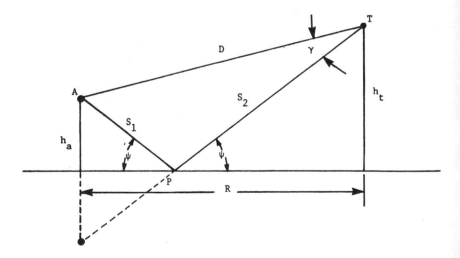

Figure 12-2 Ground plane geometry. Radar antenna is at point A, radar target at point T, and the specular point is P

If the bistatic angle in the vertical plane is small, the target return is insensitive to the difference between the angles of arrival of the direct and indirect waves illuminating the target. Because the radar cross section is proportional to power, the square root of the return is proportional to the received field strengths. Therefore, we write

$$\sqrt{\sigma} = \sqrt{\sigma_0} \left[e^{i2kD} + 2\rho e^{ik(D+I)} + \rho^2 e^{i2kI} \right] \tag{12-1}$$

where

σ_0 = free space radar cross section of the target,
ρ = effective reflection coefficient of the ground plane,
D = length of the direct path, and
$I = S_1 + S_2$ = length of the indirect path.

The first term in Eq. (12-1) is the free-space return of the target, as would be measured in the absence of the ground plane. The third term is the image return, because as far as the radar is concerned, this return appears to emanate from a mirror image of the target below the ground surface. Because the image return is generated by two bounces of the wave off the ground (one on transmission and one on reception, both along the path $S_1 + S_2$), this term is multiplied by the square of the ground reflection coefficient. The second term is called a diplane return and involves only a single bounce off the ground. However, this term is multiplied by 2 because there are two propagation paths, one being the reverse of the other.

We are interested only in the amplitude of the return, hence the exponential term e^{i2kD} can be removed from within the brackets. Then, upon squaring Eq. (12-1) to find the measured radar cross section,

$$\frac{\sigma}{\sigma_0} = \left| 1 + \rho e^{ik(I-D)} \right|^4 \tag{12-2}$$

For a perfectly conducting ground plane, $\rho = -1$ and Eq. (12-2) reduces to

$$\frac{\sigma}{\sigma_0} = \{ 2 \sin \left[(kI - kD)/2 \right] \}^4 \tag{12-3}$$

If the target and antenna heights are chosen to place the target at one of the maxima of Eq. (12-3), the received power will be 16 times stronger than if the target were measured in a free space environment. This enhancement is one of the ground plane's most powerful attractions. However, it places constraints on the antenna and target heights because the target must be placed at a maximum in the vertical interference pattern created by the ground plane. Almost invariably, the first maximum above the ground plane is the one used.

An illustrative example will show the application of Eq. (12-3) to a measurement situation. The maxima of Eq. (12-3) occur when the electrical angle $\{k(I-D)/2\}$ is an odd multiple of $\pi/2$ radians, or when

$$k\,(I\text{-}D)/2 = m\,\pi/2 \tag{12-4}$$

where m is an odd integer. Therefore, the path length difference δ must be

$$\delta = I - D = m\,\lambda/2 \tag{12-5}$$

From the geometry of Fig. 12-1,

$$D = [(h_t + h_a)^2 + R^2]^{1/2} \simeq R + \frac{(h_t - h_a)^2}{2R} \tag{12-6}$$

$$I = [(h_t + h_a)^2 + R^2]^{1/2} \simeq R + \frac{(h_t + h_a)^2}{2R} \tag{12-7}$$

hence, the path length difference is

$$\delta = I - D \simeq 2\,\frac{h_a\,h_t}{R} \tag{12-8}$$

For $m = 1$ (the first maximum), the requirement of Eq. (12-4) is therefore approximately satisfied by

$$h_a\,h_t = \frac{R\lambda}{4} \tag{12-9}$$

Equation (12-9) is a rule of thumb that is commonly used in the operation of ground plane ranges.

Selection of appropriate antenna and target heights depends on several factors, such as the maximum antenna height available, the maximum safe target height, whether an existing support column can be used and the capabilities of target handling equipment. Suppose that the target height h_t has been chosen, along with the antenna height h_a, so as to satisfy the requirement of Eq. (12-9). We may probe the received field intensity as a function of height h in the target zone by replacing h_t in Eq. (12-8) by h, and inserting this in Eq. (12-3). The resulting received signal as a function of height is

$$\frac{\sigma}{\sigma_0} = \left[2\sin\left(\frac{\pi}{2}\cdot\frac{h}{h_t}\right)\right]^4 \tag{12-10}$$

A plot of this function is shown in Fig. 12-3.

In these derivations, we have assumed that the ground plane is perfectly reflecting, implying that the voltage reflection coefficient is $\rho = -1$. Thus, the amplitude of this perfect reflection is unity and its phase angle is precisely 180 degrees. In practice, however, the amplitude is scarcely ever unity, nor is the phase angle 180 degrees. So, how do these deviations from perfection alter the ground plane performance?

Figure 12-4 shows the effect of reduced voltage reflection coefficient amplitudes, but with the 180 degree phase angle unchanged. A 10 percent reduction in the amplitude of the voltage reflection coefficient reduces the effectiveness of the ground plane by about 0.9 dB. The result of voltage reflection coefficient phase angles other than 180 degrees is shown in Fig. 12-5. Note that a change in the relative phase angle merely shifts the vertical interference pattern upward, with no change in the amplitude of the peak value.

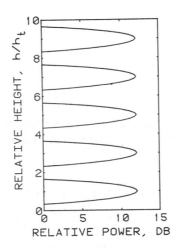

Figure 12-3 Received power as a function of vertical height of a point scatterer

However, if it were not known that the phase angle is not 180 degrees, the target would not be centered at the peak value of the first lobe. This is why it is essential that the net incident field structure in the target zone be probed (measured) prior to an RCS measurement program. The ground plane does not always behave as assumed.

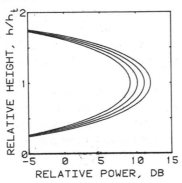

Figure 12-4 Effect of imperfect reflection in the ground plane enhancement. Traces are for voltage reflection coefficients of -0.7, -0.8, -0.9, and -1.0

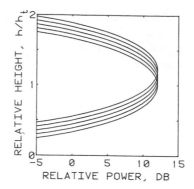

Figure 12-5 Effect of the phase angle of the voltage reflection coefficient. The amplitude was fixed at $|\rho| = 1$ and the phase angles are 140°, 150°, 160°, 170°, and 180°

12.4 EFFECT OF THE ANTENNA PATTERN

Thus far, it has been assumed that the antenna has omnidirectional properties. This assumption makes it easy to study the effects of imperfections in the ground plane. In a practical measurement situation, on the other hand, the antenna must be chosen so as to maximize system sensitivity and to properly illuminate the target. As we shall see, increasing the gain of the antenna is a two-edged sword, and the directivity of the antenna further degrades the performance of the ground plane.

Whether or not the ground plane is used in RCS measurements, one of the first things to decide is the antenna size. Although the antenna should be as large as practicable for adequate sensitivity, it cannot be so large that the target is not uniformly illuminated from one side to the other. That is to say, the antenna beam must be broad enough to adequately illuminate the target, implying that there is an upper limit on the antenna size that may be used. System sensitivity requirements, on the other hand, impose a lower limit on antenna size, and consequently we are often faced with a set of conflicting requirements.

In order to size the antenna for proper target illumination, we must make some assumptions about the antenna radiation pattern. A reasonable assumption is that the antenna aperture is illuminated with a cosinusoidal amplitude distribution and that the antenna radiation pattern is a figure of revolution about the boresight axis. The angular field pattern of such an antenna can be expressed as

$$f = \frac{\cos\left(\frac{\pi}{2}w\right)}{1 - w^2} \tag{12-11}$$

where

$$w = 2 (d/\lambda) \sin\phi \tag{12-12}$$

and d is the antenna diameter and ϕ is the off-boresight angle. The radiated power pattern is the square of Eq. (12-11) and the gain of the antenna is

$$G = \eta \left[\frac{1}{2} kd \right]^2 \tag{12-13}$$

where η is the antenna efficiency. The first sidelobes of this antenna pattern are 23.5 dB below the level of the main lobe.

A rule of thumb for accurate measurements is that the amplitude of the incident field not taper more than 0.5 dB from the center of the target to its lateral extremes. Therefore, we must find the angular separation between the center of the target and its extreme lateral dimension, insert this value for ϕ in Eq. (12-11), and solve Eq. (12-11) for the antenna diameter for $f = 0.944$ [20 $\log_{10}(0.944) = -0.5$ dB].

Unfortunately, Eq. (12-11) is a transcendental equation and its solution for w (or d) in terms of f is not directly possible. However, a method of successive approximations, or graphical methods can be used. The solutions for several values of f are listed in Table 12-1.

Table 12-1
Solution of Equation (12-2)

f, field taper, dB	w
−0.10	0.22178
−0.25	0.35026
−0.50	0.49438
−1.0	0.69642
−2.0	0.97709
−3.0	1.18703
−4.0	1.35940
−5.0	1.50713
−∞	3.00000
half-power	1.18896

Using the rule of thumb that the taper not exceed 0.5 dB, we select the value $w = 0.49438$ from Table 12-1. This can be inserted in Eq. (12-12) for a known value of ϕ and wavelength λ to find the antenna size d. If the transverse target dimension is L, and the range to the target is R, then

$$\sin\phi = \frac{L}{(4R^2 + L^2)^{1/2}} \tag{12-14}$$

Inserting this in Eq. (12-12) and solving for d,

$$d = \frac{\lambda w}{2L} (4R^2 + L^2)^{1/2} \simeq R\lambda w / L \tag{12-15}$$

By way of illustration, let us use the 0.5 dB taper requirement ($w = 0.49438$) for a 10ft target measured at the far-field distance at a frequency of 10 GHz. At this frequency, the wavelength is very nearly 0.1ft, hence, the far field distance is

$$R = 2 \frac{L^2}{\lambda} = \frac{(2)(100)}{0.1} = 2000 \text{ ft} \tag{12-16}$$

Substituting this and the other parameters into Eq. (12-16), the maximum antenna size is

$$d = \frac{(2000)(0.1)(0.49438)}{10} = 9.9 \text{ ft}$$

In fact, this is a rather large antenna to use at 10 GHz, and unless system sensitivity requires it, a 5ft antenna would be easier to use and align on the target. Moreover, the measurements probably could be carried out at a range of 1000ft without serious degradation in the measured data.

Equation (12-1) was written under the assumption that the illuminating antenna has an omnidirectional pattern, when in fact this is not the case. Let us therefore return to Eq. (12-1) and examine the effects of the antenna pattern on the vertical field profile in the vicinity of the target.

From the antenna view point, energy is transmitted and received along two principal directions, one being along the direct path and the other along the indirect path. Intuitively, one feels that the antenna boresight direction should be aligned along the bisector of these two directions, so that the energy received via either path has the same amplitude. On the other hand, the ground plane is not likely to be a perfect reflector, and somewhat less energy will be received along the indirect path than along the direct path. Consequently, the antenna might have to be aimed slightly below the bisecting angle of the two directions to equalize the energy received via each path.

Therefore, we shall depress the antenna some angle ϕ_a below the bisector of the angle between the direct and indirect propagation paths, as suggested in Fig. 12-6. Let the angle between those two paths be 2β. Now let us weight each of the three terms in Eq. (12-1) according to the angles by which the transmitted and received paths lie off the boresight axis of the antenna. When this weighting is done, the more exact form of Eq. (12-1) is

$$\sqrt{\sigma} = \sqrt{\sigma_0} \, [f^2 (\beta + \phi_a) \, e^{i2kD}$$
$$+ 2\rho f (\beta + \phi_a) f (\beta - \phi_a) \, e^{ik(D+I)} \tag{12-17}$$
$$+ \rho^2 f^2 (\beta - \phi_a) \, e^{i2kI}]$$

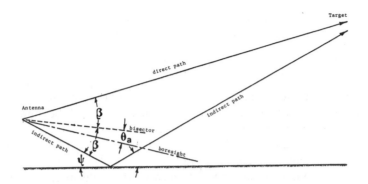

Figure 12-6 Aiming the antenna below the bisector of the angle between direct and indirect paths can equalize the power received along the two paths

Each of the pattern factors is raised to the power of 2, because the received signal intensity for each term must be multiplied by the antenna field pattern value upon transmission, and once again upon reception. Note that the transmitted and received directions are different for the second term.

As before, we seek the magnitude of the bracketed term, and the expression itself is a perfect square. Therefore, Eq. (12-17) can be simplified to

$$\frac{\sigma}{\sigma_0} = |f(\beta + \phi_a) + \rho f(\beta - \phi_a)\, e^{ik(I - D)}|^4 \tag{12-18}$$

The first term in Eq. (12-18) represents the total signal received via the direct path, and the second represents the total signal received via the indirect path. As before, these two components alternately reinforce and cancel each other as a function of height in the target region. If we seek to equalize the power received via both directions, the boresight direction of the antenna should be aligned at an angle ϕ_a such that

$$f(\beta + \phi_a) = -\rho f(\beta - \phi_a) \tag{12-19}$$

For a perfectly conducting ground plane, this implies that $\phi_a = 0$, confirming the initial notion that the antenna should be aligned along the bisector of the angle between the direct and indirect paths.

What if the reflection coefficient is $\rho \neq -1$? Equation (12-19) then becomes a transcendental equation, which, as discussed earlier, can be solved by approximate or graphical methods. Without presenting the details of the analysis, Fig. 12-7 shows a collection of solutions. We will work a specific example to illustrate the use of the chart.

The abscissa w_0 in Fig. 12-7 is

$$w_0 = 2\,\frac{d}{\lambda}\,\sin 2\beta \simeq 4\beta d/\lambda \tag{12-20}$$

This represents the electrical width of the antenna pattern subtended between the direct and indirect path directions. The ordinate w_a is

$$w_a = 2 \ \frac{d}{\lambda} \ \sin \phi_a \simeq 2 \ \phi_a \ d/\lambda \tag{12-21}$$

and represents the electrical distance by which the antenna boresight should be depressed below the bisector of the angle between the direct and indirect path directions. In order to use Fig. 12-7, we must compute w_0, enter the chart for the effective ground reflection coefficient, and read w_a. This value of w_a allows us to determine ϕ_a from Eq. (12-21).

Assume now that a 10ft target will be measured at a distance of 1000ft at a wavelength of 0.1ft using a 4ft diameter paraboloidal antenna. A 12ft support column from a previous test program will be used. To center the target on the first lobe in the vertical interference pattern created by the ground plane, we calculate a required antenna height of 2.08ft from Eq. (12-9). Luckily, this slightly exceeds the radius of our 4ft antenna, otherwise a shallow trench would have to be dug out to allow the antenna to be placed at the correct height.

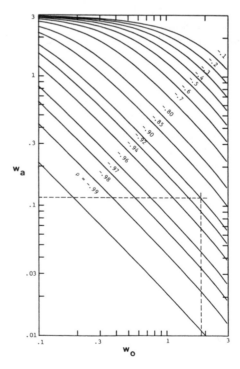

Figure 12-7 Antenna positioning chart (dashed line is for example in text)

For this combination of antenna height, target height, and range, the angles of the direct and indirect paths, as measured from either the antenna boresight axis or the local horizontal, are small enough that their sines can be replaced by the angles themselves (expressed in radians). At a range of 1000ft, a target 12ft above the ground subtends an angle of 0.57 degree (0.01 radian) above the local horizontal at the antenna, and its image subtends an angle of 0.81 degree (0.014 radian) below the local horizontal. Therefore, $\beta = 0.024$ radian and $w_0 = 1.92$. From other measurements on the range at this frequency, let us assume that the effective reflection coefficient of the ground is -0.9.

Entering the chart at the bottom at a value $w_0 = 1.92$, we trace a vertical line until it intersects the trace for $\rho = -0.9$. Then, moving horizontally to the left scale, we read a value $w_a = 0.11$ which can be solved for the angle ϕ_a. Using $w_a = 0.11$, we compute $\phi_a = 0.08$ degree or 0.0014 radian. Thus, the antenna boresight should be depressed 0.08 degree below the bisector, or about 0.20 degree below the horizontal. This is such a small angle that the antenna boresight may be aligned along the local horizontal for all practical purposes.

Even for a perfectly conducting ground plane, the directivity of the antenna radiation pattern results in less than the theoretical 12 dB enhancement. Furthermore, the location of the first lobe in the vertical interference pattern is somewhat lower than predicted when the antenna height is chosen in accordance with Eq. (12-9). This is illustrated in Fig. 12-8.

Figure 12-8 shows three vertical field profiles for a perfectly conducting ground plane. Trace *a* is the profile obtained when omnidirectional transmitting and receiving antennas are used. The relative power attains a peak value 12 dB above the free-space case, and this peak occurs at the desired height $h = h_r$. However, when a 4ft antenna is used, for a target height of 12ft, a wavelength of 0.1ft and a range of 1000ft, the vertical profile attains a peak value of only 8.7 dB above the free-space case. This is indicated by trace *b*, for the optimum antenna pointing direction selected above with the aid of Fig. 12-7.

Trace *c* shows the vertical field profile when the antenna is boresighted along either the direct or indirect path, and the peak value is barely 5.6 dB above the free-space value of the returned signal. Clearly, even for a perfectly conducting ground plane, the directivity of the antenna pattern reduces the theoretical 12 dB enhancement normally thought to be available from a perfect ground plane. Moreover, as suggested by trace *b*, the net received signal peak does not occur at $h = h_r$, but closer to 85% of that value. Therefore, the antenna directivity forces us to modify the way that the antenna height is chosen.

Because the first lobe in the vertical field profile is lower than desired when the antenna height is chosen according to Eq. (12-9), we must select a lower antenna height. Examples of different height selections are shown in Fig. 12-9.

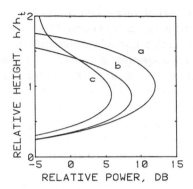

Figure 12-8 Vertical field profiles for a perfectly conducting ground plane. Trace *a* is for an omnidirectional antenna; trace *b* is for a directional antenna aligned along the local horizontal; trace *c* is for a directional antenna aimed along the direct or indirect path

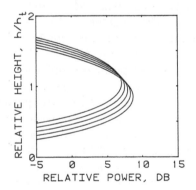

Figure 12-9 Vertical field profiles for different height settings of a directional antenna. The four traces are for 80%, 90%, 100%, and 110% of the value given by Eq. (12-9)

The uppermost curve (for an antenna height at 80% of the "normal" value) is centered very closely at the desired height $h = h_r$. However, the peak value is 7.3 dB, some 4.7 dB less than the theoretical 12 dB enhancement ordinarily associated with the perfect ground plane. Thus, each adjustment in the instrumentation system, whether it be in the antenna height, antenna directionality, or the ground reflection coefficient, degrades the performance of the ground plane range.

12.5 GROUND REFLECTION COEFFICIENT

The reflection from the ground depends on the type of soil, its dampness, and its roughness. The surface roughness diffuses energy in all directions, with the diffusion being greater for greater roughness. The diffused energy reduces the amount of energy reflected in the specular direction, thus the ground plane enhancement becomes less significant the rougher the ground. Vegetation can increase the apparent roughness and absorb some of the incident energy.

The two effects can be accommodated by an effective reflection coefficient that contains two factors independent of each other. It may be written as

$$\rho = \Gamma \rho_s \qquad (12\text{-}22)$$

where Γ is the classical Fresnel reflection coefficient associated with a perfectly flat dielectric interface and where ρ_s accounts for the reduction in reflection due to surface roughness.

We have seen that the Fresnel reflection coefficients in Ch. 3 were derived by invoking the boundary conditions at the dielectric interface. These conditions require that the total tangential electric and magnetic fields be continuous at the boundary. For typical soils, the magnetic permeability is essentially the same as that of free space; hence, the Fresnel reflection coefficients can be expressed entirely in terms of the permittivity of the two media on either side of the interface.

Because one of the media is air, whose permittivity is essentially that of free space, it is convenient to normalize with respect to the free-space permittivity, $\epsilon_0 = 8.85 \times 10^{12}$ Farad per meter. Then, the normalized permittivity becomes the dielectric constant of the ground. As shown earlier, the dielectric constant is, in general, a complex number which can be written as

$$\epsilon_r = \epsilon_r' + i\,\epsilon_r'' \qquad (12\text{-}23)$$

For the geometry of Fig. 12-6, the Fresnel reflection coefficients can be written as

$$\Gamma_h = \frac{\sin \Psi - (\epsilon_r - \cos^2 \Psi)^{1/2}}{\sin \Psi + (\epsilon_r - \cos^2 \Psi)^{1/2}} \qquad (12\text{-}24)$$

$$\Gamma_v = \frac{\epsilon_r \sin \Psi - (\epsilon_r - \cos^2 \Psi)^{1/2}}{\epsilon_r \sin \Psi + (\epsilon_r - \cos^2 \Psi)^{1/2}} \qquad (12\text{-}25)$$

where the subscripts h and v refer to horizontal and vertical incident polarizations, respectively, and Ψ is the grazing angle. If we know what the dielectric constant is, we can calculate the reflection coefficients for a variety of incidence angles Ψ.

Not many soils on outdoor ranges have been measured, but we do have some values for the soil at RATSCAT. The measurements were carried out by the Physical Science Laboratory (PSL) of New Mexico State University [2] at a frequency of 1.5 GHz. It should be noted that the soil at RATSCAT is primarily gypsum, a form of calcium sulfate whose properties are different from those of most soils. PSL fitted a fourth order polynomial through the measured data, and the results are shown in Fig. 12-10.

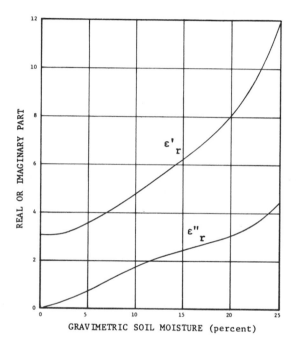

Figure 12-10 PSL's fitted polynominal to measurements of the complex dielectric constant of RATSCAT soil at 1.5 GHz

For the RATSCAT soil, a dielectric constant of $\epsilon_r = 5.5 + i2$ might be typical for a soil moisture content of 12%. Using this value in Eqs. (12-24) and (12-25), we can plot the amplitude and phase of the Fresnel reflection coefficients, as in Fig. 12-11. Note that the amplitude and phase for vertical polarization drop much faster with increasing grazing angle than for horizontal polarization. Nevertheless, the grazing angle seldom exceeds one degree for measurements made on an outdoor range and, hence, the reflection coefficients remain above 0.95 in amplitude and 179 degrees in phase. Thus, at least for these conditions, the ground plane is not far from ideal.

Because the reflection coefficients differ for horizontal and vertical polarization, ground plane measurements made for circular polarization are subject to error. The net incident wave at the target is, in general, elliptically polarized, and the ellipticity can be accentuated by the second ground reflection back to the instrumentation radar. There is no easy way to avoid this. The effect becomes more pronounced as the frequency decreases and can be particularly troublesome for frequencies in the UHF band.

It is possible for the reflection coefficient to be precisely zero for certain combinations for grazing angle and dielectric constant. This occurs for vertical polarization and all of the energy is propagated into the ground. It corresponds to the case when the numerator of Eq. (12-25) vanishes, and the angle at which it occurs is called the Brewster angle. The geometry and dielectric constants of ground plane ranges are such that this rarely occurs, however.

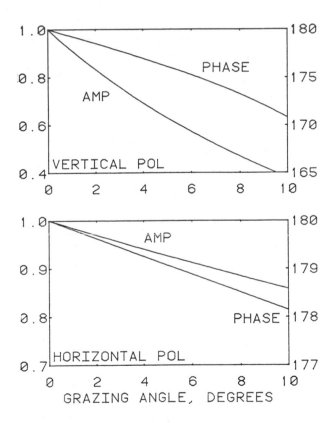

Figure 12-11 Plots of the amplitude and phase of the Fresnel reflection coefficients for small grazing angles; $\epsilon_r = 5.5 + i2$. Note the difference in the scales of the two plots

The surface roughness of the ground plays an important role because it reduces the effective reflection coefficient. The effect of surface roughness can be estimated by use of the factor ρ_s,

$$\rho_s = \exp\left[- 2\left(k\sigma_h \sin \Psi\right)^2\right] \qquad (12\text{-}26)$$

where σ_h is the RMS surface height variation. The roughness factor ρ_s which should be used in Eq. (12-22) accounts for the dispersion of energy in directions other than the specular direction, thereby decreasing the energy along the specular. Ground vegetation will modulate the ground plane reflection because brush and grass blades are moved by the wind, and there is hardly ever a perfectly calm day. Vegetation will also reduce the reflected signal and, hence, a good ground plane should be barren.

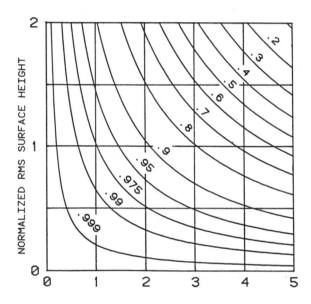

Figure 12-12 The surface roughness factor ρ_s for small grazing angles

The properties of ρ_s are illustrated in Fig. 12-12 for small grazing angles. The grazing angle for typical outdoor operation seldom exceeds five degrees, and one degree or less is quite common. If $\sigma_h = 2\lambda$ is established as an upper limit on the tolerable RMS surface roughness (which corresponds to $\rho_s = 0.91$ at $\Psi = 1$ degree), the ground plane can be quite rough and still give good performance at low frequencies. For example, this translates to 16in at 1.5

GHz, which is quite rough. On the other hand, the requirement becomes a fraction of an inch at millimeter wavelengths, and unless the ground plane is a concrete or asphalt runway, maintaining such smoothness would be difficult. Thus, for millimeter wavelengths it may be necessary to suppress ground reflections using such techniques as high gain antennas (to minimize ground illumination), large antenna and target heights, and perhaps radar fences erected between the radar and the target.

12.6 PASSIVE CLUTTER AND MULTIPATH REDUCTION

Clutter and multipath can introduce measurement errors unless they can be suppressed or exploited. As mentioned above, a ground plane range actually exploits the multipath effect by providing a smooth surface between the radar and target. However, other multipath effects are undesirable, and examples are nearby buildings, trees, and shrubbery. Multipath returns differ from clutter returns in that clutter or background signals arise from scatterers located at, or near, the target distance. The target positioner is a clutter source unless adequately shielded and, of course, a finite contribution comes from the target support column, even though it may be made from "invisible" plastic foam, which, as we have seen, is not invisible. Multipath contributions may arise from obstacles considerably farther away, as may be seen from the following analysis.

Surfaces of equal time-delay are ellipsoids of revolution, as suggested by Fig. 12-13, which shows a slice of the ellipsoid along its major axis. The radar and test target lie at the foci, and the axis of revolution coincides with the line of sight. The indirect path $S_1 + S_2$ is constant for any point on this surface, hence the path length difference δ is constant. The maximum path length difference must be greater than $c\tau$, where c is the speed of light and τ is the transmitted pulse width. If the path length difference exceeds this value, the multipath return will not be detected. In fact, depending on the type of pulse detection system used, the path length difference should be a little greater than $c\tau$.

If S is the distance between the radar and the target, the total length of the ellipsoid is

$$l = S + c\tau \tag{12-27}$$

and its width is

$$w = c\tau \left(1 + 2 \ \frac{S}{c\tau}\right)^{1/2} \tag{12-28}$$

By way of example, if $\tau = 0.2\mu s$, then $c\tau = 197$ft; if the target range is $S = 1400$ft, then

$$l = 1400 + 197 = 1577 \text{ ft}$$

$$w = 197 \left(1 + \frac{2800}{197}\right)^{1/2} = 768 \text{ ft}$$

Hence, the clear (obstacle-free) width of this test range should be about half its length. Multipath returns can be generated by obstacles or scatterers which share the range gate with the target.

Clutter can be reduced by illuminating as little ground area as possible in the vicinity of the target. The ground itself should be smooth and bare. Target support structures should be carefully designed for minimal return, and target rotators should be well shielded. If necessary, the rotator and the surrounding ground should be covered with radar absorbing material to minimize interactions between the target and the ground.

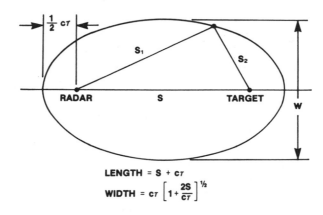

Figure 12-13 Ellipsoid of constant delay

The use of radar clutter fences can reduce the specular return from the ground in the event the ground plane effect is undesired, as might be the case at millimeter wavelengths. However, PSL concluded in its study of the RAT-SCAT ground plane range [2] that clutter fences offer no distinct advantages. This is because the top edge of the fence diffracts incident energy toward the target as well as back toward the radar, and edge diffraction is very hard to suppress. One way to reduce the diffraction is to tighten up the transmitted beam and to increase the target height to minimize the illumination of the edge — but this is precisely what would be done to eliminate the specular reflection from the ground. Consequently, the cost and effort of deploying radar clutter fences seem hard to justify. However, if the ground has vegetation which cannot be eliminated or controlled, the use of fences should not necessarily be ruled out.

12.7 DEFEATING THE GROUND PLANE

There are instances when it is more advantageous to defeat the ground plane effect than to try to exploit it. There may be several reasons, one of which is to avoid vertical antenna height adjustments. As we saw earlier in this chapter, the antenna height, target height, range, and wavelength all influence the height of the first lobe in the vertical interference pattern. In order to exploit the ground plane effect, the target must be centered in this lobe.

In most measurement situations, the range and target height remain fixed throughout a measurement program. If the frequency were also fixed, the antenna height could be set and adjusted, and left at that value for the duration of the test program. However, this is not always typical. In many programs, measurements must be made at several frequencies within a given band, requiring that the antenna height be adjusted and set for each frequency. This can be avoided if the ground plane effect is defeated.

The effect can be worse for short-pulse (wideband) measurements. The height of the first lobe of the incident field profile varies directly with the radar frequency. Therefore, if the bandwidth of the transmitted pulse is, let us say, 15 percent of the center frequency, the lobe jiggles up and down as much as 15 percent from the intended height over the frequencies in the pulse. If the target subtends a significant fraction of this first lobe, the target illumination will be seriously degraded.

Moreover, at millimeter wavelengths, particularly at 95 GHz, it may be difficult to create a ground plane smooth enough to support the forward ground reflection. For example, if a surface roughness height of 2λ is imposed, corresponding to $\rho_s = 0.91$ at $\Psi = 1$ degree, the ground plane roughness cannot exceed 0.25in. It is doubtful that a ground plane could be maintained to this degree of smoothness and, hence, at millimeter wavelengths, ground reflections may have to be suppressed rather than exploited.

One way to suppress the ground reflection has already been mentioned, radar fences, and a typical arrangement is shown in Fig. 12-14. The purpose of the fences is to eliminate the reflection from the specular point on the ground, and a single fence is hardly ever sufficient; multiple fences are often installed to suppress diffuse scattering toward the target from the ground. The fence must be made of metallic sheets or screens, and may be covered on their front sides with absorber to reduce back-scattering to the radar. The upper edges may be serrated to suppress diffraction off the edge toward the target, but the effectiveness of serrations is questionable. The Boeing Company in Seattle has installed radar fences on its 4000ft outdoor range to suppress the ground reflection.

Another technique is to construct a berm running between the radar and the target. The berm should be in the shape of an inverted *vee*, as shown in Fig. 12-15. The berm can be a mound of earth, but it should be asphalted or

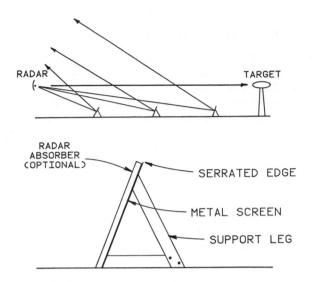

Figure 12-14 Radar fences

cemented to retard erosion. The apex of the berm should be sharp, rather than rounded, to minimize reflection off the edge toward the target. The slanted sides should be reasonably flat to deflect energy well out of the target zone.

There seem to be no established design values for this kind of berm. The Teledyne Micronetics Company in San Diego has operated an outdoor measurement range for a number of years with an asphalted berm about 4ft tall and 20ft wide. The apex of this berm does not have a sharp edge.

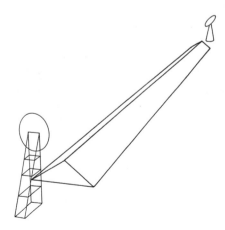

Figure 12-15 A berm constructed between the radar and the target suppresses the ground reflection

12.8 OTHER SITING CONSIDERATIONS

Most radar reflectivity measurements are performed in connection with Department of Defense programs, and many of the targets, as well as the data collected, constitute classified material or information. Moreover, test targets and data may be regarded as proprietary even if not classified, and, hence, a reflectivity range must ensure that the target is protected. These requirements are not always specific, but, in general, direct access to the site must be subject to rigid control. This could require security fences, guards, and patrols.

Physical access is not the only aspect of control. Remote observation of measurement activities is a possibility that must be considered, and this often requires that the site be well removed from urban areas. In addition, measurement activity is subject to observation from satellites, and operations may have to be suspended during the transits of certain space objects. Government agencies provide information on these satellite passes to authorized personnel.

The government also has restrictions on the power radiated by RF systems. FCC regulations are complicated and change from year to year. Therefore, range operators should become acquainted with frequency and power limitations. The nearest local FCC office should be identified and consulted whenever there is a question on operating restrictions.

Favorable environmental conditions are important, but they are not always within management's means to acquire. Local corporate real estate may have to be used for a reflectivity range, even if local weather is poor for outdoor work. Management must also reconcile the cost of maintaining a distant range in a favorable climate, as well as the travel costs of range users back and forth to the home office. The southwestern United States is a good place for an outdoor range because of the possibility of year-round operation. General weather conditions are good there because of the scant rainfall. However, the wind tend to be strong in the flat lands and is the most common cause for suspending measurement operations. Vegetation tends to be sparse or easily cleared, but the summer heat tends to be hard on personnel.

Southeastern sites also tend to have favorable year-round weather, with calmer winds but more frequent rain. The average annual precipitation in Georgia, for example, is 55in, which is almost exclusively rain (as opposed to snow) in the southern two-thirds of the state. The northeast and midwest have more moderate rainfall but much harsher winters. The Pacific northwest is calm and moderate, but wet. Further inland, the northwest climate is drier, but winters are very cold. Thus, site selection involves many factors, not the least of which are the climate and cost of operation.

12.9 EXAMPLES OF PAST AND EXISTING RANGES

Ranges in the Late 1960s

Many large corporations operate their own reflectivity ranges in support of internal development projects or government contracts for weapons systems research and development. Examples are the facilities of large aerospace companies such as McDonnell Douglas, Boeing, and General Dynamics. A large number of indoor ranges are in operation, some of them fairly small and many of them operated by academic institutions such as Georgia Tech, the University of Michigan, and Ohio State University. A few are government owned and operated, such as the Navy facility at Point Mugu, California, and the Air Force RATSCAT range in New Mexico. A few modern ranges are special-purpose facilities, not necessarily used to measure the free-space radar cross section of test objects.

It is interesting to note the similarities and differences between large outdoor ranges of 10 to 15 years ago. In a 33-month study of ranges conducted by the University of Michigan [3,4] in the late 1960s, five ranges were selected in an evaluation of large target capabilities. Of the five, two have gone out of business, and like the manufacturers of radar absorbent materials, we might surmise that the commercial market for radar reflectivity measurements is very competitive. The five ranges participating in the evaluation are listed in Table 12-2.

Of the five ranges listed, Conductron and Radiation Services ceased operations in the 1970s. All except Micronetics (now Teledyne Micronetics) used the ground plane in their measurements; the Micronetics range is unusual, as we have seen, in that an asphalted berm in the shape of an inverted *vee* runs from the transmitter to the target turntable. The purpose of this berm is to deflect ground reflected waves away from the target and away from the receiver, and it is therefore intended to provide free-space target illumination conditions. Micronetics was in the forefront of data recording in those days, being the only range to take advantage of the high-speed features of magnetic tape as a recording medium. The other four recorded data on paper tape and punched cards, which was a very slow process. Very few modern ranges, if any, record data on these media now.

Conductron and Micronetics both made measurements at much less than the far field range for the largest target and highest frequencies in the test program. The remaining three ranges also used less than the far field range, although not as foreshortened as the ranges used by Conduction and Micronetics. The data collected by Conductron suffered severe near field distortion, with the broadside lobe of a metal cylinder being split into a pair of lobes with a shallow null between them, and with the measured amplitude being several dB below the true far-field value. This was a consequence of using a CW

Table 12-2
Reflectivity Ranges Evaluated in the Late 1960s

Site	Transmitter	Max. Range Used	Geometry	Type	Data Recorded Dynamic Range	Digital Equipment
Conductron Corp.	CW	200	Ground Plane	Amplitude	40 dB	Paper Tape
Radiation Services	Pulse	1000	Ground Plane	Amplitude	40 dB	Punch Cards
General Dynamics	Pulse	1800	Ground Plane	Amplitude Phase	50 dB 360°	Paper Tape
RATSCAT	Pulse	1200	Ground Plane	Amplitude Phase (L-Band)	50 dB 360°	Paper Tape
Micronetics	Pulse	600	Berm	Amplitude	40 dB	Magnetic Tape

instrumentation system that radiated very low power, probably much less than a watt. The other ranges used more conventional pulsed systems, with pulse widths ranging from 0.1 to 1 μs.

Only two ranges (RATSCAT and General Dynamics) were able to measure the phase of the backscattered signal. General Dynamics, it might be noted, had the prime contract to build and instrument the RATSCAT range. Hence, it is not unexpected that both ranges had the same capability for coherence. General Dynamics used the longest range in the measurement of the test objects, which was a collection of precision cylinders, the largest being 32ft long and 5ft in diameter. Of particular interest is that the now-defunct Radiation Services range turned in the most accurate results for the lowest cost in the least time, a tribute to the company and its range crew. Unfortunately, not all of these virtues are apparent in Table 12-3, which is a summary of how the five ranges were rated on their measurements of the precision cylinders.

Present-Day Outdoor Ranges

Three of the outdoor ranges in Table 12-3 exist today, and several others have been built since the Michigan survey was conducted. A list of the larger outdoor present-day RCS ranges appears in Table 12-4, and even that list is incomplete.

The column headings are self-explanatory, but a few comments are in order. First, the figures listed in the fourth column (*Target Size* or *Weight*) are not strictly comparable from one range to another. In some instances, the weight capacity listed is the specification given by the manufacturer of the target rotator or turntable, and in other instances it indicates what the range operators consider typical. It is likely that none of the ranges has ever mounted target heavy enough to load the turntable to its rated capacity.

Similarly, the target size (if given) is not necessarily an indication of the maximum size for which accurate data can be collected. For example, a modest 6ft target would have to be measured at a range of 2500ft at 35 GHz in order to satisfy the far-field requirement, and this distance quadruples for every doubling of the target size.

Not included in Table 12-4 are the polarizations available for measurement, the transmitted power or the capability for coherence (phase as well as amplitude). All the ranges routinely transmit and receive linear polarizations (horizontal and vertical), and can provide circular polarization on request. Most of the ranges routinely transmit at least 1 kW, but can generate higher power at heavily used frequencies, such as in X-band. Some of the ranges have coherent systems, but the demand for coherency is not great. Most, if not all, the ranges have data processing equipment to reduce, manipulate and display the information according to the user's requirements.

Table 12-3
Range Ratings

Range	Facilities Techniques, Procedures	Cylinder Data Accuracy		Relative Cost
		Direct Polarization	Cross Polarization	
Conductron Corp.	B –	D	C	1.12
Radiation Services	B	B	C	0.47
General Dynamics, Ft. Worth	B +	C	D	1.32
RATSCAT	B	C	C	GFE
Micronetics	B –	D	D	1.09

Table 12-4
Summary of Range Capabilities

Range and Location	Type of Range	Pits or Locations	Target Size or Weight	Target Support System	Frequency Coverage	Remarks
RATSCAT; 12 miles west of Holloman AFB, N.M.	Ground plane (graded alkali)	9 (6 active); 458', 1100', 1158', 2458', 5600', 7500'	Up to 50,000 lb. and 80' capacity; a gutted F-102 has been measured	Expanable bead foam, metal pylons	30 MHz to 24 GHz continuous; spot frequencies at 35, 52, 70 and 95 GHz	Can handle and measure larger targets at longer ranges than any other facility.
GD/FW Fort Worth, Texas	Ground plane	3; maximum range 1800'; operation vans (no fixed range)	20 to 30 ft.	Expandable bead foam	1 to 16.5 GHz continuous	
Boeing; Kent, Washington	Free space (gravel) clutter fences, asphalt berms)	5 distributed over 2 ranges; 150', 300', 650', 942', 4000'	10,000 lb. capacity	Expandable bead foam (shaped)	1 to 18GHz continuous; 35 GHz planned	Routine use of shaped columns (where applicable) to reduce background returns.
Martin Marietta; 5 miles west of Orlando Airport	Ground plane (asphalted)	3; 200m, 400m, 800m	20,000 lb. capacity; 2,000 lb. and 16 ft. are typical	Expandable bead foam, metal pylons	500 MHz to 18 GHz; spot frequency at 35 GHz	New range designed and operated by experienced personnel.
Teledyne Micronetics; San Diego, California	Free space (asphalted berms)	8 distributed over 3 ranges; 125', 250', 300' 600', 650', 1000' (2)	10,000 lb. capacity	Expandable bead foam	140 MHz to 18 GHz continuous; others at (in GHz) 22-26, 35-40, 54-56, 95	Most experience in speed and quality of recording data.
Rockwell (Tulsa): Verdigris Test Site; 19 miles east of Tulsa Airport	Ground plane (asphalt and grass)	3; 300', 600' 1500', (possible use of 2900')	2,000 lb. typical an F-16 without wings or tail has been measured	Polyurethane foam	500 MHz to 1 GHz, 2.86 GHz, 8.5-9.8 GHz, 15-17.5 GHz, 35.9 GHz	Coherence not possible (all RF sources are magnetrons).
Rockwell (Los Angeles); "Weedpatch" Test Site; 75 miles northwest of LA	Ground plane	2; 185m to 300m, 1200m	1600 kg, 18m	Expandable bead foam, polyurethane foam	S, C, X, Ku, K, Ka bands	
McDonnell Douglas; Grey Butte Site; 25 miles east of Palmdale	Ground plane (asphalted)	5 distributed over 2 ranges; 1350' 1650', 2530', 2930', 3830'	up to 2500 lb.	Polyurethane foam, metal pylons	In MHz: 150-250; 500-1350; in GHz: 2.6-12.4, 15.5-17.5, 24, 35	Range sees heavy use by other contractors because of skilled staff and fast turn-around.
Northrop; Antelope Valley CA	Ground plane (asphalted)	3; 1500', 3000' (2)		Metal pylons	2 to 18 GHz	Relatively new range.

RATSCAT is probably the most impressive outdoor reflectivity range in current operation. Located on several square miles of alkali flats on the White Sands Missile Range in south-central New Mexico, the site is remote and secure. RATSCAT was originally configured with three concrete bunkers called "pits" at distances from 458 to 2358ft from a central operations building. Each pit contained a target rotator up to 17ft in diameter, and was serviced by a large mobile structure that protected the pit when not in use, and which came complete with target hoists and catwalks. Other pits were constructed later, among them a heavy duty pit with a 40ft diameter rotator rated at 50,000 pounds capacity.

As measurement requirements changed and technology advanced, some pits were abondoned, such as Pit 1, with a range of only 458ft, and Pit 6, which housed the heavy duty rotator, and new ones were constructed, such as Pit 7. Pit 7 is not a "pit" at all, but a 75ft pylon planted in the desert floor like a gaint sundial gnomon (see Ch. 11). The heavy-duty Pit 6 at a range of 5600ft will soon be replaced by a modern heavy duty rotator 10,000ft from the operations buildings, already known as Pit 8. Nevertheless, the most impressive improvement to date is the RAMS, an acronym for RATSCAT Advanced Measurement Site.

RAMS is not located at the main site, but some 25 miles away near the foothills of the San Andreas mountain range. It will be distinguished by a pylon at least 95ft tall mounted on a massive elevator that can be lowered into a deep silo drilled out of the bedrock. The lower half of the deployed target-support pylon will be shielded from the radars by a sharp change in the slope of the ground a few hundred feet in front of the structure, thereby reducing residual reflections from the pylon. Unlike the uneven alkali desert floor at the main site, which requires periodic grading to smooth out low wind-blown dunes, RAMS is paved for improved ground plane performance and low maintenance.

RATSCAT was upgraded in 1981 with the installation of ARMS, an acronym for Advanced Radar Measurement System. ARMS is a collection of four separate radar systems, all operating simultaneously, that reduces target exposure and measurement time. Although this capability comes at a high price (each radar is equipped with its own chart recorder, for example) the measurement time is cut to nearly a quarter of the time previously required. RAMS will be even more impressive: each radar will be capable of frequency stepped operation, hence radar imagery may be generated for each radar band by the appropriate processing of the collected data. The stepped-frequency capability may also be useful for other purpose that are unforeseen at the moment. It remains to be seen, however, if this complex radar instrumentation system can be kept "on the air" without subsequent modification. It may prove more difficult than imagined to operate up to six sophisticated radars simultaneously and in sychronism.

Not surprisngly, the original RATSCAT site was patterned after the General Dynamics range at Fort Worth (GD/FW), and was operated for the first years by General Dynamics personnel [5]. The GD/FW range was constructed during the development of the TFX-111, which later became operational as the F-111. At first, the range instrumentation was clustered in vans, so that the radar could be operated anywhere on the range, but a more permanent installation has been constructed since then. One of the more interesting features of the operation of the GD/FW range is the temporary shelter covering the target zone.

This shelter is a large, pressurized air bag over 100ft long inflated by a small, low pressure, high volume air pump. The incident signal from the radar and the return signal from the target both pass through the side of the shelter. There is a reflection from the near and far sides of the shelter, but the radiated pulse width and range gate settings are such that these reflections do not contaminate the desired echo signals. Moreover, the incident wave impinges on the surfaces of the structure at such angles that the reflected energy tends to be deflected in directions other than back to radar. General Dynamics range personnel report that the inflatable shelter does not seriously perturb the uniformity of the incident wave in the target zone, and the shelter denies visual access to sensitive targets. Thus, measurements ordinarily permitted only during the night without the air bag may be performed during the daylight hours.

Boeing has operated a pair of indoor chambers and a small outdoor range for many years, and recently installed a 4000ft outdoor range. Presumably, the new range was built for development work and testing in support of Beoing's ALCM contract. Unlike the RATSCAT and GD/FW ranges, which utilize the ground plane effect, the Boeing ranges were designed to reduce ground reflections. The older, shorter range was constructed with an asphalted berm running down range; the berm is interrupted at several places by flat pads where a rotator can be installed.

The newer range is essentially a gravel runway along which several transverse clutter fences have been erected. Real estate in the area is limited, and the range had to be fitted into an awkward corner of Boeing property. The range has been mistaken for a landing strip on at least one occasion.

Like most military aircraft companies, Boeing has been active in RCSR technology for several years. Success in this technology means that target support reflections can become significant sources of error in RCS measurements. One way to reduce reflections is to design the support column for low return over some azimuth sector, which is itself a form of RCSR. Boeing range operators routinely use support columns that have a rhombic profile in the horizontal plane (see Ch. 11). The flat sides of these prismatic columns are large sctterers when presented at normal incidence to the instrumentation radar, but presumably the target can be remounted at a different azimuth with respect to the column, and the low RCS sector shifted to a different aspect angle range. A complete pattern can be reconstructed by splicing together the more accurate portions of the partial patterns.

The Martin-Marietta Aerospace Company constructed a new RCS test range at its Orlando facility in the early 1980s. The range is paved with asphalt to stabilize the ground plane and is equipped with modern digital equipment for data acquisition, reduction and display. Like the RATSCAT ARMS, up to four radars may be operated simultaneously to minimize the measurement time. This reduces exposure of the target to unauthorized observation as well as increasing the number of patterns that may be recorded during a given test session. Martin-Marietta tooks steps in 1984 to acquire the stepped frequency capability implemented at RAMS, which will significantly enhance its diagnostic data collection system.

Like the older, shorter Boeing range, the Micronetics range uses asphalted berms to suppress ground reflections. Maximum range is limited to about 1000ft, however, which precludes the accurate measurement of large targets at high frequencies. The range has the capability for coherent RCS data collection, and the Micronetics engineers once claimed they could improve data accuracy by subtracting the column return from the total measured return. This implies that two complex (phase and amplitude) patterns must be recorded: target plus column and column alone. The phasor subtraction of the second data set from the first assumes that the two sources of signal return are independent and decoupled. The extent to which this assumption is valid has not been tested anywhere, not even at RATSCAT.

Micronetics also had an "overwater" range which had been used in testing submarine periscopes. The range was essentially a very large swimming pool fitted with a hydraulic target support system. The target could be raised or lowered to the desired exposure height, rotated, and tilted. Because of the shallowness, size, and location of the pool, however, the water surface was much smoother than that typically found at sea. Nevertheless, the "overwater" range was a unique testing facility of significant diagnostic value. The pool was drained in the fall of 1982 and Micronetics planned to fill it in and convert it to a more conventional range.

Rockwell International (Tulsa Division) uses magnetrons exclusively for RF sources. Although this is an economical way to generate high power at modest cost, it precludes any kind of coherent measurement. Rockwell Tulsa uses polyurethane exclusively for its target support columns. Polyurethane foams have the smallest "volume" return. However, polyurethane generates toxic fumes when set afire, and the foamed version is not as strong as the polystyrene foams.

Rockwell International (North American Aviation Division) has an outdoor range about 75 miles northwest of Los Angeles. The site is called "the Weedpatch" and is being upgraded in support in support of the B-1 bomber project at Rockwell.

The Grey Butte test facility owned and operated by McDonnell Douglas is popular, at least if judged by the number of contractors who use it. It seems to be favored because of the experience of the personnel manning the site, its location, the speed and convenience with which a measurement program can be completed, and the quality of the data. Presumably, it is much cheaper to use than the RATSCAT range. Grey Butte was the only range, other than RATSCAT, that used the metal pylon target support system, but the newer ranges are also exploiting the advantages of the metal pylon.

The Northrop Aircraft Company constructed a new test range in California's Antelope Valley in the early 1980s. Two asphalted ranges, one of them 3000ft long, were installed at a slight angle to each other. The range features modern instrumentation and data processing equipment, target support fixtures, and target handling equipment. The Lockheed Aircraft Company is reported to have constructed a new test range, but no further information is available.

The information listed in Table 12-4 shows the variability as well as the similarity between outdoor RCS ranges, but the table does not include more subjective factors such as the personnel who maintain and operate the ranges. RATSCAT is familiar to most RCS analysts and specialists because it is a national range, and because major weapons systems must be tested there. However, excessive range time and cost have impelled major users, such as Lockheed and Northrop, to build their own facilities.

In the past, range operators were willing to furnish statistics and information on the capabilities of their facilities. This is no longer the case. Because of the competitive nature of their facilities, corporations no longer release technical data. Thus, although there may be additional outdoor RCS ranges that could be added to Table 12-4, we have no information to include in the table.

12.10 SUMMARY

In this chapter we have discussed some features peculiar to the operation of outdoor RCS ranges. Because of the long target ranges, relatively high transmitter power (kilowatts) must be developed, making it necessary to use pulsed radar instrumentation systems. A typical, amplitude-only pulsed radar was described.

Most, but not all, outdoor ranges exploit the signal enhancement due to the ground plane effect. However, the signal enhancement is always less than ideal, and the degradation is due to the directivity of the radar antennas used, as well as imperfections in the ground plane itself. The use of the ground plane is not without some disadvantages, and some outdoor ranges choose to eliminate the ground plane by the use of berms or radar fences.

The chapter concluded with a discussion of examples of past and existing outdoor ranges. Over the last ten years, some have gone out of business, but new and better ranges have been and are being built. Information concerning the newest ranges is difficult to obtain, hence the list of ranges and capabilities is incomplete.

REFERENCES

1. M.I. Skolnik, ed., *Radar Handbook*, New York: McGraw-Hill, 1970.
2. T.F. Bush, ed., "Evaluation of the RATSCAT Ground Plane Range," Report No. PE00911, Physical Science Laboratory, New Mexico State University, Las Cruces, NM, August 1978.
3. R.E. Hiatt, E.F. Knott, and T.M. Smith, "Evaluation of Selected Radar Cross Section Measurement Ranges," Part I, University of Michigan Radiation Laboratory, Report No. 7462-1-F(I), March 1969.
4. R.E. Hiatt, E.F. Knott, and T.M. Smith, "Evaluation of Selected Radar Cross Section Measurement Ranges, Part II: Cylinder Tests and Range Evaluation Procedures," University of Michigan Radiation Laboratory, Report No. 7462-1-F(IIA), March 1969.
5. H.C. Marlow, D.C. Watson, C.H. Van Hoozer, and C.C. Freeny, "The RATSCAT Cross Section Facility," *Proc. IEEE*, Vol. 53, August 1965, pp. 946-954.

CHAPTER 13

INDOOR RCS RANGES

E. F. Knott

13.1 OVERVIEW

Indoor RCS measurement ranges offer advantages not possible on outdoor ranges. Probably the single most important disadvantage of making measurements outdoors is the weather. Measurements cannot be made in the rain because of possible water damage to the target, clutter returns from the raindrops, the collection of moisture on target support columns, and the sheer discomfort of personnel. Even in the absence of rain, winds threaten to blow expensive target models off their support columns. Although light guy lines can be used to stabilize the target on its support, the guy lines contribute to the radar return and are not safe in winds of 15 mph or more.

Moreover, outdoor measurements are subject to overhead observation by satellite, and some targets are particularly sensitive. Although the transit hours of these satellites are known, and the target may be dismounted from its column or otherwise shielded, this interrupts the measurements, often consuming several hours. In fact, satellite transits are among the reasons that the silo design discussed briefly in Ch. 12 was implemented at the RATSCAT advanced measurement site. Although night operations may prevent observation within the visible spectrum, they do not prevent observation within the infrared bands.

An indoor RCS range eliminates these hazards, which collectively account for an estimated 35% loss of range time at outdoor ranges such as RATSCAT.

The target is completely shielded from the weather and unauthorized observation, and personnel can work in a comfortable environment. The chamber can be fitted with hoists to facilitate target installation and removal, and the enclosure itself prevents radar energy from escaping into the environment. The target may also be stabilized by light guy lines.

Nevertheless, the indoor RCS measurement facility is not without its disadvantages. Its use must be restricted to relatively small targets, and a great deal of care must be taken to eliminate or reduce the effects of reflections from its walls. Although the wall reflections can be reduced with absorbing materials and further suppressed by the use of suitable instrumentation, the operation of an indoor anechoic chamber represents a considerable investment of time and money.

13.2 CHAMBER ABSORBING MATERIALS

The primary problem facing early experimenters was how to suppress reflections from objects in a test laboratory. Measurements of interest then were antenna impedances and antenna patterns. It is well known that the impedance of an antenna can be measured indoors if the antenna is well separated from any indoor obstacles, but the measurement of its radiation pattern is quite another matter. Consequently, experimenters draped indoor objects with "space cloth," a carbon-impregnated fabric intended to soak up radar waves.

Research on these kinds of materials was pursued by Winfield W. Salisbury, who filed a patent on an "Absorbent Body for Electromagnetic Waves" in 1943 [1]. In his patent, Salisbury made this observation:

It is often desirable to prevent the reflection of electromagnetic radiation from surfaces or objects. For instance, when the field characteristics of an electromagnetic radiator, such as a dipole antenna or a waveguide are to be tested, it is more convenient to make such tests in the laboratory than in the open air where they are usually made, because of the necessity of eliminating reflections. It is also often desirable to provide a dummy antenna whereby a transmitter may be operated for test purpose and the radiation entirely absorbed. It may also be highly advantageous to protect an airplane or ship or other object against radar echo detection systems of the enemy by treating the surface thereof or a large portion of it so as to prevent reflection of electromagnetic waves of the particular wavelength used. I have provided a very simple and easily constructed composite surface for absorbing electromagnetic radiation, thus accomplishing these and other desirable results.

Salisbury's invention was obviously of strategic value, as attested by the fact that it took nine years for the patent to be awarded after its filing in 1943.

The principles of the now famous Salisbury screen were discussed in Ch. 8, and we will discuss them no further. Salisbury's absorber, on the other hand, could be manufactured by using "space cloth" on panels of plywood backed by metal foil or sheets. Thus, an enclosure could be erected in a laboratory to provide a relatively low reflecting environment for antenna testing.

Neher applied Salisbury's absorber design to the interior surface of an open wooden pyramid [2]. The apex of the pyramid faced away from the source of illumination, and the test antenna was placed in the aperture of the open base, as shown in Fig. 13-1. Thus, energy that propagated past the test antenna was absorbed in the tapered interior walls of the pyramid.

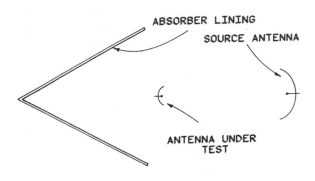

Figure 13-1 Neher's non-reflecting background device

These crude, but pioneering, attempts to build a reflection-free test environment for antenna work were later greatly refined and eventually used for RCS measurements. In the meantime, broadband materials had been developed based on the coating of the fibers of non-conducting fibrous mats with a mixture of finely divided carbon embedded in a rubber matrix [3]. These mats were produced commercially on a large scale and ranged in thickness from 1in to 4 or 6in. The thinner mats were flexible, and all of them could be stapled to portable panels of plywood or onto the walls of an anechoic chamber. The mats are still marketed commercially, attesting to the popularity and success of the design.

As good as these improvements in absorbers proved to be, the reflections from interior chamber walls were still too high for RCS measurements. A spongy pyramidal kind of absorber appeared on the market which could be improved almost without limit by making the pyramids taller and taller. The pyramidal concept appears to have been originated by Tiley [4], who suggested that the absorber be cast in molds into which is a "high loss mixture" poured. The mixture, Tiley went on to say, "may include a finely divided carbon in the form of lamp black, graphite, deflocculated graphite in water, or other resistant material retained in position by a suitable binder, such as plaster of paris, a synthetic resin, or other dielectric material."

Although the pyramidal shape was recommended by Tiley, the binders he recommended were too heavy for practical applications. Sponge rubber proved to be better, and for many years, the B.F. Goodrich Sponge Products Division in Shelton, Connecticut, was a major producer of that material. Goodrich engineers devised a rather simple, but effective, fabrication technique.

First, large slabs of polyurethane were foamed from rubber. This material has an open cell structure, meaning that the interior is filled with myriad interconnected cells and channels. This is the property that allows a compressed sponge to soak up water when released underwater. The foam rubber is easily cut with a hot wire, although the resulting fumes are toxic, and workmen must wear masks and the shop must be well ventilated.

The hot wire is stretched across a fixture that oscillates up and down over a belt or a moving table. The blank slabs are fed under the wire at constant speed, whereupon the oscillating wire carves out a saw-tooth wedge design. The slabs may be rotated 90 degrees and fed through the oscillating hot wire a second time to form the pyramids. The result is a monolithic foam blank which must be impregnated with the lossy substance. However, this procedure is used only for pyramids up to about a foot or two in height; taller pyramids require a different fabrication technique.

The lossy substance is a form of carbon, and getting it into the foam requires compressing the foam block, then releasing it in a suspension of graphite in a large dip tank. The blank may be compressed and released several times. A final compression drives out most of the suspension fluid, leaving the fine carbon particles trapped behind in the foam structure. After the pyramidal block has dried out, it may be painted with a form of low reflectivity paint. The paint helps protect the absorber, and it also improves the visual lighting in an anechoic chamber. (It can be depressing working in the totally black environment of an anechoic chamber lined with unpainted absorber.)

These pyramidal foam blocks typically have square bases, 2ft along a side. High frequency pyramidal absorbers can be as thin as 3 or 4in, while very low frequency pyramids as tall as 12ft have been manufactured (see Fig. 13-2). They can be fastened to the walls of the chamber with special adhesives (a permanent installation), or by means of mating fasteners glued to the wall and to the rear of the absorber panel.

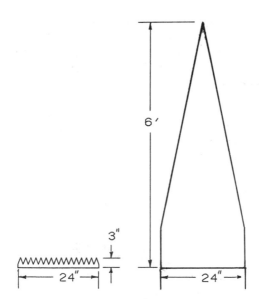

Figure 13-2 Some extremes in pyramidal absorbers

Typically, the chamber must satisfy stray radiation requirements, implying that chamber doors must be fitted with RF gaskets and that the chamber be completely enclosed within a metal shell or screen. Rigid specifications for such RF isolation require continuous soldered seams between metal sheets or screens, and the shielding of power lines, cables, and communication lines entering the chamber. The design of anechoic chambers is complicated by requirements for lighting, heating, cooling, ventilation, and strict adherence to fire safety codes. We will not concern ourselves with those requirements here.

The pyramidal design is a geometrical way of matching the impedance "experienced" by the incident wave just outside the tips of the absorber to the short-circuit (zero impedance) value of the metal bulkhead at the base of the absorber. A heuristic way to see this is to examine the impedance of a very thin transverse layer through the pyramid as function of the depth inward from the tip.

Referring to Fig. 13-3, the area of the pyramid sliced by a transverse plane is

$$A = \left(\frac{z}{h}\right)^2 A_0 \qquad (13\text{-}1)$$

where h is the height of the pyramid, z is the position of the slice inward from the tip, and A_0 is the area of the base. Assuming the net admittance associated with that particular location is weighted according to the area of the absorber and the remaining area not covered by the absorber,

$$Y = \frac{Y_0(A_0 - A) + GA}{A_0} \tag{13-2}$$

where Y is the net admittance, Y_0 is the admittance of free space, and G is the conductance of a thin slice of the absorber.

Inserting Eq. (13-1) into Eq. (13-2),

$$Y = \left[1 - \frac{z}{h}\right]^2 Y_0 + \left(\frac{z}{h}\right)^2 G \tag{13-3}$$

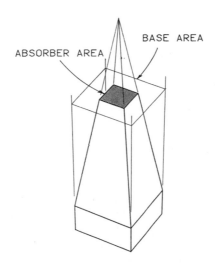

Figure 13-3 Geometry for examining the impedance taper along a lossy pyramid

Outside the pyramid and beyond the tip, the admittance takes on the free space value. As the wave propagates inward, the admittance rises with the square of the distance. Finally, at the base of the pyramid ($z = h$) the admittance is simply G, the value of the material conductance. Presumably, the conductivity of the carbon-loaded absorber is high enough that G is sensibly the value of the metal backing plate. Figure 13-4 shows a generic plot of the normalized impedance, which is the reciprocal of Eq. (13-3).

Whatever the actual variation in impedance, the incident field appears to decay exponentially once inside the tips. Figure 13-5 shows the results of the fields probed between the pyramids of an 18in pyramidal absorber. The frequency used in these measurements was 3 GHz, and the abscissa is the distance x/λ as measured *outward* from the trough at the bottoms of the pyramids. The field strength remains essentially at the incident field strength level for a distance of approximately 1.5λ inside the tips. Then it decays at a rate of about 6 dB per wavelength for this particular sample.

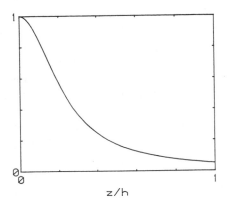

z/h

Figure 13-4 Normalized impedance as a function of distance inward from pyramidal tips

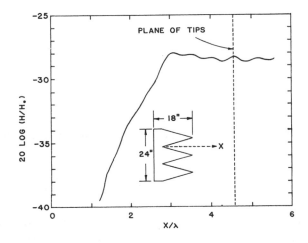

Figure 13-5 Measured field intensity between the pyramids of a pyramidal absorber

The very deep pyramidal absorbers (2ft or more) are intended for the lower frequencies of operation in anechoic chambers. It is impractical to evaluate these materials using the techniques mentioned in earlier chapters of these notes. Instead, the absorbers are mounted on metal plates and inserted into large waveguides and the VSWR is measured. This technique is described by Hiatt, *et al.* [5], and was discussed in Ch. 10.

13.3 CHAMBER CONFIGURATIONS

Once experimenters learned the importance of reducing extraneous reflections, true anechoic chambers were constructed, as opposed to partitioning off part of a laboratory with movable panels. At first, anechoic chambers were rectangular, because this was the shape of the room made available. Later, the concept of the tapered chamber was introduced to suppress the specular wall reflections.

As in a ground plane range, an anechoic chamber is subject to multipath effects, although its walls may be covered with absorbing materials. This is illustrated in Fig. 13-6; there are four paths like these (floor, ceiling, and two side walls), hence, the absorber placed at these positions should be made of good quality materials.

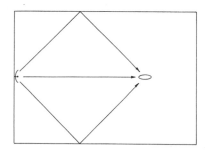

Figure 13-6 A rectangular anechoic chamber suffers multipath illumination due to wall reflections

Among the early attempts to suppress the wall reflections was a fluted design attributable to the Emerson and Cuming Company. Instead of being flat, the walls and ceiling were brought to longitudinal cusps along the midpoints of the chamber walls, as shown in Fig. 13-7. These cusps have precisely the same purpose as the berms described in Ch. 12, namely, to force wall reflections to propagate in some other direction than through the target zone. The size of the "quiet zone" of minimum perturbation due to wall reflections depends on the depth of the cusps relative to the transverse width of the walls.

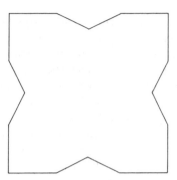

Figure 13-7 Fluted wall design is intended to deflect wall reflections away
from target zone. This is a transverse section through the chamber

Probably the best geometry for an anechoic chamber is the tapered design
illustrated in Fig. 13-8. The taper effectively removes the sidewall regions
where specular reflections can occur; hence, chambers with this design are
superior to the conventional rectangular chamber. The tapered concept was
first described by Emerson and Sefton [6], and King, *et al.* [7], built small
models of rectangular and tapered rooms and probed the internal fields using
a small horn as a transmitter. King and his colleagues found that the multipath
fields were essentially eliminated, although they also discovered that the decay
in field strength with propagation distance deviated from the normal free
space value. Thus, absolute antenna gain measurements made in a tapered
anechoic chamber might be in error unless a known gain standard antenna can
be used for calibration.

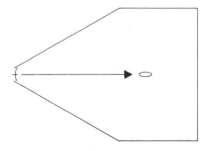

Figure 13-8 Tapered anechoic chamber has little, if any, sidewall reflections

Later it was observed that the improper decay in field strength with distance was in fact due to multipath reflections from near the apex of the tapered chamber. These arise because the illuminating antenna is not a point source that can be placed at the very apex of the chamber. Because the wall reflections lie so close to the antenna, an interference pattern is created along the chamber axis. The period of this pattern is so long that it cannot be measured in the chamber, yet its influence was to alter the apparent rate of decay of the radiated field. Nevertheless, tapered chambers are superior to rectangular chambers for radar cross section measurements, especially if the measurements must be made at low frequencies for which high gain antennas (to reduce sidewall illumination) cannot be used.

Upon investigation of the capabilities of tapered chambers, Dybdal and Yowell [8] reported that, with careful attention given to the absorbent material, reflectivity measurements can be made at frequencies as low as 120 MHz and as high as 93 GHz in a tapered anechoic chamber. At millimeter wavelengths (about one-eighth inch at 93 GHz), the sharp tips on the pyramidal absorbing material in the chamber must be maintained, otherwise the effectiveness of the material is degraded. Apparently, the paint ordinarily applied to absorbent materials cannot be used at these very short radar wavelengths, and the rounding or breaking of the tips due to handling or abuse during installation, must be avoided to preserve the high performance.

For many years, the largest tapered anechoic chamber was at the NASA Manned Spacecraft Center in Houston, Texas. It is 175ft long and has a transverse cross section of 55 X 55ft. The rear wall of this huge chamber is hung on rails and can be slid out of the way so that large test objects can be brought into the chamber, or so that tests can be made on objects up to a quarter of a mile away. Such large chambers require tall support columns (in this case, up to 25ft tall), and special attention must be given to target handling and target mounting problems. Recently, the NASA chamber was surpassed by Emerson & Cuming's design of the largest chamber ever built, which was purchased by the Japanese Space Agency [9].

The "compact range" represents an attempt to increase the size of the targets that can be measured indoors. With limited indoor area, it is difficult to obtain acceptably flat incident phase fronts, particularly at the higher frequencies. Two different techniques are available: lenses and reflectors. In either case, the device (lens or reflector) is used to collimate the beam radiated by the antenna. Within certain limitations, these devices straighten out the incident phase fronts, making it possible to conduct measurements indoors within a fraction of the distance normally required to satisfy the far-field range criterion. In one case, energy passes through the device, and in the other it is reflected. Thus, the lens must be stationed between the source antenna and the target, while the target and the source antenna must be placed on the same side of a reflector device.

Kraus[10] discusses frequency-sensitive metallic lenses and nearly frequency-independent foamed plastic lenses. Because indoor RCS ranges must be operated over a range of frequencies, the foamed plastic configurations are more desirable. The use of foamed plastics, instead of more dense materials, minimizes the reflections from the lens, because it is reflection from the target on the other side of the lens that must be measured, and we do not want the measurements contaminated by contributions from the lens.

Assuming that one face of the lens can be flat (for ease of machining, for example), the profile of the convex side is a hyperboloid of revolution. As shown in Fig. 13-9, the convex side faces the source antenna and the flat side faces the target. The convex side takes on a different profile if the lens is reversed, with the flat side toward the antenna. The hyperbolic profile depends on the focal distance and the index of refraction of the lens material.

The distance F from the source to the apex of the conventional lens is the focal length. A ray striking the lens surface is refracted (bent) at the surface, whence it travels parallel to the lens axis toward the rear face of the lens. As it turns out, this parallel path occurs when the total electrical path from the source to the rear face is constant for all rays emanating from the source. Consequently, all rays arrive with the same relative phase angle, and the phase of the transmitted wave is constant over the rear face. Thus, the lens flattens out the incident phase fronts that would have been spherical in the absence of the lens.

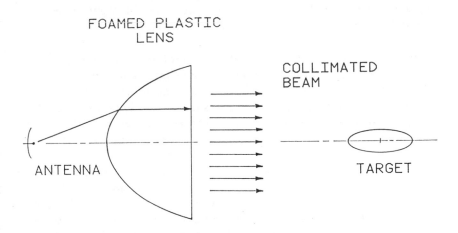

Figure 13-9 Compact range using a microwave lens

Foam plastic lenses have been successfully built and used in the past. Mentzer [11] used a styrofoam lens 1.1m in diameter at X-band frequencies. Olver and Saleeb [12] even introduced loss in a polyurethane lens to correct for the amplitude taper, but this lossy lens was only 0.43m in diameter.

A much more ambitious design was undertaken by Plessy, Inc., [13]. The lens was about 8ft in diameter, 6ft thick, and had to be fabricated from 17 separate blocks of urethane foam. The lens was intended for a compact range only 38ft long at the Naval Ship Systems Engineering Station in Philadelphia. The lens performance was within specifications for frequencies up to about 4 GHz, but failed to meet the specifications at higher frequencies. The performance failure was attributed to the interfaces between the foam blocks used to construct the lens. Plessy explored the possibility of building another lens from a single block of material, but the foam fabricator used by Plessy could not guarantee the required uniformity in dielectric constant for a monolithic (one-piece) lens, and Plessy eventually abandoned the concept.

The Boeing Military Airplane Company has built a unique millimeter wave measurement facility for model work [14], and is exploring the use of the lens for flattening out the incident phase fronts. The range from antenna to target is approximately 100ft, but at the highest frequency of interest (95 GHz), this distance limits the model size to about 12in. However, if a suitable lens can be designed and built, targets as large as 3ft can be measured. Unfortunately, the status of the lens study is not known.

As with anechoic chambers, early compact ranges exploiting paraboloidal reflectors were first used for antenna measurements [15,16]. The configuration of the paraboloidal reflector compact range is shown in Fig. 13-10. Only a segment of the paraboloid of revolution is used, and this segment lies to one side of the axis of revolution. Consequently, the reflector is called an offset paraboloid.

A feed horn is placed at the focus of the paraboloid and is aimed toward the reflector. A property of this reflecting surface is that all paths from the focus to a plane transverse to the reflector axis are equal in length. Thus, all rays reflected off the surface are in phase over such planes, and the reflected beam is collimated. Energy reflected by the target traverses these paths in the opposite direction, and converges back at the feed horn, where it is collected and delivered to a radar receiver.

Typically, a shield made of radar absorbing material is constructed about the feed horn to prevent stray target reflections from being received directly. The compact range performance is degraded by diffraction from the edges of the reflector, but it is claimed that edge serrations reduce the effect. Scientific Atlanta, Inc., markets a commercial compact range capable of measuring targets up to 6 or 8ft in size. Insofar as we know, this is the only company that markets compact ranges.

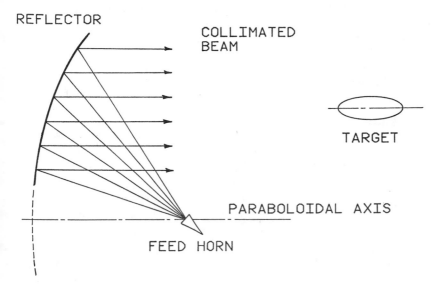

REFLECTOR

COLLIMATED
BEAM

TARGET

PARABOLOIDAL AXIS

FEED HORN

Figure 13-10 Compact range using an offset paraboloidal reflector

13.4 INSTRUMENTATION

Because measurements are made at relatively close ranges in indoor anechoic chambers, ordinary pulse radars cannot be used without special design
modifications. This is because the direct transmitter-to-receiver leakage, the
target return, and reflections from the chamber walls are all received within a
very short time of one another. Although the receiver can be protected against
the direct transmitter-to-receiver leakage, it remains "blinded" for a significant fraction of a microsecond thereafter, unless the receiver is designed specifically for close range measurements. Moreover, even if the receiver can
recover its full sensitivity by the time the target echo arrives, reflections from
the chamber walls arrive within an even shorter time afterward.

Although instrumentation systems have been built that have short recovery
times, a less expensive alternative to the pulsed system is the continuous wave
(CW) system. However, a CW instrumentation radar cannot use the time
delay characteristics of the return signals to isolate desired from undesired
returns. How can we avoid contamination of the desired target signal? Figure
13-11 shows the answer, which is a nulling loop.

The schematic diagram of Fig. 13-11 is one form of a CW cancellation
system. A sample of the transmitted signal is routed through a precision phase
shifter and a precision attenuator, thereby allowing the phase and amplitude

of this sampled signal to be varied. The signal sample is injected into the receiving line, and added to the signals received from the target and the chamber walls.

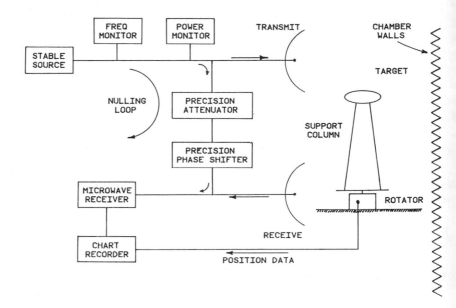

Figure 13-11 CW cancellation system shown schematically

The purpose of the cancellation signal is to "buck out" the direct leakage from the transmitting to the receiving antenna, the residual return from the chamber walls, and returns from the target support column. Hence, the circuit must be adjusted without the target in place. Without the target, the sampled signal is adjusted to obtain zero signal into the receiver. This tuning procedure is best accomplished by alternately adjusting the attenuator and the phase shifter by small amounts. When an acceptably small residual signal is indicated on the receiver, the target may be installed and measured; such measurements should be preceded and followed by the substitution of a known calibrated standard, such as a sphere, thick cylinder, or corner reflector.

The CW cancellation scheme is extremely sensitive to small changes in the transmitted frequency and, almost invariably, the RF source is phase-locked to a harmonic frequency of a crystal oscillator housed in a constant temperature oven. Experimenters have found that submerging the RF source (usually a klystron oscillator tube) in a constant temperature oil bath helps stabilize the

ystem. In addition, the phase of the signals within the waveguide runs de-
ends on the ambient room temperature because this controls the physical
imensions of the waveguide. The extreme sensitivity of the system also makes
t susceptible to normal noise and vibration in a laboratory environment.
Nevertheless, despite the sensitive characteristics of this system to noise,
emperature, vibration, and frequency stability, it has been used successfully
or many years in many laboratories, yielding high quality data when carefully
sed.

Many times a single antenna is used for transmission and reception instead
of the two antennas shown in Fig. 13-12. In this case, a hybrid *tee* is used to
prevent the transmitted signal from leaking into the receiver, and to tune out
he unwanted chamber echoes as well. However, the circuit in Fig. 13-12
works differently from that in Fig. 13-11.

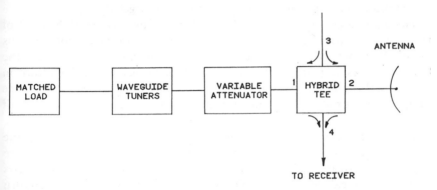

Figure 13-12 CW cancellation with a single antenna for transmission and
reception

As in the cancellation scheme shown in Fig. 13-11, the one shown in Fig.
13-12 must be tuned without any target or calibration standard on the support
column. The variable attenuator controls the amount of signal reaching the
waveguide tuners and the matched load at the left side of the diagram. As
before, signal cancellation is achieved by adjusting the attenuator and the
tuners for a null indication (low signal) at the receiver.

With this condition has been achieved, half the input power delivered to the
hybrid *tee* is routed to the antenna, and half is routed to the tuning network.
The reflections from the tuning network propagate back toward the hybrid *tee*
from the left, and reflections from the antenna, the chamber walls and the
target support column from the right. If the tuning network has been carefully

adjusted, the two sets of reflections are of equal amplitude, but 180 degrees out of phase. Thus, when they combine in the receiving arm, they cancel each other.

When the target is installed, only the reflections attributed to the target are fed to the receiver. However, if the effective reflection coefficient of any component or signal change over so slightly, the previous cancellation condition is lost, and the target measurements may be in error. Thus, it is essential to remove the target immediately after the measurement to verify that the receiver null condition has been maintained.

CW cancellation systems have been assembled which can maintain this careful adjustment for up to 30 minutes. They require careful assembly and operation, and the indoor temperature must remain very steady. Conversely, some systems are incapable of supporting the null condition for more than a few minutes at best. With a good system, a residual null signal can be maintained more than 30 or 40 dB below the received target signal.

There are cases when a good null indication is a poor gauge of the value of the measured target data. This occurs for targets whose radar cross sections are small and whose physical sizes are not so small, and is due to the way the target shadows the rear wall. The target changes the incident field structure on the rear wall, and hence the wall reflections that were tuned out in the absence of the target are not at all the same as those that are received in the presence of the target.

The perturbation of the fields incident on the rear wall are due to the forward scatter of the target, and this scattering is proportional to the square of the geometrical area of the target. Consequently, the perturbation grows worse with increasing target size. This would not be serious if the target return also increased with size, but there are some targets whose radar returns are more or less independent of size, at least for certain aspect angle regions.

Solomon reported investigations of this effect that were carried out at the Cornell Aeronautical Laboratories (C.A.L.) in the early 1960s [17]. (Solomon was a magazine editor and did not carry out the work; instead, he interviewed engineers at C.A.L.) Chamber background levels were measured indirectly by swinging small spheres back and forth within the antenna beam. As such, the sphere return goes in and out of phase with the residual chamber echo, (even after precise tuning), and the amplitude of the signal excursion can be used to deduce the background contribution.

According to data presented by Solomon, a 10 dB change in the electrical size of the sphere produced a 24.1 dB change in the background contribution. Actual measurements are not given, so there is no way to evaluate possible errors. If the chamber contribution is due to the forward scatter of the target, the background should increase 20 dB (instead of 24 dB) for every 10 dB change in sphere size. Whatever the actual numbers, it is clear that forward target scattering can be a serious source of measurement error.

Because the error is not apparent during a routine measurement, it is possible for an experimenter to be unaware of it. One way to check it is to have the target rotator mounted on a carriage that can be moved slightly toward or away from the radar. The total return will undergo a cyclic variation as the target signal alternately cancels and reinforces the contribution from the rear wall. The amplitude of this variation should remain less than ±0.5 dB for high quality measurements. The signal variation completes a cycle in a distance of λ/2; hence, a few inches of motion is all that is required to assess the error. The target must remain stationary (except for the carriage motion) during this test, of course, otherwise normal aspect angle pattern variations would be interpreted as resulting from background signals.

Several approaches have been attempted to suppress the contribution from the rear wall, but few have been successful. One is to "tune out" the contribution with movable panels of absorbing material, but this is difficult, if not impossible, to implement. Another is to hinge or pivot the rear wall to allow it to be swung to a position that places a null in its own scattering pattern precisely toward the instrumentation radar. However, as we have seen, that pattern depends on the distribution of the incident fields over the wall, and the distribution changes when the target is installed. Thus, it is doubtful if the positionable rear wall ever worked as its inventors intended.

Another approach is to remove the rear wall entirely and to provide an outdoor background. This was implemented in the huge NASA chamber at the Manned Spacecraft Center in Houston, but was not intended to suppress the rear wall return. Instead, it was intended to allow pattern measurements of antennas mounted on spacecraft and illuminated from outdoors. No data are available to assess whether the absence of a rear wall improved the measurement environment.

Probably the most effective technique to suppress rear wall contributions is to purchase the highest performance absorbing materials available, and to install them on the rear wall. If funds are limited, then the experimenter must decide for himself (or his company must make the decision) how valuable the additional measurement accuracy may be. When solutions to measurement problems are expressed in dollars per decibel, the problems sometimes go away automatically.

These arguments hold for a CW cancellation system, but there is another solution if the instrumentation can be redesigned. A particularly attractive measurement system is the FM/CW system, in which the transmitted frequency is linearly swept over a range of frequencies. The FM/CW system, which is capable of providing range discrimination, is shown in Fig. 13-13.

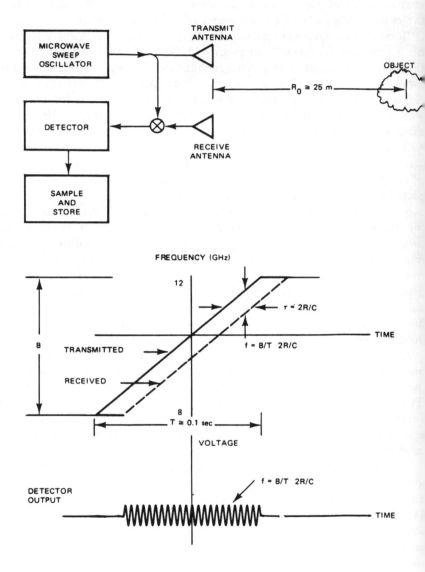

Figure 13-13 FM/CW instrumentation system

This system radiates a signal whose radar frequency is swept linearly over a range of frequencies. This can be done continuously using a microwave sweep oscillator, as shown in the diagram, or with a frequency synthesizer, whose output is a "comb" of discrete frequencies. A sample of the transmitted signal is used as a local oscillator signal, and is mixed with the target signal collected by the receiving antenna. Because the target return is delayed by the amount $\tau = 2R/c$, where R is the range and c is the speed of light, the frequency of the transmitter signal sample is slightly higher than that of the target return signal at the time of detection. Thus, the frequency of the detected signal is proportional to the range between the target and the antennas. Consequently, contributions from the rear wall of the chamber can be eliminated by passing the detected signal through a filter that suppresses all frequencies above a certain threshold which corresponds to the rear wall return.

The system shown in Fig. 13-13 happens to operate at X-band frequencies with a bandwidth $\beta = 4$ GHz (from 8 GHz at the beginning of the frequency sweep to 12 GHz at the end of the sweep), but the system can be designed and built for any radar band. The sweep time T is shown to be about 0.1 sec, but this can vary from system to system, depending on the particular components used. The transmitted frequency at any time t after the initiation of the sweep, therefore, is

$$f_t = f_0 + \frac{t\beta}{T} \tag{13-4}$$

where f_0 is the lower frequency of the swept band. The frequency of the signal received from a scatterer at a range R is

$$f_r = f_0 + \frac{(t-\tau)\beta}{T} \tag{13-5}$$

The frequency of the detected signal is the difference between the two, or

$$F = f_t - f_r = \frac{\tau\beta}{\tau} = 2\frac{\beta R}{cT} \tag{13-6}$$

Thus, for a single point scatterer, the detected signal frequency is constant, as suggested at the bottom of Fig. 13-13.

A typical target, however, has several scatterers, and sometimes several dozen; consequently, the detected signal covers a spectrum of frequencies. This spectrum allows the individual scatterers to be sorted according to range. The resolution in range is inversely proportional to the bandwidth

$$\Delta R \simeq \frac{c}{2\beta} \tag{13-7}$$

Thus, with the system of Fig. 13-13, we should be able to resolve scatterers that are only 1.5in apart.

In addition to resolving scatterers in range, the FM/CW system is also capable of resolving them in cross range, and this information can be used to create a target "image," as discussed in Ch. 11. Cross range resolution is achieved by rotating the target through a small angle $\Delta\theta$ in azimuth. The rotation imparts a Doppler shift to the target return signal, and the frequency shift serves to locate the scatterer in cross range. For example, zero Doppler shift places the scatterer along the line of sight between the radar and the center of rotation; a positive shift means the scatterer is moving toward the radar, thereby placing the scatterer on one side of the axis of rotation, while a negative shift indicates motion away from the radar, and the scatterer lies on the other side of the axis of rotation. This is shown in Fig. 13-14.

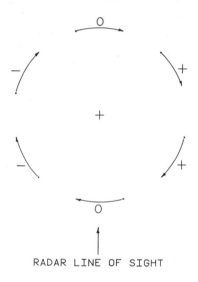

RADAR LINE OF SIGHT

Figure 13-14 Doppler shift can be negative, positive, or zero depending on the location and motion of a scatterer relative to the center of rotation

For a scatterer moving through a circular arc, the Doppler frequency shift is

$$f_d = 2\Omega\, x/\lambda \qquad\qquad (13.8)$$

where Ω is the angular speed of rotation, and x is the cross range location of the scatterer, which can be either positive or negative. For cross range discrimination between scatterers, the received data are processed over an aspect angle window $\Delta\theta$, where $\Delta\theta$ is typically several degrees. The resolution in cross range is approximately

$$\Delta x \simeq \frac{\lambda}{2\Delta\theta} \tag{13-9}$$

Thus, the range and cross range resolutions can be made comparable by appropriate selection of the system bandwidth and the aspect angle window.

There is no Doppler shift associated with the rear wall return, hence, radar imagery may contain a "ridge" of apparent scatterers along the radar line of sight. However, most of these contributions can be removed by the use of filters, as mentioned above, associated with the processing of the received data. (For further information, see reference [18]).

13.5 EXAMPLES OF SOME INDOOR RCS RANGES

The world is dotted by several dozen indoor radar reflectivity ranges. Many of them share common methods of instrumentation and calibration, and many have similar sensitivities and equipment. Rather than cataloging the capabilities and features of these ranges, which do not differ in large measure from one facility to the next within a given class, it is of interest to examine the features of some of the more unusual ranges.

A selection of such ranges is presented below. The list is not intended to be exhaustive, nor to endorse the particular virtues of the facilities discussed. It was assembled only with the idea of demonstrating the variety of approaches for determining radar reflectivity characteristics.

MACROSCOPE

MACROSCOPE is an unusual facility operated by the US Army Mobility Equipment Research and Development Command at Ft. Belvoir, Virginia. This CW range uses a homodyne detection scheme in which a sample of the transmitter signal is used as the local oscillator signal. The frequency of operation is about 98.6 GHz, and the facility is used primarily in diagnostic studies of the returns from scale models of ships and military vehicles. The transmitted signal is amplitude modulated at a rate of 100 kHz.

The range can be operated in one of two methods. In one mode, the target is illuminated by a broad beam, and is oscillated back and forth at a rate of a few hertz, thereby shifting the frequency of scattered wave. This allows for discrimination against background return. A second detector extracts the amplitude of the shifted signal from the 100 kHz modulation. In the other mode of operation, a microwave lens is installed and the target (on its ground plane) is scanned in range and cross range through the focus of the lens. The resolution of the system is purported to be about 0.5 in in range and cross range. A signal processing system displays the imagery on a television screen, and the image reveals the strong scattering centers on the target as bright spots. The computer used with the facility and the software developed for it allows close studies to be made of small areas on the target. Although the amplitudes of the various contributions to the radar echo can be obtained in a cross sense, the facility is most useful in the location of dominant target scatterers.

In its early configuration, the system oscillated the target and ground plane together, but currently only the target is moved. The data are calibrated in amplitude by replacing the target with a metallic calibration sphere. Although an imperfectly conducting ground plane could be used, or a rough one to simulate the reflection from the sea or terrain, this has not been done. The operators of the range have not been required to perform this simulation, and they contend that the diagnostic purposes of the facility are served just as well with a "shiny" ground plane because the target images in the ground plane are brighter. A brief description of the facility is given by Paddison, *et al.* [19], in their paper on ship cross section.

EMI Modeling Facility

EMI Electronics Limited has developed a radar modeling capability at the UK National Radio Modeling Facility. Emphasis at this facility is on the development of instrumentation systems, and the collection and interpretation of radar scattering data at frequencies up to 2000 GHz (2 THz). Because virtually all of the measurements and testing are performed on scale models, EMI has fabricated many models over the years, ranging from missiles and shells to ships and aircraft. EMI has also developed state of the art components, such as RF sources and detection systems.

All the measurements are conducted indoors. As of 1978, nine different radar systems were operable in conjunction with seven different model support systems. Unlike most indoor facilities, this one makes limited use of radar absorbing material and relies instead on range gating to eliminate background reflections. As many as seven different systems can be operated simultaneously, although it is doubtful that this often occurs.

Targets are usually suspended from overhead carriage by means of thin electrical lines, although tapered support columns may also be used. The line of sight is inclined such that the support lines are not viewed at normal incidence. For some measurement systems, the target is slowly rotated by the support carriage while being viewed by a stationary radar. For others, the target is held stationary while the instrumentation is rolled past the target on circular or straight tracks. Obviously, the tracks must be very smooth and precisely laid for coherent measurements at these frequencies, which EMI has the capability of performing.

A succinct survey of the modeling facility, including photographs and diagrams, is given by Cram and Woolcock [20].

The University of Michigan

The Radiation Laboratory at the University of Michigan operates a unique surface field measurement range in which the induced fields on a target are probed. The facility cannot be strictly classified as a reflectivity range, but it is sufficiently unusual to warrant brief discussion. Developed in the mid-1960s in support of US Air Force studies of the scattering from re-entry bodies, the facility is currently being used to investigate surface field as might be induced by the electromagnetic pulses generated by nuclear blasts.

The range is housed in an indoor tapered anechoic chamber of more or less conventional design, except that a hole has been cut in the ceiling of the target end of the chamber. The test object is placed under this hole on a foamed plastic column in the desired orientation and remains stationary while the surface fields are probed. An extremely broadband transmitting antenna has been developed and installed at the transmitting end (the apex) of the chamber, and the frequency capability of the system is continuous from 100 MHz to 5 GHz.

Two kinds of surface probing can be performed. A "free space" unbalanced loop probe, typically from 3 to 10mm in diameter, can be moved over the surface of the target, thereby sampling the total tangential magnetic field. The probe loop is formed at the end of a small diameter, semi-rigid coaxial line which projects downward from the hole in the chamber ceiling. The probe and its lead are supported by a low-reflectivity support structure attached to a carriage above the ceiling, and the carriage is driven by small, remotely controlled dc motors. Because the metallic probe lead projects vertically into the chamber, the incident wave is always chosen horizontally polarized to minimize interactions with the probe lead and reduce stray pickup.

In the other mode of operation, small loop probes or monopoles are attached to the target surface at selected stations and their leads brought outside the target at some point of geometric symmetry. The leads are run vertically up to the hole in the ceiling, as with the movable probe, to minimize stray induced fields. The fixed probes are oriented to sample the tangential magnetic field or the normal electric field, depending on the data to be collected.

The RF sources are a pair of sweep oscillators that are switched into or out of operation by computer control. These sources can cover the frequency range from 10 MHz to 18 GHz, but the entire range is not used. A set of four power amplifiers, each covering a portion of the entire system, boosts the 120mW source output to 1W of input to the broadband antenna. The amplifiers are switched in and out of operation by the control computer as needed, depending on the particular frequency demanded.

The computer is programmable, and the user can specify the frequencies and frequency intervals to be used in the measurements. The data are collected at a sequence of discrete frequencies, rather than continuously over the entire spectrum. The probe output signals are detected and amplified by a system of preamplifiers, and a microwave network analyzer extracts the phase and amplitudes of the signals at each frequency. These are then converted to digital form by A/D converters, and stored or written on tape by the control computer for subsequent analysis and processing.

A brief description of the facility in its early configuration is given Knott, *et al.* [21], and a discussion of later improvements may be found in Liepa [22, 23].

Georgia Tech

Georgia Tech Research Institute operates a compact range that uses an offset paraboloidal reflector. The reflector in Georgia Tech's original compact range was a 10ft diameter dish, and only the upper half of it was used. While this facility was being used for RCS measurements in connection with sponsored research, Scientific-Atlanta developed a commercial compact range that uses a 12 × 16ft reflector. In the late seventies, Scientific-Atlanta donated its prototype reflector to Georgia Tech, and that reflector is now in use in Georgia Tech's compact range.

The compact range is housed in a room approximately 25ft square and 14ft tall. The room is partially lined with absorbing material, and the rear wall behind the target is covered with high quality pyramidal material. The compact range has been instrumented for three radar bands of operation (S-band, X-band, and K-band), and the CW cancellation system was used. More recently, a stepped-frequency instrumentation system has been installed, covering frequencies from 2 to 18 GHz. This system operates much like the one installed at Ohio State University in 1983 [24]. The extreme bandwidth of the system gives range resolutions of better than 1cm. The compact range has a quiet zone approximately 4ft wide, and extending from the rear wall to near the focus of the reflector.

No photos of Georgia Tech's compact range are available, but its physical arrangement is similar to that of Scientific-Atlanta's Model 5750 compact range shown in Fig. 13-15. This photo was taken from the rear wall of the facility and shows the serrated edges of the reflector. The prototype reflector in Georgia Tech's compact range does not have these serrations. The reflector is illuminated by a horn on the far side of the target support column. Because of the support column and the use of absorber panels to shield the horn from stray reflections due to the target, the horn cannot be seen in the photo.

Boeing Millimeter Modeling Facility

The Boeing Military Airplane Company has installed a unique millimeter wave measurement facility. First described at the 1980 Radar Camouflage Symposium [14], this facility is intended for scale model measurements. Currently instrumented for three millimeter bands (35 GHz, 60 GHz, and 95

GHz), the conventional CW cancellation system is used. The RF sources are phase-locked to highly stable, low frequency crystal oscillators, and interestingly enough, the system is tuned from a remote control panel in the control room.

Figure 13-15 Scientific-Atlanta Model 5750 compact range

Figure 13-16 is an artist's sketch of the facility, which is comprised of four major features or systems. The anechoic chamber itself has a 16 × 16ft square cross section and is 130ft long. The chamber is lined with a pyramidal type of carbon loaded urethane foam which is only 2in thick. This material was tested panel-by-panel, and exhibits a reflectivity of –50 dB or less. Unlike most pyramidal absorbers, it is not painted, and the interior of the chamber is very dark. The facility is equipped with a fire sprinkler system and has passed the required fire inspection.

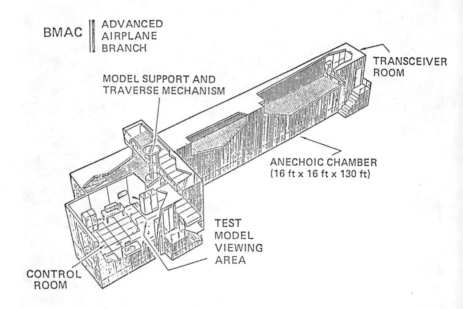

Figure 13-16 Boeing's modeling facility

The RF portions of the radar instrumentation system are housed entirely in the transceiver room at the far end of the facility. The antennas are aimed down the longitudinal chamber axis through a port in the center of the wall separating the room from the anechoic chamber. The instrumentation is turned on and initially adjusted manually, whereafter subsequent adjustments — including the nulling of residual chamber reflections — are made by remote control. The received signal is detected and amplified in the transceiver room, and the video output signal is routed to the control room for plotting and digital recording. System tuning, data recording and processing, and target control are all performed in the control room.

A video camera and monitor system provide the operator with a continuous view of the target attitude and motion. Perhaps the most unusual feature of the entire facility is the method of mounting, supporting, and removing the test target. The chamber is equipped with a pair of rotators slaved together, one in the ceiling and one in the floor. Mounted around the circumference of each rotator is a set of small reels driven by stepping motors. Support lines wound on the reels run from the ceiling rotator to the floor rotator, and the target is suspended in a harness attached to these lines. The target attitude can be controlled by reeling line out or in on selected reels, which can be accomplished under computer control once the target has been installed in the support harness.

Both rotators are mounted on a traversing mechanism that translates the rotators out of the anechoic chamber and into a work station at one end of the control room. The floor of the work station is several feet higher than the floor of the chamber, which places the model at a convenient working height for personnel. The traversing mechanism moves the target out of the chamber in a matter of minutes, making calibration simple and easy to perform with a test object of known return, such as a sphere.

The target-to-antenna spacing is approximately 100ft, thereby restricting far-field measurements to targets of about 12in in size. Boeing is investigating the possibility of using a lens to collimate the beam and thereby increase the allowable target size. Preliminary studies have suggested that a 10ft diameter lens will provide far-field conditions for targets up to 4ft in size, but reflections from the lens will contaminate the measurements unless reflections can be removed. One possible method is the use of synthesized wideband pulses, as discussed in Ch. 15.

13.6 SUMMARY

In this chapter we have discussed the salient features of indoor RCS ranges. Those features include the absorbing material used to line the interior of the chamber, the chamber geometry, and the instrumentation needed to make the measurements. The pyramidal absorbers must be chosen with regard to the desired performance at the lowest expected measurement frequency. The lower this frequency, the thicker and more expensive the material will be.

The shape of the chamber can be optimized; the conventional rectangular shape is not as good as a taper provided at the transmitting end of the chamber. The rear wall contributes errors to the measurements, even when the CW cancellation system is employed to "buck out" the rear wall return. This is due to forward scattering from the target and is important for targets whose projected area is much larger than the target RCS.

Special instrumentation — specifically an FM/CW radar — can greatly suppress rear wall contributions, and can provide radar image generation as well. Such images are useful in diagnostic tests and RCS reduction studies. Finally, at the end of the chapter we described a few unusual facilities and how they can be used.

REFERENCES

1. W.W. Salisbury, "Absorbent Body for Electromagnetic Waves," U.S. Patent 2,599,944; granted 10 June 1952 (filed 11 May 1943).

2. L.K. Neher, "Non-Reflecting Background for Testing Microwave Equipment," U.S. Patent 2,656, 535; granted 20 October 1953 (filed 6 August 1945).

3. H.A. Tanner, "Fibrous Microwave Absorber," U.S. Patent 2,977,591; granted 28 March 1961 (filed 17 September 1952).

4. J.W. Tiley, "Radar Wave Absorption Devices," U.S. Patent 2,464,006; granted 8 March 1949 (filed 28 April 1944).

5. R.E. Hiatt, E.F. Knott and T.B.A. Senior, "A Study of VHF Absorbers and Anechoic Rooms," University of Michigan Radiation Laboratory Report No. 5391-1-F, February 1963.

6. W.H. Emerson and H.B. Sefton, Jr., "An Improved Design for Indoor Ranges," *Proc. IEEE*, Vol. 53, August 1965, pp. 1079-1081.

7. H.E. King, F.I. Shimabukuro, and J.L. Wong, "Characteristics of a Tapered Anechoic Chamber," *IEEE Trans. Antennas Propag.*, Vol. AP-15, May 1967, pp. 488-490.

8. R.B. Dybdal and C.O. Yowell, "VHF to EHF Performance of a 90-Foot Quasi-Tapered Anechoic Chamber," *IEEE Trans. Antennas Propag.*, Vol. AP-21, August 1965, pp. 1079-1081.

9. Private Communication with Mr. William F. Emerson, Emerson & Cumming, Inc., 5 November 1984.

10. John D. Kraus, *Antennas*, New York: McGraw-Hill, 1950, pp. 382-390.

11. J.R. Mentzer, "The Use of Dielectric Lenses in Reflection Measurements," *Proc. IRE*, Vol. 41, February 1953, pp. 252-256.

12. A.D. Olver and A.A. Saleeb, "Lens-Type Compact Antenna Range," *Electron. Lett.*, Vol. 15, No. 14, 5 July 1979, pp. 409-410.

13. Private Communication with Leland Hemming on 21 January 1981. Mr. Hemming is the Technical Director of Advanced Electromagnetics, Inc., of San Diego. At the time of the communication, he was with Plessy Inc., Materials Division, San Diego.

14. H.S. Burke, T.G. Dalby, W.P. Hansen, Jr., and M.C. Vincent, "A Millimeter Wave Scattering Facility," paper given at the 1980 Radar Camoutlage Symposium, Orlando, FL, 18-20 November, 1980; Technical Report No. AFWAL-TR-81-1015, Avionics Laboratory, Wright-Patterson AFB, OH, pp. 327-336.

15. R.C. Johnson, H.A. Ecker, and R.A. Moore, "Compact Range Techniques and Measurements," *IEEE Trans. Antennas Propag.*, Vol. AP-17, September 1969, pp. 568-576.

16. R.C. Johnson, "Antenna Range for Providing a Plane Wave for Antenna Measurements," U.S. Patent 3,302,205; granted 31 Jan. 1967.

17. Leslie Solomon, "Radar Cross Section Measurements: How Accurate Are They?" *Electronics*, Vol. 35, July 20, 1962, pp. 48-52.

18. Dean L. Mensa, *High Resolution Radar Imaging*, Dedham, MA: Artech House, 1981.

19. F.C. Paddison, C.A. Shipley, A.L. Maffett, and M.H. Dawson, "Radar Cross Section of Ships," *IEEE Trans. Aerospace Electron. Syst.*, Vol. AES-14, January 1978, pp. 27-34.

20. L.A. Cram and S.C. Woolcock, "Review of Two Decades of Experience Between 30 GHz and 900 GHz in the Development of Model Radar Systems," *Millimeter and Submillimeter Wave Propagation and Circuits*, AGARD (Advisory Group of Aerospace Research and Development) Conference Proceedings No. 245, September 1978.

21. E.F. Knott, V.V. Liepa, and T.B.A. Senior, "A Surface Field Measurement Facility," *Proc. IEEE*, Vol. 53, August 1965, pp. 1105-1107.

22. V.V. Liepa, "Sweep Frequency Surface Field Measurements," Sensor and Simulation Note 210, U.S. Air Force Weapons Laboratory, July 1975.

23. V.V. Liepa, "Measurements of Surface Fields on Scale Model E-4B Aircraft," University of Michigan Radiation Laboratory Report No. 016708-1-F, October 1979.

24. E.K. Walton and J.D. Young, "The Ohio State University Compact Range Cross-Section Measurement Range," *IEEE Trans. Antennas Propag.*, Vol. AP-32, No. 11, November 1984, pp. 1218-1223. (This paper contains a photograph of the compact range).

For background information, further reading is recommended in:

W.H. Emerson, "Electromagnetic Wave Absorbers and Anechoic Chambers Through the Years," *IEEE Trans. Antennas Propag.*, Vol. AP-21, July 1973, pp. 484-490.

14.1 INTRODUCTION

The objective of this chapter is to present the considerations required in estimating the RCS of complex targets, and the very important topic of data reduction. The topics to be covered are:

- High frequency scattering by a complex target. This is a discussion of the seven high frequency scattering mechanisms, and phasor addition of the scattering from multiple scattering centers
- Simple estimation techniques for obtaining RCS estimates for a hypothetical missile target
- Interpretations of cylinder-ogive scattering patterns in terms of physical mechanisms, with comparisons to simple estimates
- RCS data reduction, including discussions·of user needs, types of reduced data, and RCS data format examples

14.2 HIGH FREQUENCY SCATTERING BY A COMPLEX TARGET

High frequency scattering occurs when the wavelength of the incident radiation is much smaller than the scattering body itself and of the various scattering center components comprising the body. This condition is usually met for most targets of interest in the microwave region. *The Radar Cross Section Handbook*, Vol. 2 [1], lists seven mechanisms which can combine to form the total RCS signature of a complex body. They are

1.   Specular scattering;
2.   Scattering from surface discontinuities such as edges, corners, and tips;
3.   Scattering from surface derivative discontinuities;
4.   Creeping waves or shadow boundary scattering;
5.   Traveling wave scattering;
6.   Scattering from concave regions such as ducts, dihedrals, and trihedrals;
7.   Interaction scattering such as multipath or other multiple bounces which occur due to collocated scattering centers.

We begin with a review of the hierarchy of scattering shapes and the formula for their peak scattering magnitudes shown in Table 6-1.

The intensity of the specular scattering depends on the radii of curvature at the specular point where the angle of incidence is equal to the angle of reflection. For the backscatter case, the specular points are those locations on the scattering body where the local surface normal points are directed back at the radar. The specular RCS of doubly curved surfaces is independent of frequency and proportional to the product of the two radii of curvature at the specular point. For singly curved surfaces, such as that of a cylinder, the RCS is linearly proportional to frequency, and varies as the square of the length and as the first power of the radius. For flat surfaces, for which both radii of curvature are infinite, the RCS varies as the square of the frequency and as the square of the area. Specular scattering is typically independent of polarization.

Scattering from surface discontinuities such as edges is a diffraction phenomenon and depends on polarization. In contrast to surface scattering, for which the specular direction is a single direction, the specular direction for edge scattered rays lie along the generators of a forward cone centered on the edge. For backscatter, the peak return occurs when the edge is perpendicular to the radar line of sight. The maximum return from leading edges occurs when the incident electric field is parallel to the edge, and from trailing edges when the electric field is perpendicular to the edge. The magnitude of this peak is independent of frequency and varies as the square of the edge length l:

$$\sigma \simeq \frac{l^2}{\pi} \qquad\qquad (14\text{-}1)$$

A 1ft long knife edge (0.305m), for example, has a peak specular RCS of 0.03m^2, or −15.3 dBsm.

Scattering due to creeping waves may be important for simple targets, but is not believed to be a significant mechanism for complex targets not designed for low cross section. The creeping wave return has the same polarization dependence as traveling waves.

Scattering due to traveling waves (TW) is a polarization-dependent phenomenon, which occurs only when there is a component of the incident electric field along the surface in the direction of propagation. This phenomenon occurs only near grazing incidence, which is near nose-on for a long, thin body. The TW return is caused by a reflection of the forward surface wave at the rear termination of the scattering body, and depends on the nature of the discontinuity there. Although the TW phenomenon is understood qualitatively, quantitative theoretical prescriptions for the magnitude are generally lacking. The angular location of the first peak of the pattern is given by Eq. (5-87).

For concave, re-entrant, duct-like targets, there are no routine theoretical treatments available. The scattering is due to a multiplicity of interior interactions among internal surfaces, some of which eventually come back out of the duct. The scattering is generally a high return mechanism.

Interactions between isolated scatterers can generally be ignored, but may be important for scatterers that are close to one another. Additional reflections may increase the combined RCS above that of each element taken separately, or one scatterer may shield another, thereby reducing the net RCS.

14.2.1 Phasor Addition

The RCS of a collection of scattering centers is the coherent sum of the contributions of the individual scattering centers. We shall examine the effects of phasor addition on the peak values of the RCS as well as on the average.

For our purpose, multiple scattering is the net return from a collection of scattering centers, each of which acts independently of all others. We assume that there are no mutual interactive reflections. We further assume that the incident field on each scatterer is the field impinging on the collection, thereby ignoring multiple bounces and shadowing effects. The objective of this exercise is to illustrate what happens when the returns from N scatterers are added in and out of phase.

As pointed out in Ch. 7, the net return from a collection of N scatterers is given by

$$\sigma = \left| \sum_{n=1}^{N} \sqrt{\sigma_n} \, e^{i2kR_n} \right|^2 \tag{14-2}$$

where σ_n is the complex RCS of each scatterer and depends on the viewing angle, relative position, polarization, and frequency. Here, $2R_n$ is the round trip distance from the radar to the nth scattering center and the summation is a complex addition of electric field phasors. The square of this sum is the net power scattered back to the radar.

To illustrate phasor addition, consider N point scatterers, each with $\sigma_n = 1m^2$, randomly distributed within a circle of radius 10m. The monostatic RCS as a function of aspect angle is shown in Fig. 14-1 for 1, 2, and 5 point scatterers. The plot shows that the RCS from any one scatterer is independent of aspect, while the return from the collection is very aspect dependent. This is due to changes in their relative positions along the radar line of sight as the aspect angle changes.

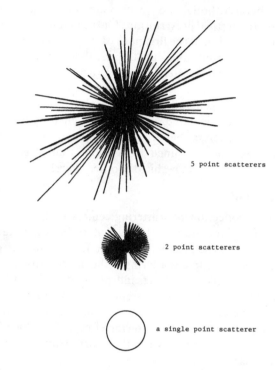

5 point scatterers

2 point scatterers

a single point scatterer

Figure 14-1 The phasor addition of the returns of independent scatterers can produce complicated scattering patterns

Consider now two scattering centers, each with an amplitude of one square meter. The total cross section is then given by Eq. (14-2) as

$$\sigma = |\sqrt{\sigma_1}\, e^{i2\vec{k}\cdot\vec{r}_1} + \sqrt{\sigma_2}\, e^{i2\vec{k}\cdot\vec{r}_2}|^2 = \sigma_1 + \sigma_2 + 2\sqrt{\sigma_1\sigma_2}\cos[2\vec{k}\cdot(\vec{r}_1 - \vec{r}_2)] \quad (14\text{-}3)$$

The following should be observed:

(1)   The maximum RCS is 4m². This is known as the coherent maximum, and is given by

$$\sigma_{max} = (\sqrt{\sigma_1} + \sqrt{\sigma_2})^2 \quad (14\text{-}4)$$

which, for N scattering centers, is

$$\sigma_{max} = |\sum_{n=1}^{N} \sqrt{\sigma_n}|^2$$

(2)   The minimum RCS is zero for the pair because each has the same amplitude. This is known as the coherent minimum. If the two amplitudes are different, the net echo RCS is given by:

$$\sigma_{min} = |\sqrt{\sigma_1} - \sqrt{\sigma_2}|^2 \qquad (14\text{-}5)$$

(3)   The average value over a large sector in aspect angle (say 10 lobe widths or more), is about $2m^2$, which is simply the sum of the individual RCS values of the pair. This is known as the noncoherent sum, and is given by

$$\sigma_{NC} = \sigma_1 + \sigma_2 \qquad (14\text{-}6)$$

which for N scatterers becomes:

$$\sigma_{NC} = \sum_{n=1}^{N} \sigma_n \qquad (14\text{-}7)$$

Note that if the two scattering centers have the same value, the coherent maximum is twice (or 3 dB higher than) the noncoherent sum.

(4)   The summation described above must be performed in square meters (not dB) because we are adding scattered electric fields, not scattered power.

14.3 RCS ESTIMATION FOR A HYPOTHETICAL TARGET

An important, but often neglected, task in RCS measurements is to estimate the expected RCS levels before the target is measured. This establishes some guidelines for the measurements, however coarse the estimates may be, for gauging the worth of the data. A more precise estimate can be obtained, of course, by mathematically modeling the target with one of the various computer codes available. However, as we will show, highly refined and extensive target modeling is not always necessary. For purposes of illustration, we will demonstrate elementary prediction techniques with the hypothetical missile geometry shown in Fig. 14-2, with specified dimensions and frequency.

The scattering in the broadside region will be dominated by specular scattering from the body. For a singly curved cylindrical body, the peak RCS is given by

$$\sigma = kal^2 = 53 \text{ m}^2 = 17.3 \text{ dBsm} \qquad (14\text{-}8)$$

where a is the cylinder radius and l the length. This RCS value increases by 3 dB for each doubling of the frequency.

For a doubly curved surface missile body, such as shown in Fig. 14-2, the specular cross section is given by the geometric optics formula

$$\sigma_{max} = \pi a_1 a_2 = 2.2 \text{ m}^2 = 3.4 \text{ dBsm} \qquad (14\text{-}9)$$

where a_1 and a_2 are the principal radii of curvature at the specular point. The specular point typically first appears on a body like that in Fig. 14-2 in the 60 to 70 degrees region. The RCS increases more or less uniformly as the specular point slides aft along the body, along which the radius of curvature increases.

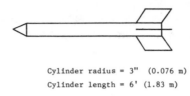

Cylinder radius = 3" (0.076 m)
Cylinder length = 6' (1.83 m)

A. A cylindrical body

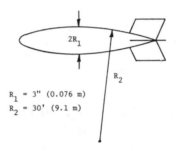

R_1 = 3" (0.076 m)
R_2 = 30' (9.1 m)

B. Revolved doubly curved surface

Figure 14-2 Hypothetical missile geometries. Calculations in text assume a frequency of 10 GHz, for which the wavelengths is 0.03 m

The specular scattering from doubly curved surfaces varies smoothly (without rapid oscillations) because the variation in the principal radii of curvature is a smoothly varying function of the aspect angle. This is different from the specular scattering from a cylinder or flat plate, for which the specular peak may be so narrow that when we move slightly off the specular direction, the RCS contains rapid oscillations due to the sidelobes.

In addition to body scattering, the return from fins presented normal to the radar can also contribute. For a zero degree roll, this requires the fins to be oriented perpendicular to the line of sight. Because this is not the usual missile fin orientation, this scattering mechanism is not applicable.

Scattering mechanisms contributing to the echo in the nose-on sector are more difficult to estimate. This is because the nose-on sector is usually dominated by non-specular scattering. The nose-on sector is usually taken to be that angular region within 60 degrees to either side of nose-on. The non-specular RCS pattern in this region has been described by some as the "fuzz ball region," indicating the difficulty they experience in understanding the RCS mechanisms contributing to the RCS for these aspect angles. The RCS in the nose-on sector is very sensitive to frontal geometry. For example, if a seeker radome is present, the nose structure is similar to a re-entrant duct for which the return is controlled by internal geometric structures behind the radome. If there is no radome, and if the nose has a sharp tip with a very low return, the nose-on sector can be dominated by other mechanisms, such as rear corner curved edge diffraction, straight edge diffraction from fins, and traveling wave contributions. In the absence of a radome, and in the presence of a blunt nose, specular scattering is possible in the nose-on region. An example is a hemispherical nose configuration.

When a radome is present, the return is extremely difficult to predict. This is true even if the geometric structure behind the radome is known, because the internal geometry behind the radome may not be simple. However, an upper bound can be obtained by modeling the radome as a trihedral corner reflector. For this purpose, assume a reflector length of 4in and an incident frequency of 10 GHz. The equation for the maximum return from a thrihedral suggests

$$\sigma = \frac{12\pi a^4}{\lambda^2} = 4.5 \text{ m}^2 = 6.5 \text{ dBsm} \tag{14-10}$$

This large return, of course, is to be expected because the trihedral reflects incident EM energy back to the radar over a rather wide angle, while the actual radome geometries backscatter less incident energy.

The diffraction from rear edges viewed in the nose-on region can be estimated from the maximum return from a straight edge [Eq. (14-1)]. We use half the perimeter as the edge length l, accounting for the fact that only half the perimeter is illuminated by an incident electrical field parallel to the edge. Equation (14-1) then yields

$$\sigma \simeq \frac{l^2}{\pi} = \frac{(0.5 \times 2\pi a)^2}{\pi} = \pi a^2 = 0.018 \text{ m}^2 = -17.4 \text{ dBsm} \tag{14-11}$$

where a is the radius of the rear edge discontinuity.

Two important features of this result should be noted: it is independent of frequency and depends only on the radius of the discontinuity at the rear. While this formula is similar to that for the RCS of a sphere, the underlying scattering mechanisms are fundamentally different. In one case, we have diffraction from an edge at normal incidence, and in the other we have a surface specular geometric optics return. At angles well away from nose-on incidence, for which none of the rear edge contributions are any longer in phase, the rear edge return will fall off as suggested by $[(\sin(x)/x)]^2$. Off axis, only one flash point on the rear edge will be visible to the radar.

Diffraction due to the straight leading and trailing edges of control surfaces has a maximum value given by Eq. (14-1). As stated earlier, this result does not depend on frequency, and it occurs only when the edge is viewed in the specular direction. As suggested earlier, an edge 1ft long has a maximum specular return of –15.7 dBsm. When straight edges are viewed in non-specular directions, the return can be expected to vary as $[(\sin x)/x)]^2$. Edge scattering is polarization dependent: front edges have maximum return when the incident electric field is parallel to the edge; and rear edges have maximum return when the electric field is perpendicular to the edge. Because the edges of most fins are presented obliquely to the radar in the nose-on sector, fin edge scattering in the nose-on sector is not usually a dominant mechanism.

The traveling wave return is polarization dependent. It occurs only when there is a component of the incident electric field along the body surface and in the plane of incidence. The TW return is due to a reflection of the induced forward surface wave at the rear termination of the body. This reflected current wave radiates back toward the radar like an end-fed long-wire antenna. The peak of the traveling wave return, if present at all, occurs in the nose-on region at an angle given by Eq. (5-87). Thus, as the body becomes electrically longer, the TW peak moves closer to nose-on incidence. Unfortunately, the magnitude of the peak is not obtainable from any simple formula.

Specular scattering may be an important mechanism if a radome is not present and the nose geometry is not a sharp tip. For example, a spherical metallic nose cap with a three-inch radius would have an RCS of

$$\sigma = \pi a^2 = 0.018 \text{ m}^2 = -17.4 \text{ dBsm} \tag{14-12}$$

The echo areas of other frontal geometries can be estimated using formulas related to those in the Table 6-1.

The returns in the broadside aspect sectors are due to specular scattering mechanisms and, hence, can be estimated fairly accurately. The forward sector returns, however, are generally due to nonspecular mechanisms; hence, it is much more difficult to identify the scattering mechanisms or to estimate the RCS. Forward sector mechanisms include returns due to the seeker/radome, rear edges, straight edges, surface traveling waves, and specular contributions from the nose.

14.4 CYLINDER-OGIVE SCATTERING PATTERNS

An example of the measured RCS of a very simple missile-like geometry is shown in Fig. 14-3 from reference [2]. This geometry is a cylinder 36in long attached to an ogival nose 15.75in long, for a total length of 51.75in. The diameter is 5.75in. The body was measured at 2.93, 5.4, and 9.5 GHz, and the

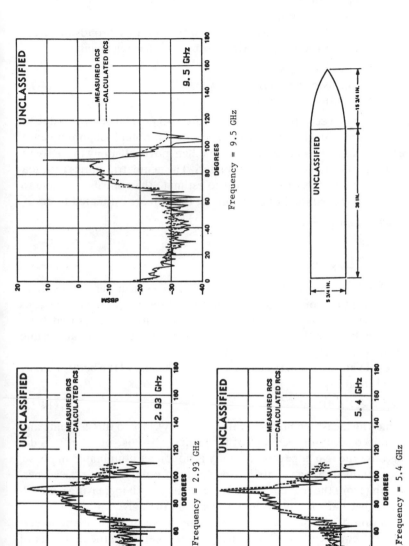

Figure 14-3 Comparison of measured and predicted results for a cylinder with an ogive nose, horizontal incident polarization [2]

incident and received polarizations were horizontal (i.e., in the plane perpendicular to the axis of rotation).

The broadside peak at 90 degrees is due to the specular broadside return from the cylinder and is well approximated by the theoretical values of

$$\sigma = kal^2 = \begin{cases} 5.7 \text{ dBsm at } 2.93 \text{ GHz} \\ 8.4 \text{ dBsm at } 5.4 \text{ GHz} \\ 10.9 \text{ dBsm at } 9.5 \text{ GHz} \end{cases} \tag{14-13}$$

The broadside region echo from about 70 degrees to near 90 degrees is due to the specular return from the ogival nose. The specular point first pops up onto the ogive tip at an aspect near 70 degrees, and then moves aft along the ogive to the point where the ogive joins the cylinder. The geometric optics formula for doubly curved surfaces can be used to estimate the contribution due to the ogive. The principal radii of curvature at the join are estimated to be

$$R_1 = 2.87 \text{ in and } R_2 = 36 \text{ in}$$

so that

$$\sigma = \pi R_1 R_2 = -6.8 \text{ dBsm} \tag{14-14}$$

Note that R_2 is large and unchanging, while R_1 ranges from zero at the tip to the body radius at the join as the specular point moves from the tip towards the join. The estimate based on these radii agree with the measured data, provided we realize that the ogive and the cylinder sidelobe returns add in and out of phase with changing aspect angle.

The nose-on value is due to the circular edge at the rear of the body and may be estimated from Eq. (14-11):

$$\sigma = \pi R_1^2 = -17.8 \text{ dBsm} \tag{14-15}$$

We note that the measured nose-on values are very close to this estimate and that they indeed seem independent of frequency. Off nose-on, the RCS decreases because the rear edge discontinuity is no longer illuminated in phase, and there remains only one contributing specular point on the rear corner. The scattering from the rear circular edge decreases with increasing azimuth angle until a specular flash point first appears on the ogive tip.

An indication of the presence of a traveling wave is given at the lowest measurement frequency, 2.93 GHz. The angular location of the first peak of that pattern is predicted by Eq. (5-87) to be

$$\theta = 49 \sqrt{\lambda/l} = 14° \tag{14-16}$$

At the higher frequencies, there is no indication of a TW return, supporting the notion that the TW return decreases with frequency.

Figure 14-3 shows the predictions of the MISCAT II code which, except for the possible TW lobe, gives a reasonable prediction. Figure 14-4 shows an example of a Georgia Tech prediction for the same geometry using the CROSS RCS code. Again the results compare reasonably well with the measured values.

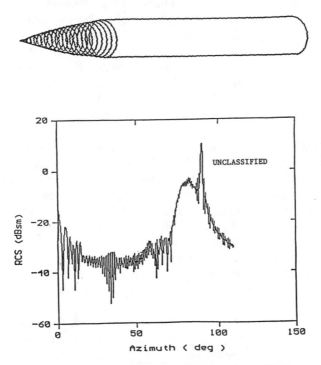

Figure 14-4 Predicted RCS of the cylinder-ogive for 9.5 GHz as obtained from Georgia Tech's CROSS model

14.5 DATA REDUCTION AND PRESENTATION

An important element of RCS analysis is what to do with raw measured data, and how to reduce and present the data. Although there are no standards for doing this, a possible approach is outlined in this section.

For typical test programs, raw data are obtained for several targets or configurations for multiple pitch angles, roll angles, frequencies, and polarizations. Typical data sets can number over 700 individual runs, with perhaps 50 or more runs over the test matrix for just one target or configuration. Each raw data set is in the form of (angle, RCS) data pairs, typically obtained every 0.1 degree for either a partial or complete azimuthal cut. Data reduction includes not only each raw data set, but also summaries for each target in the test matrix.

Data reduction can take many forms, and while we cannot possibly cover them all, it is worthwhile mentioning how data are smoothed. Most targets of interest have dozens or even hundreds of scattering elements, and we have seen that the RCS pattern with even a handful of scatterers scintillates rapidly with changing aspect angle. At the higher frequencies, the individual lobes in the pattern may be spaced less than 0.1 degree apart, and a measured pattern can often consist of a band of ink across the chart paper. While the RCS specialist may be interested in the details of such lobe structures, managers and radar system designers need characterizations of the data in a much broader sense. Averages, medians, and standard deviations of individual test runs are more meaningful, as well as the statistics formed over the parameters in the test matrix.

In forming an average or median, we must decide how many contiguous data points will be used. Because the RCS pattern is usually sampled at a fixed angular rate, often at intervals of 0.1 degree, the decision amounts to selecting an angular "window" over which the averaging will be performed. Typically, the window ranges in width from one to ten degrees, depending on the characteristics of the RCS fluctuations. A scientific criterion may be used in the selection, such as ensuring that at least three or four lobes be included in the window, or we may select the width on the basis of previous experience with similar data sets. In one case, the selection is based on a scientific objective, while in the other it is simply because "that's the way we've always done it."

Once the angular window width has been selected, we must choose the "slide," or the amount by which the window will be indexed across the RCS pattern. The slide should never be greater than the width of the window, for this will create gaps in the pattern over which no averaging is performed. The slide can be as small as one datum, for which one new datum is added to the collection to be averaged while the oldest datum is dropped from the list, or it can be as large as the window itself, for which a complete set of new data points are fetched and averaged. Obviously, smaller slides generate finer patterns, but require more processing time. Slides of one and two degrees are common, and slides are often set to the window width in order to generate preliminary "quick look" RCS plots. These plots are usually only for the purpose of preliminary assessment of a test run just completed during measurements, with the expectation that finer processing will be performed later.

The raw (but calibrated) data are invariably stored on the recording medium in decibel values, and depending on the kind of processing to be performed, we may need to take the antilog of each datum prior to further processing. This is particularly true of the square meter mean (average), because the logarithm of the square meter average of a collection of numbers is not the same as the average of their logarithms. On the other hand, the logarithm of the median value of a collection of numbers is the same as the median of their logarithms; hence, the antilog need not be taken prior to subsequent processing. In the discussion below, "linear space" refers to the antilogs of the data values, while "log space" refers to the decibel (logarithmic) form of the recorded data.

Each reduced data run may take several forms: smoothed data over a specified window and slide for three percentile levels; sector data over three specified angular regions for median, mean, and standard deviation in dB (log) space; mean and standard deviation in square meter (linear) space; and probability density function (PDF) and cumulative distribution function (CDF) for each of the sectors. Figure 14-5 is an example of raw data (in this case, a metallic cylinder), and Fig. 14-6 is an example of the reduced data. The test matrix summary may have the following forms:

- A matrix "road map" showing run number, pitch, roll, frequency, polarization, and operator comments and the front sector median, mean, and standard deviation
- A data file for use in plotting a target response as a function of the test matrix variables such as RCS *versus* pitch angle, frequency, polarization, and roll angle

An example of a matrix road map is shown in Fig. 14-7. Test matrix summary plots, as might be obtained from the summary data, are shown in Fig. 14-8.

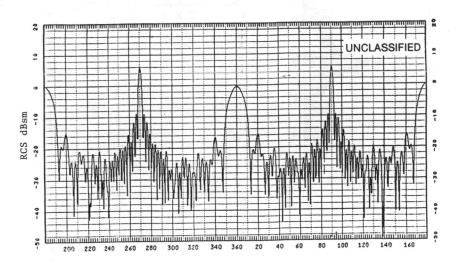

Figure 14-5. An example of raw RCS data. The target was metallic cylinder of 5.5 in diameter and 28 in length measured at 5.4 GHz

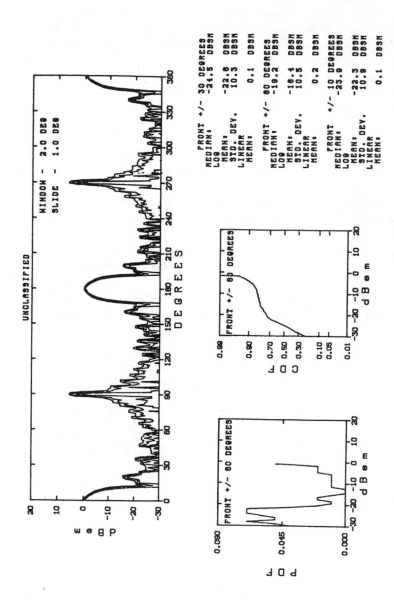

Figure 14-6 The reduced data for the cylinder of Fig. 4-5

SYSTEM: EXAMPLE

PITCH	ROLL	FREQ	POL.	RUN #	MEDIAN	MEAN	STD
(degrees)		(MHz)			RCS(dBsm)	RCS(dBsm)	(dBsm)
−20	0	3200	V	1000	0.00	0.00	0.00
−20	0	3200	H	1000	0.00	0.00	0.00
−20	0	6500	V	1000	0.00	0.00	0.00
−20	0	6500	H	1000	0.00	0.00	0.00
−20	0	9700	V	1000	0.00	0.00	0.00
−20	0	9700	H	1000	0.00	0.00	0.00
−20	0	16000	V	1000	0.00	0.00	0.00
−20	0	16000	H	1000	0.00	0.00	0.00
−10	0	3200	V	1000	0.00	0.00	0.00
−10	0	3200	H	1000	0.00	0.00	0.00
−10	0	6500	V	1000	0.00	0.00	0.00
−10	0	6500	H	1000	0.00	0.00	0.00
−10	0	9700	V	1000	0.00	0.00	0.00
−10	0	9700	H	1000	0.00	0.00	0.00
−10	0	16000	V	1000	0.00	0.00	0.00
−10	0	16000	H	1000	0.00	0.00	0.00
0	0	3200	V	1000	0.00	0.00	0.00
0	0	3200	H	1000	0.00	0.00	0.00
0	0	6500	V	1000	0.00	0.00	0.00
0	0	6500	H	1000	0.00	0.00	0.00
0	0	9700	V	1000	0.00	0.00	0.00
0	0	9700	H	1000	0.00	0.00	0.00
0	0	16000	V	1000	0.00	0.00	0.00
0	0	16000	H	1000	0.00	0.00	0.00
10	0	3200	V	1000	0.00	0.00	0.00
10	0	3200	H	1000	0.00	0.00	0.00
10	0	6500	V	1000	0.00	0.00	0.00
10	0	6500	H	1000	0.00	0.00	0.00
10	0	9700	V	1000	0.00	0.00	0.00
10	0	9700	H	1000	0.00	0.00	0.00
10	0	16000	V	1000	0.00	0.00	0.00
10	0	16000	H	1000	0.00	0.00	0.00
20	0	3200	V	1000	0.00	0.00	0.00
20	0	3200	H	1000	0.00	0.00	0.00
20	0	6500	V	1000	0.00	0.00	0.00
20	0	6500	H	1000	0.00	0.00	0.00
20	0	9700	V	1000	0.00	0.00	0.00
20	0	9700	H	1000	0.00	0.00	0.00
20	0	16000	V	1000	0.00	0.00	0.00
20	0	16000	H	1000	0.00	0.00	0.00

PROJ #0000   15-MAR-84

Figure 14-7 Test matrix "roadmap" example

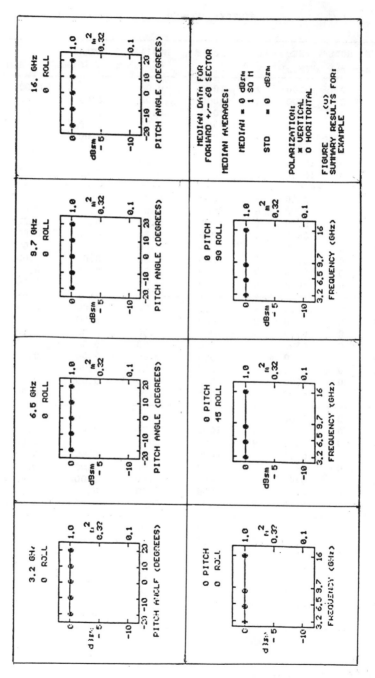

Figure 14-8 Example of format of summary results

14.5.1 RCS Data Requirements

The raw data typically are in the form of the cross section in dBsm every 0.1 degree, or 3600 data points per 360 degree run. The objective of data reduction is to present this data in a format more meaningful to various users of the data. Thus, the display format and reduction technique must be selected with the end-user in mind. A data base objective, for example, could have a variety of purposes:

- Threat effectiveness models
- Determining impact of component signatures on total vehicle signatures
- Radar cross section reduction (RCSR)
- Developing design requirements
- Specification requirements

The following discussion presents the rationale for the various reduction formats along with examples.

One purpose for measuring RCS is to determine the target detectability by a given radar. Detection, however, is a statistical process which involves the signal level at the radar receiver. Most detection analyses use a statistical model for target fluctuations and a statistical description for unwanted contaminating signals such as clutter, multipath, and noise. The typical track radar antenna has high gain (a narrow pencil beam, typically 1 to 2 degrees wide) in one or both angular directions. The radar receiver responds to a target return averaged over this beamwidth window, and is one reason why raw data are typically smoothed over narrow windows.

The RCS of airborne targets is typically of interest over a limited sector or cone in space for which a threat is present. For aircraft, this threat sector is a forward cone in both the yaw and pitch planes of the target. Targets can be characterized over given threat sectors by median data or RCS distribution functions, which include medians, averages, and standard deviations. The average median is defined as the average over the test matrix (frequency, polarization, pitch, and roll) of each sector median.

RCSR analysts are usually interested in raw or nearly raw RCS data. This is because they are concerned with the scattering mechanisms responsible for the return in some localized sector for which they may be able to suggest methods for reduction. Data of interest to project managers and operations analysts, on the other hand, are for encounter scenarios or engagement computer codes. Generally, they seek only one or two numbers, rather than dozens or hundreds of numbers, characterizing a given target over a sector in space. These are the mean and standard deviation or median.

From the above discussion, it is evident that the types of RCS data required by various members of the user community are:

- Raw data
- Smoothed data
- Statistical data
- Single number, median, mean, and standard deviation characterizations

A final consideration in any data reduction exercise is the large volume of data to be processed. For example, for 700 runs if one page were devoted to presenting each reduced RCS pattern, it would require 700 pages to display the raw data patterns. For two pages per run, 1400 pages would result, and so on. It is clear that data reduction should be limited to one page per run, and it should place as much information as possible onto this single page format.

14.5.2 Reduced Data Format Example

In the one-page reduced data example shown in Fig. 14-6, smoothed data are presented in the form of the 10, 50, and 90 percentile points within a two-degree window for every 1 degree slide over the 360 degree rotation. Each window has 21 data points, and the decibel difference between the 10% and 90% curves is an indication of the "band of ink" that would be observed on the raw data plots. If the 10% and 90% points are close to each other, then there is little fluctuation in the raw data, while widely separated 10% and 90% points indicate large fluctuations. If those data were distributed log normally, then the 90% point would be 1.29 standard deviations above the median and the 10% point would be 1.29 standard deviations below the median. The 90% curve tends to follow the peaks of the raw data while the 10% curve tends to follow the nulls or valleys. As an example, the raw cylinder data in Fig. 14-5, and reduced in Fig. 14-6, show that the specular end-on return near zero degrees has a small deviation because of the 10% and 90% points are very close to each other. The separation of the 10% and 90% curves in the regions near 60 degrees and 120 degrees indicates large fluctuations in the raw data. It is instructive to compare these raw data with the smoothed 10%, 50%, and 90% data.

Statistical distributions are presented for the forward ±60 degrees sector in the form of two plots: the probability density function (PDF) and the cumulative distribution function (CDF) plotted on a normal probability scale. The PDF is a measure of the probability of occurrence that the cross section is a given level, σ. It is a mathematical statement that $P(\sigma)\,d\sigma$ is the probability that σ lies between σ and $\sigma + d\sigma$. As such, $P(\sigma)$ is a number between zero and unity. An examination of the PDF curve can show how well a given data set conforms to one of the standard distribution functions (log normal, Rayleigh, Weibull, *et cetera*), used in target modeling analysis. The PDF curve is obtained from the raw data points in the sector by sorting these values into bins ordered from the lowest to the highest RCS level measured.

The cumulative distribution function (CDF) is a measure of the fraction of the total data values which are below a given RCS value. It is obtained by integrating the PDF curve. The fraction of data values that lie between σ_1 and σ_2 is given by

$$\int_{\sigma_1}^{\sigma_2} P(\sigma)\, d\sigma \qquad (14\text{-}17)$$

The fraction of values which lie below a given value, yielded by Eq. (14-17) with $\sigma_1 = -\infty$,

$$\mathrm{CDF}\,(\sigma) = \int_{-\infty}^{\sigma} P(\sigma)\, d\sigma \qquad (14\text{-}18)$$

is the area under the PDF curve from $-\infty$ to σ. The total fraction of data below $\sigma = \infty$ is, by definition, unity:

$$\mathrm{CDF}\,(\infty) = \int_{-\infty}^{\infty} P(\sigma)\, d\sigma = 1 \qquad (14\text{-}19)$$

The median (50 percentile level) σ_m is that value of σ for which CDF = 0.5:

$$\mathrm{CDF}\,(\sigma_m) = 0.5 = \int_{-\infty}^{\sigma_m} P(\sigma)\, d\sigma \qquad (14\text{-}20)$$

The CDF is plotted on a normal probability scale *versus* σ in dBsm. This scale is chosen so that a log normal (Gaussian in dB space) probability distribution would be a straight line whose slope is proportional to the standard deviation. A steep slope indicates a small standard deviation while a shallow slope indicates a large standard deviation. The CDF curve in Fig. 14-6 for the ±60 degree forward sector of the cylinder shows that 90% of the data were less than –3.5 dBsm and 30% of the data less than –28 dBsm. In addition, the slope of the CDF curve for the cylinder is not a straight line, hence the distribution is not log normal. This can also be seen on the PDF curve, where a bimodal distribution is observed. The reader is encouraged to compare the forward ±60 degree sector raw data, smooth data, PDF, and CDF curves.

Single-number statistical characterizations of the data are given for three angular sectors. These data are presented in three blocks on the right-hand side of the page (Fig. 14-6), showing the median, the logarithmic average and standard deviation, and the square meter average expressed in dBsm. The median has the same value whether computed in log or in linear space.

Additional characteristics of the raw and reduced cylinder data are the following. The end-on aspects lie at 0 degrees and 180 degrees, and broadside incidence is at 90 degrees and 270 degrees. The broadside RCS is given by $\sigma = ka l^2 = 6$ dBsm, where a is the cylinder radius, and l is the length. The end-on RCS is given by the flat plate formula, $\sigma = 4\pi A^2/\lambda^2 = 0$ dBsm, where A is the area of the flat end. The smoothed data in the end-on region does not fluctuate; hence, the smoothed data for the 10%, 50%, and 90% points all lie almost one atop the other. In the broadside regions, where fluctuations are present in the raw data, the 10% curve is somewhat below the 50% curve, which in turn is barely below the 90% curve.

The probability distribution function for the ±60 degree forward sector shows a bimodal distribution of data, with one peak near –23 dBsm and a smaller peak near –2 dBsm. This second peak is due to the end-on return, which is about 0 dBsm. The cumulative distribution function is not a straight line because the PDF is not a log normal distribution. Note that the CDF 50% point corresponds to the median for the ±60 degrees forward sector, that 30% of the sector data are below –28 dBsm, and that 90% of the data are below –4 dBsm. The sector statistics for the forward and rear sectors produce nearly the same values. The forward ±30 degrees sector has higher medians and means than does the forward ±60 degrees, due to the large end-on return.

14.5.3 Artificial Data

Artificially generated log normal data are shown in Fig. 14-9 and 14-10 for a median of –10 dBsm and standard deviations of 5 dB and 10 dB, respectively. These are useful in understanding the statistics of a known distribution, and they serve as a check for data reduction software. Raw data for 3600 data points were generated with a Gaussian distribution in dB space, which has the probability distribution:

$$P(\sigma) = \frac{1}{\sqrt{2\pi}\,\sigma_s}\, \exp\left[-\frac{1}{2}\left(\frac{\sigma - \sigma_m}{\sigma_s}\right)^2\right] \tag{14-21}$$

where σ_m is the median or center of the distribution and σ_s is the standard deviation. Numerical values were generated using a uniform random number generator and the following formula to convert the uniform random deviate into a Gaussian random deviate

$$\sigma = \sigma_m + \sqrt{2}\,\sigma_s\,\sqrt{-\log_e u_2}\,\cos\,(2\pi u_1) \tag{14-22}$$

where σ_m is the distribution median, σ_s is the Gaussian standard deviation, and u_1 and u_2 are two distinct uniform random numbers between 0 and 1.

The probability distribution curves do have a log normal or Gaussian bell shape centered at the median of –10 dB. The plot with a standard deviation of 5 dB is narrower than that of the 10 dB distributions. The peaks of the distributions correspond to

$$\frac{1}{\sqrt{2\pi}\,\sigma_s} \tag{14-23}$$

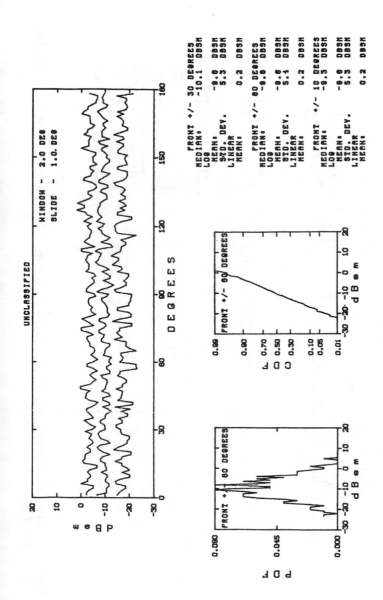

Figure 14-9 Example of reduced data for an artificially generated log normal data set with a median of –10 dBsm and a standard deviation of 5 dB

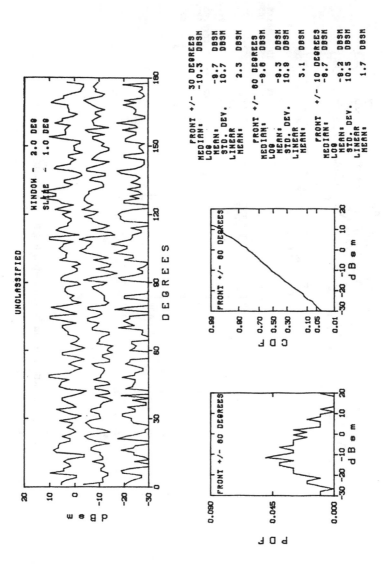

Figure 14-10 Example of reduced data for an artificially generated log normal data set with a median of -10 dBsm and a standard deviation of 10 dB

which is 0.08 and 0.04 for σ_s = 5 and 10 dB, respectively. For a Gaussian distribution, the cumulative distribution is

$$CDF = \int_{-\infty}^{\sigma} P(\sigma) \, d\sigma = \frac{1}{\sqrt{2\pi} \, \sigma_s} \int_{-\infty}^{\sigma} \exp \left\{ - \frac{1}{2} \left[(\sigma - \sigma_m)/\sigma_s \right]^2 \right\} d\sigma$$

$$(14\text{-}24)$$

which is closely related to the error function. Tables of this function are available, for example, in the Chemical Rubber Company (CRC) Standard Math Tables [3]. The 0.5 (50 percentile) point corresponds to the median of the distribution ($\sigma = \sigma_s$), while the 0.1 (10 percentile) and 0.9 (90 percentile) points are 1.29 standard deviations below and above the mean, respectively. The 0.84 (84 percentile) and 0.16 (16 percentile) points are respectively one standard deviation above and below the mean, so that 68% of the data values fall within ±1 standard deviation of the mean

The CDF results in Figs. 14-9 and 14-10 show that the CDF is a straight line, as indeed is required when plotting CDF for a Gaussian distribution on a normal probability scale. In addition, the slope for the σ_s = 10 dB case is less than that for the σ_s = 5 dB case. The 50% points on each CDF curve correspond to −10 dBsm, the median and mean of the distribution. The 10% and 90% points are indeed 1.29 standard deviations above and below the median.

The sector statistical data summary produces nearly the same values for each parameter because these artificially generated Gaussian data have no angular variation. The medians are all approximately −10 dBsm, and the standard deviations are approximately 5 dB or 10 dB.

The smoothed data over the 2° window show a random variation. In each case, the 50 percentile (center) curve has an average value of −10 dB, which was the median of the distribution. In Fig. 14-9, for which σ_s = 5 dB, the 10th and 90th percentile curves are centered at 1.29 · 5 = 6.45 dB above and below the −10 dB median, that is, at −16.5 dB and −3.6 dB. In Fig. 14-10, for which the standard deviation was 10 dB, the 10 and 90 percentile curves are centered at −22.9 dB and + 2.9 dB, respectively, which is 1.29 × 10 = 12.9 dB below and above the median.

14.5.4 Data Processing Approach

Reduced statistical data fall into three distinct classes:
- The sector means and standard deviations in log and linear space
- The probability density function, the cumulative distribution function, and the median computation
- The azimuthal smoothed data for the 10, 50, and 90 percentile of data within a specified window and slide

The following paragraphs more fully describe each area.

Sector Statistics

The sector means and standard deviations are computed in log (dB) and in linear space for each of the angular sectors. Each angle-RCS data pair is examined to determine if it occurs within one of the sectors. If so, then it is included in the appropriate sector calculation. The arithmetic average is computed using the usual definition:

$$\bar{\sigma} = \frac{1}{N} \sum_{i=1}^{N} \sigma_i \tag{14-25}$$

where N is the total number of occurrences of σ_i within a defined sector. The linear space average is determined from the decibel σ values by first taking the antilog

$$\sigma \, (\text{m}^2) = 10^{(0.1 \times \sigma_{dB})} \tag{14-26}$$

and then repeating Eq. (14-25) using σ with units of m².

The difference between the average in linear space and the average in log space is equivalent to the difference between an arithmetic average

$$\bar{x}_A = \frac{1}{N} \sum_{i=1}^{N} x_i \tag{14-27}$$

and a geometric average

$$\bar{x}_g = \left[\prod_{i=1}^{N} x_i \right]^{1/N} \tag{14-28}$$

which can be represented in log (dB) space as

$$10 \log_{10} \bar{x}_g = \frac{10}{N} \sum_{i=1}^{N} \log_{10} x_i \tag{14-29}$$

The arithmetic average tends to weight large values of x_i more than the geometric average does. The arithmetic average of σ in dB space is equivalent to the geometric average of σ computed in square meter space.

The standard deviation for each sector is defined as [3]:

$$\text{STD} = \left[\frac{\sum_{i=1}^{N} (\sigma_i - \bar{\sigma})^2}{N-1} \right]^{1/2} \tag{14-30}$$

where $\bar{\sigma}$ is the mean over the sector. This is computed assuming that N is a large number, so that

$$\frac{1}{N-1} \simeq \frac{1}{N}$$

Equation (14-30) is then expanded as

$$STD = \left[\frac{1}{N} \sum (\sigma_i^2 - 2\sigma_i \bar{\sigma} + \bar{\sigma}^2) \right]^{1/2}$$

$$= \left[\frac{1}{N} \sum \sigma_i^2 - 2\bar{\sigma} \frac{1}{N} \sum \sigma_i + \frac{1}{N} \sum \bar{\sigma}^2 \right]^{1/2} \tag{14-31}$$

$$= [\bar{\sigma^2} - 2\bar{\sigma}\bar{\sigma} + \bar{\sigma}^2]^{1/2}$$

$$= [\bar{\sigma^2} - (\bar{\sigma})^2]^{1/2}$$

so that the standard deviation is the square root of the average of the square minus the square of the average. The software computes $\bar{\sigma^2}$ along with $\bar{\sigma}$, from which the STD is then computed. The variance, which is the square of the standard deviation, is not explicitly computed.

PDF, CDF, Median

The probability density function (PDF), cumulative distribution function (CDF), and sector median were computed as follows. The PDF and CDF were computed and plotted for the forward ±60 degrees sector. The median was computed for all three sectors. The variable definitions needed to compute these quantities from the raw data are given in Table 14-1.

Table 14-1
Data Reduction Variables for PDF, CDF, and Median

Variable	Definition	Units
$\sigma_{max/min}$	largest/smallest value of cross section occurring within sector	dBsm
N	total number of bins (= 51)	—
Δ	bin width = $(\sigma_{max} - \sigma_{min})/N$	dBsm
I_n	number of RCS data occurrences within the nth bin	—
$J_n = \sum\limits_{i=i}^{n} I_i$	total data count up to and including nth bin	—
$J_{tot} = J_N = \sum\limits_{i=1}^{N} I_i$	total data count in all N bins (= 1201 for front +/− 60° sector)	—

The probability density function $P(\sigma)$ is defined as the probability that the cross section σ lies between σ_0 and $\sigma_0 + d\sigma$, that is,

$$\text{Probability of } \sigma_0 \leq \sigma \leq \sigma_0 + d\sigma = \int_{\sigma_0}^{\sigma_0 + d\sigma} P(\sigma)d\sigma = P(\sigma)d\sigma = F(\sigma) \tag{14-32}$$

where $P(\sigma)$ within the integral is the probability density function. $F(\sigma)$ is a positive quantity whose value is less than or equal to unity. The probability that σ lies within the range $-\infty \leq \sigma \leq \infty$ is unity, which requires that

$$\int_{-\infty}^{\infty} P(\sigma')\, d\sigma' = 1 \qquad (14\text{-}33)$$

The probability density function $P(\sigma)$ has the units of inverse cross section.

The PDF is computed as a discrete histogram for N bins. The fundamental definition for $P(\sigma) \to P(n\Delta)$ is that the probability of $\sigma = n\Delta$ is, where n is the bin number with $1 \leq n \leq N$ and Δ is the bin width,

$$\int_{(n-0.5)\,\Delta}^{(n+0.5)\,\Delta} P(\sigma')\,d\sigma'$$

$$\qquad (14\text{-}34)$$

$$= P(n\Delta)\,\Delta = \frac{\text{number of occurrences in } n\text{th bin}}{\text{total number of occurrences}} = \frac{I_n}{J_{tot}}$$

From this definition, we see that the density function is thus defined as

$$P(n\Delta) = \frac{I_n}{J_{tot}\,\Delta} \qquad (14\text{-}35)$$

Each raw RCS datum within each sector is examined to determine which bin it belongs in, and the appropriate bin counter I_n is then incremented. After all the data are sorted, the density function histogram is computed for $n = 1$ to N using Eq. (14-35).

The cumulative distribution function (CDF) is defined as the probability that the cross section is less than or equal to some value σ:

$$\text{CDF}(\sigma) = \int_{-\infty}^{\sigma} P(\sigma')\,d\sigma' \qquad (14\text{-}36)$$

which is the area under the PDF curve from $-\infty$ to σ. The total area under the PDF curve is unity because CDF $(\sigma = \infty) = 1$. The discrete CDF is obtained from the PDF histogram as the area under that histogram:

$$\text{CDF}(n\Delta) = \sum_{i=1}^{n} P(i\Delta)\Delta \qquad (14\text{-}37)$$

$$= \sum_{i=1}^{n} \frac{I_i}{J_{tot}\,\Delta}\,\Delta$$

$$= \frac{J_n}{J_{tot}} \qquad (14\text{-}38)$$

where the defining Eq. (14-35) for PDF was used along with the defining equations for I_i, J_n, and J_{tot} from Table 14-1.

The median can be computed using one of two methods.
Median, Method 1: The median is defined as the value σ_{50} for which $CDF(\sigma_{50})$ = 0.5, that is, half of the data values occur above σ_{50} and half below. This approach to compute the median is to find the bin in which the σ_{50} value falls and to then linearly interpolate within that bin to compute the median. This approach assumes that the RCS data values are linearly or uniformly distributed within the bin which contains σ_{50}. That is, the CDF *versus* σ curve is assumed to be linear over the bin width Δ. This approach is based on the rationale that the finite angle-RCS data set represents a continuum of RCS values which describe the target. Linear interpolation is then used on the bin containing σ_{50}. This removes any bias due to quantitative error in the measurement process as might be introduced by the analog to digital converter with its finite quantum levels. This approach is to be contrasted to the *Method 2* sort approach on the finite measured data set where a sorting of data values from low to high is performed and then a middle value is chosen as the median.

The median is computed during the CDF computation. During the summation, the bin number n for which $CDF(n\Delta) \geq 0.5$ has just occurred represents the bin location of σ_{50}. At this point, σ_{50} is known only within a bin width resolution Δ. That is, $CDF[(n-1)\Delta)] < 0.5$ and $CDF(n\Delta) \geq 0.5$. Linear interpolation on the nth bin is used to compute σ_{50}:

$$\sigma_{50} = [(\sigma_{min} + (n-1)\Delta)] + (\text{slope}) \{(0.5 - CDF[(n-1)\Delta]\} \tag{14-39}$$

where σ_{min} is the lowest RCS value in the sector, $(\sigma_{min} + (n-1)\Delta)$ is the left edge of the bin containing σ_{50}, and the slope is the rate of change of RCS with CDF for the nth bin,

$$\text{slope} = \frac{\Delta}{CDF(n\Delta) - CDF((n-1)\Delta)} \tag{14-40}$$

Expression (14-40) can be further recast, using Eq. (14-39) and Table 14-1, into quantities involving the raw counts in each histogram bin:

$$CDF(n\Delta) = \frac{J_n}{J_{tot}} \tag{14-41}$$

$$\text{slope} = \frac{\Delta J_{tot}}{J_n - J_{n-1}} \tag{14-42}$$

Inserting these expressions into Eq. (14-39), we obtain

$$\sigma_{50} = (\sigma_{min} + (n-1)\Delta) + \frac{\Delta J_{tot}}{J_n - J_{n-1}} \left[0.5 - \frac{J_{n-1}}{J_{tot}} \right] \tag{14-43}$$

Now, $(J_n - J_{n-1})$ can be rewritten as

$$J_n - J_{n-1} = \sum_{i=1}^{n} I_i - \sum_{i=1}^{n-1} I_i = I_n \qquad (14\text{-}44)$$

so the final computational formula for σ_{50} becomes

$$\sigma_{50} = (\sigma_{min} + (n-1)\Delta) + \frac{\Delta}{I_n} (0.5\, J_{tot} - J_{n-1}) \qquad (14\text{-}45)$$

where n = bin number which contains σ_{50}.

Median, Method 2: This is a sort approach to find the median. The bin in which the σ_{50} value falls is determined during the CDF computation. The RCS values within this bin are then sorted to find the median. During the summation process used to compute the CDF, the bin number n for which CDF($n\Delta$) ≥ 0.5 just occurred represents the bin location of σ_{50}. At this point, σ_{50} is known only to within a bin width resolution Δ. That is, CDF$[(n-1)\Delta)] < 0.5$ and CDF($n\Delta$) ≥ 0.5. A sort algorithm is then used on the RCS values contained in the nth bin to find the median value. This differs from the procedure used in *Method 1* where a linear interpolation scheme was used. For all practical purposes, the answers obtained by either method are the same.

The choice of using interpolation or sorting to find the median depends on the user's bias. Interpolation is computationally faster and, in a statistical sense, yields a better estimate of the distribution median. The sorting approach has the advantage that for a given set of data values, the exact datum for the median can be identified (one or two values if the set has an odd or even number of values, respectively).

Smooth Data

The 360° smooth data for the 10th, 50th, and 90th percentile levels for a 2° window with a 1° slide were computed from the raw angle-RCS data pairs as follows.

The smoothed data were computed for a 2° window and a 1° slide for the 21 RCS data values, and included the two end-points. The window was centered on the slide. For example, the smoothed data for 10° azimuth included the 21 RCS data values from 9° to 11°, while the smooth data for 11° included the 21 RCS data values from 10° to 12°.

Let *NWIND* be the number RCS data values within each window. These values are then sorted from low to high. Each percentile level is chosen from this array using an index chosen as the integer part of the product (percent · *NWIND*). For example, if there were 21 (= *NWIND*) values in the sorted window array then: the second value in the array is the tenth percentile = integer (.10 $\times$ 21)); the eleventh value is the 50th percentile (= integer (0.5 $\times$ 21)); and the nineteenth value is the 90th percentile (= integer (0.9 $\times$ 21)).

14.6 SUMMARY

This chapter has discussed the considerations required when estimating the scattering from complex targets in the high frequency regime. Scattering mechanisms which apply are:

- Specular
- Discontinuities, such as edges
- Surface discontinuities
- Creeping waves
- Traveling waves
- Re-entrant or concave structures, such as ducts
- Interactions, such as multiple bounces

Phasor addition is the physical process in which the returns from individual scattering centers is summed, resulting in the typical "band of ink" seen in RCS patterns. The peaks in the RCS pattern are the result of coherent addition of the scattering centers. The noncoherent sum is the result of averaging over an angular interval sufficient to average out the peaks and nulls of the "band of ink."

RCS estimates of complex targets should precede any measurement program and often times can be accomplished by application of very simple formulas for various angular sectors as shown in the presented example.

Data reduction requirements vary with user needs. Types of reduced data include:

- Smoothed data over specified angular windows with specified slides for several percentile levels
- Sector data
- Statistical data
- Single number characterizations for median, mean and standard deviation in both log and square meter space

Measurement programs in which much data are collected require careful consideration for the format of the reduced data. The presentation format for RCS variation with test matrix variables of frequency, azimuth, pitch and roll angles, and polarization must be considered.

REFERENCES

1. G.T. Ruck, ed., *Radar Cross Section Handbook*, Vols. 1 and 2, New York, Plenum Press, 1970.
2. C.H. Hargraves, Jr., "The Application of RCS Analysis Techniques to a New Class of Body Shapes (U)," *Radar Camouflage Symposium Proceedings*, 1980.
3. CRC, *Standard Mathematical Tables*, The Chemical Rubber Company.

APPENDIX A

MATHEMATICS REVIEW

J. F. Shaeffer

A.1 VECTOR MATHEMATICS

Electromagnetic fields are vector fields and, hence, a knowledge of vector mathematics is a prerequisite for understanding EM phenomena. In addition, certain very fundamental vector notions concerning the divergence and curl of a vector field are essential to EM analysis. The power and utility of vector mathematics allows us to analyze physical phenomena independently of a specific coordinate system. Only when we are ready to generate numbers or an analysis for a specific problem do we specify the coordinate system. Then we may specifically write down vector components and vector operations in terms of that coordinate system and its basis vectors.

For normal three-dimensional space, a vector $\vec{A}$ has three scalar components, (A_x, A_y, A_z), which are required to specify its direction and magnitude. Types of vector products are dot, cross, and direct. The dot product is the projection of one vector along the direction of another:

$$\vec{A} \cdot \vec{B} = |\vec{A}| |\vec{B}| \cos \theta = A_x B_x + A_y B_y + A_z B_z \qquad \text{(A-1)}$$

where θ is the angle between the vectors. Because $\vec{A} \cdot \vec{B}$ is the projection of $\vec{A}$ along $\vec{B}$, or $\vec{B}$ along $\vec{A}$, the dot product is a scalar quantity.

The vector cross product is the area of a parallelogram formed by two vectors and is itself a vector which is perpendicular to the plane of the parallelogram:

$$\vec{C} = \vec{A} \times \vec{B} = |\vec{A}||\vec{B}| \sin \theta = \begin{vmatrix} \hat{i} & \hat{j} & \hat{k} \\ A_x & A_y & A_z \\ B_x & B_y & B_z \end{vmatrix} \tag{A-2}$$

$$= (A_y B_z - A_z B_y)\,\hat{i} + (A_z B_x - A_x B_z)\,\hat{j} + (A_x B_y - A_y B_x)\,\hat{k}$$

where $\hat{i}, \hat{j}, \hat{k}$ are unit vectors aligned along the x, y, z directions. The vector direction of the cross product follows the right-hand rule.

The direct product of two vectors is a dyadic (tensor) quantity having nine components,

$$\vec{A}\,\vec{B} = (A_x\,\hat{i} + A_y\,\hat{j} + A_z\,\hat{k})\,(B_x\,\hat{i} + B_y\,\hat{j} + B_z\,\hat{k}) \tag{A-3}$$

The types of vector triple products are the triple scalar product

$$\vec{A} \cdot (\vec{B} \times \vec{C}) \tag{A-4}$$

which is a scalar, and is the volume of a parallelepiped having sides $\vec{A}, \vec{B}$, and $\vec{C}$, and the triple vector cross product

$$\vec{A} \times (\vec{B} \times \vec{C}) = \vec{B}\,(\vec{A} \cdot \vec{C}) - \vec{C}\,(\vec{A} \cdot \vec{B}) \tag{A-5}$$

which has a vector result and can be expanded using the "bac-cab" expansion.

Vector fields can be functions of position and time. Hence, vector fields can be differentiated and integrated. The scalar line integral is

$$\int \vec{A} \cdot d\vec{l} \tag{A-6}$$

which is a measure of $\vec{A}$ along the line $\vec{l}$. The scalar surface integral is

$$\int \vec{A} \cdot d\vec{S} \tag{A-7}$$

which is a measure of the amount of flux $\vec{A}$ passing through surface $\vec{S}$.

The operator ∇ is called "del" and is defined as

$$\nabla = \frac{\partial}{\partial x}\,\hat{i} + \frac{\partial}{\partial y}\,\hat{j} + \frac{\partial}{\partial z}\,\hat{k} \tag{A-8}$$

The gradient of a scalar field is a vector which points in the direction of maximum rate of change of the scalar field

$$grad(u) = \nabla u = \frac{\partial u}{\partial x}\,\hat{i} + \frac{\partial u}{\partial y}\,\hat{j} + \frac{\partial u}{\partial z}\,\hat{k} \tag{A-9}$$

The change of u in a specific direction is du $- \nabla u \cdot d\vec{l}$. If a vector field can be represented entirely by the gradient of a scalar function u, then u is called the potential of the field. Further, the vector field is conservative. Lines perpendicular to ∇u are equal potential lines.

The divergence and curl of a vector field are two very important properties of a vector field. The divergence of the field is a measure of its source or sink; at each point in space it describes the amount of field produced or consumed. As such, the divergence of a vector field is a scalar quantity defined as

$$div \vec{A} = \nabla \cdot \vec{A} = \frac{\partial A_x}{\partial x} + \frac{\partial A_y}{\partial y} + \frac{\partial A_z}{\partial z} \tag{A-10}$$

The curl of a vector field is a measure of the magnitude and direction of the *rotation* of the vector field. Experimentally, if the vector field were that of water velocity, a small paddle wheel would be a measure of the fluid rotation. The paddle turns at points where the curl is non-zero. The vector orientation of the curl is parallel to the axis of rotation of the paddle wheel. The curl is defined formally as

$$curl \vec{A} = \nabla \times \vec{A} = \begin{vmatrix} \hat{i} & \hat{j} & \hat{k} \\ \frac{\partial}{\partial x} & \frac{\partial}{\partial y} & \frac{\partial}{\partial z} \\ A_x & A_y & A_z \end{vmatrix} \tag{A-11}$$

$$= \left(\frac{\partial A_z}{\partial y} - \frac{\partial A_y}{\partial z} \right) \hat{i} + \left(\frac{\partial A_x}{\partial z} - \frac{\partial A_z}{\partial x} \right) \hat{j} + \left(\frac{\partial A_y}{\partial x} - \frac{\partial A_x}{\partial y} \right) \hat{k}$$

On vector field maps, the divergence is different from zero only at those points where field lines originate or disappear. The curl is non-zero only at those points surrounded by closed or spiraling field lines, as shown in Fig. A-1. A vector field whose divergence is zero everywhere (no sources or sinks), is a solenoidal field. A vector field whose curl is zero everywhere is an irrotational field.

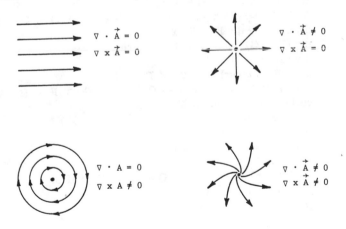

Figure A-1 Vector fields in terms of divergence and circulation

The divergence of the gradient is called the Laplacian and is

$$\nabla \cdot \nabla = \nabla^2 = \frac{\partial^2}{\partial x^2} + \frac{\partial^2}{\partial y^2} + \frac{\partial^2}{\partial z^2} \tag{A-12}$$

This operation arises from taking the curl of a curl of a vector field which may be defined as

$$\nabla^2 \vec{A} = \nabla (\nabla \cdot \vec{A}) - \nabla \times (\nabla \times \vec{A}) \tag{A-13}$$

The important vector integral relations are as follows. The amount of flux $\vec{A}$ flowing in or out of a closed surface is given by a volume integral of the source-sink function $\nabla \cdot \vec{A}$,

$$\int (\nabla \cdot \vec{A})\, dv = \oint \vec{A} \cdot d\vec{S} \qquad \text{(Gauss' theorem)} \tag{A-14}$$

where the surface S completely encloses volume v.

The field $\vec{A}$ around a closed contour $\vec{l}$ is given by a surface integral of the rotation function $\nabla \times \vec{A}$,

$$\int \nabla \times \vec{A} \cdot d\vec{S} = \oint \vec{A} \cdot d\vec{l} \qquad \text{(Stokes' theorem)} \tag{A-15}$$

where $d\vec{l}$ is the curve bounding surface S. A summary of important vector identities is given in Fig. A-2.

$$\mathbf{\nabla}(\varphi + \psi) = \mathbf{\nabla}\varphi + \mathbf{\nabla}\psi$$
$$\mathbf{\nabla}\varphi\psi = \varphi\mathbf{\nabla}\psi + \psi\mathbf{\nabla}\varphi$$
$$\text{div } (\mathbf{F} + \mathbf{G}) = \text{div } \mathbf{F} + \text{div } \mathbf{G}$$
$$\text{curl } (\mathbf{F} + \mathbf{G}) = \text{curl } \mathbf{F} + \text{curl } \mathbf{G}$$
$$\mathbf{\nabla}(\mathbf{F} \cdot \mathbf{G}) = (\mathbf{F} \cdot \mathbf{\nabla})\mathbf{G} + (\mathbf{G} \cdot \mathbf{\nabla})\mathbf{F} + \mathbf{F} \times \text{curl } \mathbf{G} + \mathbf{G} \times \text{curl }$$
$$\text{div } \varphi\mathbf{F} = \varphi \text{ div } \mathbf{F} + \mathbf{F} \cdot \mathbf{\nabla}\varphi$$
$$\text{div } (\mathbf{F} \times \mathbf{G}) = \mathbf{G} \cdot \text{curl } \mathbf{F} - \mathbf{F} \cdot \text{curl } \mathbf{G}$$
$$\text{div curl } \mathbf{F} = 0$$
$$\text{curl } \varphi\mathbf{F} = \varphi \text{ curl } \mathbf{F} + \mathbf{\nabla}\varphi \times \mathbf{F}$$
$$\text{curl } (\mathbf{F} \times \mathbf{G}) = \mathbf{F} \text{ div } \mathbf{G} - \mathbf{G} \text{ div } \mathbf{F} + (\mathbf{G} \cdot \mathbf{\nabla})\mathbf{F} - (\mathbf{F} \cdot \mathbf{\nabla})\mathbf{G}$$
$$\text{curl curl } \mathbf{F} = \text{grad div } \mathbf{F} - \mathbf{\nabla}^2\mathbf{F}$$
$$\text{curl } \mathbf{\nabla}\varphi = 0$$
$$\oint_S \mathbf{F} \cdot \mathbf{n}\, da = \int_V \text{div } \mathbf{F}\, dv$$
$$\oint_C \mathbf{F} \cdot \mathbf{dl} = \int_S \text{curl } \mathbf{F} \cdot \mathbf{n}\, da$$
$$\oint_S \varphi\mathbf{n}\, da = \int_V \mathbf{\nabla}\varphi\, dv$$
$$\oint_S \mathbf{F}(\mathbf{G} \cdot \mathbf{n})\, da = \int_V \mathbf{F} \text{ div } \mathbf{G}\, dv + \int_V (\mathbf{G} \cdot \mathbf{\nabla})\mathbf{F}\, dv$$
$$\oint_S \mathbf{n} \times \mathbf{F}\, da = \int_V \text{curl } \mathbf{F}\, dv$$
$$\oint_C \varphi\, \mathbf{dl} = \int_S \mathbf{n} \times \mathbf{\nabla}\varphi\, da$$

Figure A-2 Summary of vector identities

A.2 VECTOR FIELD SOURCES

A very important concept for the source of a vector field is the following. Any arbitrary vector field $\vec{A}$ can be expressed as the sum of two volume integrals, one related to conservative sources (usually charge), and the other related to non-conservative sources (usually current):

$$\vec{A} = -\frac{1}{4\pi} \nabla \int_{\substack{\text{all} \\ \text{space}}} (\nabla \cdot \vec{A}) \frac{e^{ikR}}{R} \, dv + \frac{1}{4\pi} \nabla \times \int_{\substack{\text{all} \\ \text{space}}} (\nabla \times \vec{A}) \frac{e^{ikR}}{R} \, dv$$

(A-16)

This is known as Poisson's theorem, Fig. A-3, and it shows that a vector field can be expressed in terms of a conservative part through a scalar potential ϕ and a non-conservative part through a vector potential $\vec{F}$ as

$$\vec{A} = -\nabla \phi + \nabla \times \vec{F}$$

(A-17)

where the scalar potential ϕ is an integral over the sources and sinks of the field $\vec{A}$, i.e., its divergence,

$$\phi = \phi_o + \frac{1}{4\pi} \int (\nabla \cdot \vec{A}) \frac{e^{ikR}}{R} \, dv \qquad \text{(scalar potential)} \quad \text{(A-18)}$$

and where the vector potential $\vec{F}$ is an integral of the rotation of the field $\vec{A}$, i.e., its curl,

$$\vec{F} = \vec{F}_0 + \frac{1}{4\pi} \int (\nabla \cdot \vec{A}) \frac{e^{ikR}}{R} \, dv \qquad \text{(vector potential)} \quad \text{(A-19)}$$

where ϕ_0 and $\vec{F}_0$ are constants. Thus, once the divergence and curl of a vector field are given by a physical theory such as Maxwell's equations for electromagnetics, elastic theory, or fluid theory, then the field $\vec{A}$ is completely specified. For electromagnetics $(\nabla \cdot \vec{A})$ is electric or magnetic charge density, while $(\nabla \times \vec{A})$ is an electric or magnetic current density.

We can further rewrite Poisson's equations by specifying

$$\nabla \cdot \vec{A} = \rho/\epsilon \qquad \text{(charge)}$$

$$\nabla \times \vec{A} = \vec{J} \qquad \text{(current)}$$

(A-20)

to obtain

$$\phi = \frac{1}{4\pi\epsilon} \int \rho \, \frac{e^{ikR}}{R} \, dv$$

$$\vec{F} = \frac{1}{4\pi} \int \vec{J} \, \frac{e^{ikR}}{R} \, dv \qquad\qquad (A\text{-}21)$$

Next we carry out the specified operations on ϕ and $\vec{F}$ to obtain

$$\vec{A} = \int \rho \, \nabla \psi \, dv - \int \vec{J} \times \nabla \psi \, dv \qquad\qquad (A\text{-}22)$$

where the Green's function is $\psi = e^{ikR}/4\pi R$. The gradient and curl operators are performed with respect to the field points, while ρ and $\vec{J}$ are functions of the source points.

Sources for Arbitrary Vector Field $\vec{A}$

Vector Fields have two parts:

(1) Conservative part
(2) Solenoidal part

$$\vec{A} = - \nabla\phi + \nabla \times \vec{F}$$

where

$$\phi = \int_V (\nabla \cdot \vec{A}) \, \frac{e^{ikR}}{4\pi R} \, dv$$

$$\vec{F} = \int_V (\nabla \times \vec{A}) \, \frac{e^{ikR}}{4\pi R} \, dv$$

The physics of the problem specify:

$$\nabla \cdot \vec{A} = \rho$$

$$\nabla \times \vec{A} = \vec{J}$$

then

$$\vec{A} = \int \rho \, \nabla\psi \, dv + \int \vec{J} \times \nabla\psi \, dv, \quad \psi = e^{ikR}/4\pi R$$

Figure A-3 Poisson's theorem

The concept of a Green's function is that it relates a source at one point in space to an observed field at another point in space. In Poisson's equations, the Green's function is simply the $e^{ikR}/4\pi R$ phase shift (time lag), and spherical spreading decay between the source and field points. In applying Poisson's theorem, we must keep in mind the distinction between field coordinates and source coordinates when performing the differentiation. Thus, for a vector field whose curl is zero we have $\nabla \cdot \vec{A} = \rho$ and

$$\vec{A} = -\nabla \phi \cdot -\nabla \int \rho \psi \, dv = \int \rho \nabla \psi \, dv \qquad (A\text{-}23)$$

where $\nabla \psi = (1 + ikR) \, e^{ikR} \, \hat{R}/4\pi R$, where $\hat{R}$ is a unit vector pointing from the source point to the field point. The gradient operation "slides" over ρ because ρ is only a function of source coordinates, while

$$R = [(x - x')^2 + (y - z')^2 + (z - z')^2]^{1/2}$$

is a function of both source and field coordinates.

For the case where the field $\vec{A}$ is solenoidal, $\nabla \cdot \vec{A} = \rho = 0$, and $\nabla \cdot \vec{A} = \vec{J}$, we have

$$\vec{A} = \nabla \times \vec{F} = \nabla \times \int \vec{J} \psi \, dv = \int \vec{J} \times \nabla \psi \, dv \qquad (A\text{-}24)$$

where again $\vec{J}$ is a function of source coordinates only.

A.3 SCALAR GREEN'S THEOREM

In field analyses involving sources of fields, we often encounter two regions in space, as shown in Fig. A-4. In general, there are field sources in both regions. However, we may know only the sources in region I, while the only knowledge of the sources in region II is the field values they create on the surface separating the two regions. Because the total field in region I is a function of sources in both regions, there are several important relationships which give the effect of region II sources in terms of fields on the surface separating the two regions.

The first relationship is Green's theorem for *scalar* fields, such as field potentials, which relates volume integrals to surface integrals:

$$\int (\phi \nabla^2 \psi - \psi \nabla^2 \phi) \, dv = \oint (\phi \nabla \psi - \psi \nabla \phi) \cdot d\vec{S} \qquad (A\text{-}25)$$

Here ϕ and ψ are scalar functions. Letting ϕ be the scalar potential and $\psi = e^{ikR}/4\pi R$ the above identity leads to

$$\phi(\vec{r}) = \int \rho \psi \, dv + \oint (-\phi \nabla \psi + \nabla \phi \, \psi) \cdot d\vec{S} \qquad (A\text{-}26)$$

450

This equation states that the scalar potential *anywhere inside* the surface S is determined by the charge distribution ρ inside S, and by the value of the potential ϕ and its normal component $\nabla \phi$ at all points on the surface S from sources external to S. For a charge-free region, the potential is governed by the surface potentials. If the region is not charge-free, and the surface S recedes to infinity, then the potential is only due to the first term; that is, it is an integral over the source ρ.

When considering high frequency radiation problems where the vector character of the electric and magnetic fields are neglected (when they are considered scalar fields), then the above representation for the scalar E and H fields is known as the scalar Huygens-Kirchhoff integral. This is typically used to obtain the field diffracted by an aperture; the fields on the surface are zero everywhere except for the aperture, and the aperture field distribution is assumed to be precisely that of the incident field. Then the field in region I is the integral given by Eq. (A-26) with the known surface field as the source function.

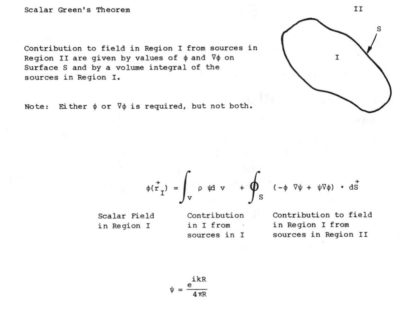

Scalar Green's Theorem

II

Contribution to field in Region I from sources in Region II are given by values of ϕ and $\nabla\phi$ on Surface S and by a volume integral of the sources in Region I.

Note: Either ϕ or $\nabla\phi$ is required, but not both.

$$\phi(\vec{r}_I) = \int_v \rho \, \psi d \, v \quad + \oint_S (-\phi \, \nabla\psi + \psi\nabla\phi) \cdot d\vec{s}$$

Scalar Field in Region I

Contribution in I from sources in I

Contribution to field in Region I from sources in Region II

$$\psi = \frac{e^{ikR}}{4\pi R}$$

Figure A-4 Scalar Green's theorem

A.4 VECTOR GREEN'S THEOREM

For vector functions, a similar representation of the fields is possible, involving an integral over the sources enclosed within surface S, and field values on S due to sources external to S. The vector analog is not usually given in most textbooks for this case. However, for application to electromagnetic fields Stratton gives the following:

$$\vec{E}(\vec{r}) = \int_{\text{volume}} [i\omega\mu\vec{J}\,\psi - \vec{M} \times \nabla\psi + \rho\nabla\psi]\,dv$$

$$-\oint_{s} [i\omega\mu(\hat{n} \times \vec{H})\psi + (\hat{n} \times \vec{E}) \times \nabla\psi + (\hat{n} \cdot \vec{E})\nabla\psi]\,dS \tag{A-27}$$

where fields on S ($\hat{n} \times \vec{H}$, $\hat{n} \times \vec{E}$, and $\hat{n} \cdot \vec{E}$), are due to external sources. This equation and its magnetic counterpart are the so-called Stratton-Chu (or Chu-Stratton) equations for the EM field. Again, the important concept is that the field interior to S is due to a contribution over sources within the volume (the first integral in Eq. (A-27)) plus a contribution from surface sources (currents and charges) from sources in region II. While we have not seen the general vector form of this equation, it is apparent that the surface terms involve both the curl of the field (surface currents, $\hat{n} \times \vec{E}$) and the divergence of the field (surface charge, $\hat{n} \cdot \vec{E}$). The general field $\vec{E}$ is due to both solenoidal and conservative surface components.

This completes the introduction to vector mathematics. References [1] through [4] are good reviews for the required vector analysis.

REFERENCES

1. O.D. Jefimenko, *Electricity and Magnetism*, New York, Meredith Publishing, 1966.

2. J.R. Reitz and F.J. Milford, *Foundations of Electromagnetic Theory*, Addison-Wesley, 1967.

3. R.F. Harrington, *Time Harmonic Electromagnetic Fields*, New York, McGraw-Hill, 1961.

4. J.A. Stratton, *Electromagnetic Theory*, New York, McGraw-Hill, 1941.

INDEX

Author Index

Corporate Index

Subject Index